Library of
Davidson College

MEDICAL IMAGES AND DISPLAYS

Two of the first color X-ray pictures (somewhat faded since 1950), each displaying and comparing three independent pieces of information at every spot. The three separation negatives were radiographs made at 40, 60, and 80 kilovolts on anodes of iron, molybdenum, and silver, respectively. A near infrared image to extend the spectral information of the mouse is not included here. (By Mackay and Collins.)

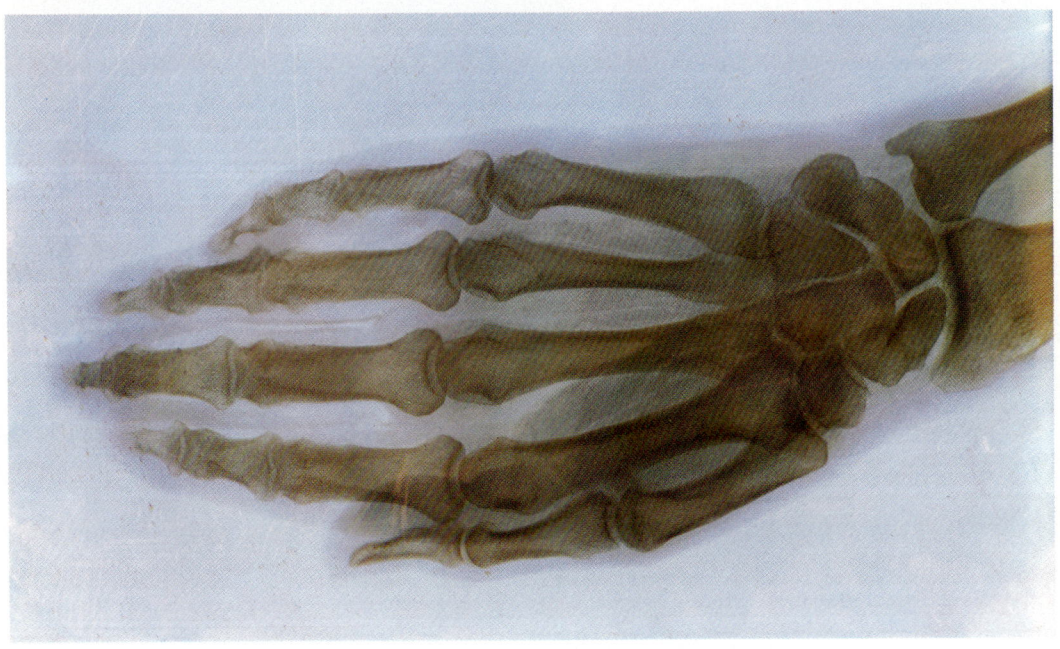

MEDICAL IMAGES AND DISPLAYS

Comparisons of Nuclear Magnetic Resonance, Ultrasound, X-Rays, and Other Modalities

R. STUART MACKAY

Boston University

A Wiley-Interscience Publication
JOHN WILEY & SONS
New York   Chichester   Brisbane   Toronto   Singapore

Copyright © 1984 by John Wiley & Sons, Inc.

All rights reserved. Published simultaneously in Canada.

Reproduction or translation of any part of this work beyond that permitted by Section 107 or 108 of the 1976 United States Copyright Act without the permission of the copyright owner is unlawful. Requests for permission or further information should be addressed to the Permissions Department, John Wiley & Sons, Inc.

Library of Congress Cataloging in Publication Data:

Mackay, R. Stuart (Ralph Stuart), 1924-
   Medical images and displays.

   "A Wiley-Interscience publication."
   Bibliography: p.
   Includes index.
   1. Diagnosis, Radioscopic. 2. Imaging systems in medicine. 3. Electroluminescent display systems. 4. Optoelectronic devices. I. Title. [DNLM: 1. Radiography. 2. Tomography. 3. Ultrasonics. 4. Nuclear magnetic resonance. 5. Technology, Radiologic. WN 200 M153m]
   RC78.M186   1984   616.07′54   84-5124
   ISBN 0-471-89617-9

Printed in the United States of America

10 9 8 7 6 5 4 3 2 1

*To my students,
whose problems and successes
have provided many years
of stimulation and pleasure*

PREFACE

This book is intended for physicians, engineers, physicists, and biologists, and hopefully will prove of interest to any others who might make or use imaging equipment. For those not routinely involved with equations, physical understanding is emphasized rather than mathematical brevity and derivations, with some formulas included for calculation and to summarize functional relationships. The concepts should be helpful also to those examining objects of wood, metal, rubber, concrete, and so on, without cutting into them, that is, to "nondestructive testing" engineers. The ideas of image processing and display are of potential value to those producing scientific images of any sort.

All possible references, or indeed most, can not be given since for each major topic there is a library of relevant books and journals. Sometimes a single reference is used to introduce the work of groups that had produced dozens of articles. Some older references are included that may be helpful to those whose knowledge is limited to a computer search of recent literature. The Introduction to the 1959 *Yearbook of Radiology* recommended reading the original of Chapter 9 of *Advances in Biological and Medical Physics,* Vol. 6, for 1958. Much of this has become incorporated into routine thinking but it can be considered a companion piece to the present much more general summary. The citations attempt to indicate the ramifications and diversity of this field.

For assistance with this project I wish to thank two of my good friends and former graduate students: Michael Reid, Associate Professor of Radiology at the University of California, and Bob Markevitch, Manager of the Systems and Analysis Section of Ampex.

R. STUART MACKAY

San Francisco, California
August 1983

CONTENTS

1. **MEDICAL IMAGES**   1

2. **FROM DIRECT CURRENT TO COSMIC RAYS: A SPECTRUM OF MODALITIES**   3
    - 2.1. Steady Current   4
    - 2.2. Low Frequencies   6
    - 2.3. Radio   8
    - 2.4. Nuclear Magnetic Resonances   9
    - 2.5. Infrared   10
    - 2.6. Visible Light   14
    - 2.7. Ultraviolet   15
    - 2.8. X-Rays   16
    - 2.9. Radioactivity and Nuclear Medicine   17
    - 2.10. Cosmic Rays   24
    - 2.11. Additional Modalities   25
        Sound and Ultrasound; Corona Discharge Photography
    - 2.12. Near Field versus Far Field   27
    - 2.13. Emanation versus Traversal and Summary   29

3. **IMAGE FORMATION**   31
    - 3.1. Scan versus Parallel Processing   31
    - 3.2. Absolute Limits   37
        Quanta, Noise, Resolution, Image Tubes, and Intensifiers; Television; Optimum Conditions
    - 3.3. Computed Images   46
    - 3.4. Equipment Evaluation   49

4. COMPUTED TOMOGRAPHY — 53
4.1. Introduction 53
4.2. Digital Computed Tomography 62

5. X-RAYS — 83
5.1. Sources 83
5.2. Interactions and What Is Sensed 86
5.3. Speed of Examination 91
5.4. Biological Hazard and Dose 94
5.5. Contrast Agents, Tracers, and Stains 97

6. ULTRASOUND — 100
6.1. Sources 100
6.2. Interactions and What Is Sensed 107
6.3. Scan Types 110
  A, B, M, Sector, and Compound Modes; Mechanical versus Electrical Real Time; Phased Array; Synthetic Aperture; Converter Tubes or Cameras
6.4. Hazard and Dose 125
6.5. Contrast Agents 127
6.6. *Post Mortem* Effects 129
6.7. Tiny Bubbles 130
  Response; Diving; Contrast Echocardiology; Tumors versus Abscesses and Other Pending Studies

7. NUCLEAR MAGNETIC RESONANCE — 138
7.1. The Effect and Its Observation 138
7.2. Speed of Image Formation 158
7.3. Hazards 161
7.4. Contrast Agents 163

8. SPECTRAL INFORMATION — 166

9. MOTION AND FLOW — 180

10. IMAGE PROCESSING — 197
10.1. Photographic Film 197
10.2. Contrast Enhancement 200
10.3. Edge Emphasis 205
10.4. Deblurring 209
10.5. Smoothing 215
10.6. Changes, Movement, and Differences 215
10.7. Quantitative Analysis 219

11. DIGITAL METHODS — 225

12. DISPLAYS   **232**
    12.1. Gray Scale 235
    12.2. Color 236
    12.3. Pseudo 3D 238
    12.4. True 3D 241
    12.5. Motion Pictures 254
    12.6. Holograms 256

References   **259**

Index   **273**

1
MEDICAL IMAGES

A medical image is the display of the distribution of some particular property within the body. Examples of properties are temperature, mobile proton density, stiffness, electron density, velocity, iodine affinity, electrical activity, and so on. The image is produced by interacting and exploring with some imaging modality to which the body is rather transparent, and which thus distributes itself in space; this distribution is converted to visibility, or to electrical form for computer processing or analysis. Intermediate distributions might be a pattern of X-rays or a distribution of high frequency sound intensities. In any case, adjacent organs or structures appear distinct and abnormalities or distortions can be seen. Each method supplies supplementary information, and no one is best in all cases, so no one imaging modality has superseded all the others. However, some modalities are more generally useful than others.

This division of the formation process into two steps, an invisible and a visible step, is useful both for analysis and for practical implementation. Thus an X-ray picture might simply be said to be a medical image formed by shining X-rays through a subject onto a film. In that case X-ray intensity or exposure time is increased until the film is suitably blackened. However, it is often better to deliver and limit patient dose by passing just enough photons through the subject to define the smallest region of interest, and to let these enter a television camera whose output can then be turned up any amount—enough to expose film, to look at, or to burn the paint off the wall if desired, all without an increase in patient dose.

Also, processing of an image before display can unveil otherwise unnoticeable detail already present. A separate chapter will be devoted to this matter.

It might be mentioned that a conversion step in some form is almost inevitable, be it by film, sound detector, or other means. Radiation that passes perfectly through tissue will not interact with the retina to be seen. (If there actually were a totally invisible man as in the mystery stories he would have to be blind: if he could see, at least his retinas would have to be noticeable to others.) In the case of X-rays, for example, they could not form useful images if the body were perfectly or uniformly transparent to them and, for the same reason, there is some small interaction between X-rays and the eye that allows them to be sensed, partly due to fluorescence and partly not. Indeed an older technique for measuring the length of the eye in the presence of an opacity required the subject to describe the circle perceived when a sheet of X-rays was shone across the eye, starting from the front and moving backward until the circle shrank to a point, but the process was inefficient and required relatively high intensities.* Even if the size of an X-ray distribution intermediate image could be reduced to enter the eye, it would be at least ineffective for the radiologist to look directly into the X-ray tube; other imaging modalities that slightly affect the eye must also be converted to a visible pattern for maximum sensitivity and minimum damage in the overall process.

Some useful images are displayed through the ears to give a sense of spatial distribution (Kay et al., 1977; Kay, 1980) or across the abdomen by vibrators, voltage stimulation electrodes, or scanned water jet (Collins and Madey, 1974). Such displays are helpful for blind guidance prostheses and some medical instruments, but, with humans, vision is a primary sensory input and it is to the eye that medical images are usually displayed. Not all visible displays communicate information equally well, and some of the relevant properties of the human eye are given in later chapters.

---

*In recent times, eye length is routinely measured before a cataract operation by timing the transit of pulses of ultrasound; length combined with an optical measurement of corneal curvature allows selection of a suitable replacement lens for implanting in the eye after the operation, and scanning to make an image also allows search for tumors and other problems at that time.

2

FROM DIRECT CURRENT TO COSMIC RAYS
A Spectrum of Modalities

Many of the modalities used in imaging involve a fluctuating electric and magnetic field that is called electromagnetic radiation. These possible modalities can be ordered for consideration according to wavelength, which is just velocity divided by the frequency of oscillations. An overall summary of this is given in Figure 2.1 which spans sizes from atomic to continental. Sharp boundaries of demarcation between regions are purposely not indicated because they blend into each other. For example, radio techniques can generate frequencies normally considered to be in the infrared region of the frequency spectrum while thermal effects generate frequencies that can be detected by radio techniques. Similarly, the longest wavelength that can be detected by the eye depends somewhat upon intensity and varies between animal species, thus making the meaning of visible light slightly vague. Two regions are shown relatively sharply and these are the wavelengths most strongly radiated by bodies at biological temperatures and the approximate wavelengths seen by the human eye. The modalities shown above in that figure can be introduced to traverse the body and form a sort of image, or in a few cases emanate directly from the body, while those below normally involve emanation from different structures within the body. The energy and source temperature scale increases to the right. Cosmic rays are included at the far right because they represent extremely high energy processes, though they usually involve the passage of charged particles.

4 FROM DIRECT CURRENT TO COSMIC RAYS

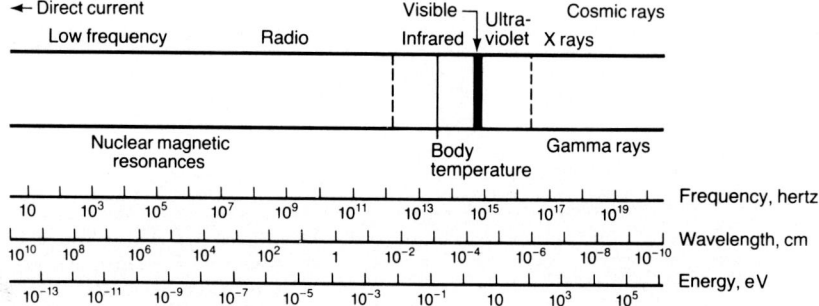

FIGURE 2.1. The scale of electromagnetic radiation frequencies and associated properties. The full visible spectrum from red to violet is ordered in the band labeled visible. At lower frequency, the wavelength for 60 Hz in power lines is about the distance across the United States. At 1 MHz the energy of each photon is small; smooth energy transfer is apparent rather than the granular nature of the radiation, and chemical bonds are not readily disrupted by application. The bottom two lines incidentally give the minimum X-ray wavelength produced by a specified applied tube voltage. The frequency and wavelength scales do not apply to ultrasound. Boundaries of modality regions are purposefully left indistinct.

It has become familiar that radiation can be regarded either as discrete quanta or as waves. For certain purposes in later chapters we will find it useful to employ a quantum or photon representation but here it is well to recall that several factors are affected by wavelength or frequency. In an object, geometrical detail finer than roughly a wavelength will not generally be detected, except in the interesting case of tiny bubbles observed with high frequency sound; resolution may be considerably worse than this, being limited by machine factors. Higher frequencies are associated with higher energies, energy being frequency times Planck's constant h; this determines potential for breaking chemical bonds, producing hereditary mutations, and so on. Ability to form a beam of radiation and its best mode of generation and detection depend on frequency and on proximity to the source; the distance where the near field becomes the far field depends upon wavelength. Penetration depends upon frequency or wavelength. There may be resonances or accentuated interactions at specific frequencies characteristic of the structure of the subject. These few introductory remarks are amplified in later chapters.

2.1. STEADY CURRENT

Unchanging electric currents or extremely slowly changing currents can flow in the body, and their presence and distribution can be significant. Their frequency of changing is essentially zero and the electric and magnetic fields associated with them decrease steadily with distance but do not change much with time. (Obviously their time is ultimately limited to the long period between conception and death.) Such currents generally are not introduced to explore the body because they are able to stimulate excitable tissue and cause problems ranging from muscle spasm to ventricular fibrillation (though some of the problem occurs during the changes of turning on and turning off the current), but small currents of this sort indicate functioning by being generated within the body itself.

Detection of steady currents by their associated voltages is very difficult by electrodes that are applied to the surface of the skin. This is because of battery-like chemical changes that may take place in the vicinity of any one electrode while something different is taking place at the other. Such problems can arise even when the amplifier has an extremely high input resistance so as to draw very little current from the subject. When well-matched pairs of electrodes are applied to separated parts of the body there often appears a small difference in potential whose origin may be unknown.

Associated with any steady current is not only a voltage but also a magnetic field. The motion of ions in tissue produces measurable magnetic fields that can be recorded. No electrodes are involved in their recording, which suggests applications ranging from clandestine observation to observation of severe burn victims, but mention is made here because of the possibility of observing the fields associated with steady currents both as to magnitude and direction. Though observations have been reported of the magnetic field associated with the activity of the heart, brain, and isolated nerves, the detected fields are small with respect to the fluctuating earth's magnetic field and so extreme precautions must be taken with regard to shielding, detector sensitivity, and techniques whereby signals from pairs of detectors cancel out the effects of large scale overall changes. Since Harvey Fishman, for his 1963 master's thesis from our Berkeley laboratory, recorded probably what were magnetic transients in isolated lobster axons using coils and a transistor amplifier, considerable progress has been made. Magnetic mapping of the human brain has now made it possible to identify the three-dimensional location of sources producing epileptic discharges (Barth et al., 1982), as well as observing auditory effects (Romani et al., 1982).

Blood flowing across an external applied magnetic field will generate small voltages and currents with which are associated secondary magnetic fields. Thus a map of magnetic fields may also, in principle, aid the study of the vascular system. These and other more immediately practical considerations are best mentioned in the next section dealing with electrical or magnetic signals that cyclically change but at a relatively slow rate.

Even though a zero frequency has associated with it an effectively infinite wavelength, magnetic fields and current paths can be shaped somewhat by suitable placement of the sources. The types of pattern generated are similar to those seen when iron filings are sprinkled in the vicinity of various combinations of permanent magnets. In the next few sections are mentioned methods in which various electric and magnetic fields are produced by configurations of coils or electrodes and how such fields are detected by similar combinations; generally a configuration that produces a certain field distribution in tissue will, if used as a

2.2. LOW FREQUENCIES

When the carrier of information has an intrinsic cyclic aspect then slower drift in amplifiers or electrodes can easily be rejected as artifact rather than being misconstrued as useful data. In this section we consider frequencies from roughly a few cycles per second to a few hundred thousand per second. Voltages within this range are generated within the body, and their distribution can be mapped, or voltages within this range can be applied to the body to map the distribution of some property. The electric currents and fields involved will spread and in no case would we expect to obtain extremely fine geometrical detail by some process resembling radio direction finding. When dealing with currents applied for exploratory purposes we will be concerned about the comfort and safety of the subject since low frequency currents are able to stimulate excitable tissue; indeed it is found that the frequency of electric current able to produce electrocution at lowest magnitudes is roughly that supplied as power to homes.

The rate of arrival of information about changes in whatever is being observed is not greater than roughly the rate of arrival of cycles. To say that the frequency response of the measuring system is about the same as the frequency of the carrier of information can be a misleading oversimplification as to what is practical. (Instrumentation examples that might appear as exceptions include the possibility of displaying dc information on an ordinary ac coupled oscilloscope by repeatedly short-circuiting its input, or the stabilization of an electromagnetic flow meter by occasionally reversing the magnetic field; in one case the frequency response apparently extends below the cyclic rate and in the other above it.)

A familiar example of a bioelectric potential is the electrocardiogram. The heart is an extended low frequency electrical source, the signal from whose parts can interfere or add in various ways nearby at the skin surface. In this near field one can attempt to calculate surface signals resulting from the somewhat nonuniform electrical conductivity of the body. When the frequency of changes in an electromagnetic field is low the wavelength is very long and it is less useful to think of penetration through the body as being like radio rays. In this case the way to sense activity mostly at a given spot is to move a detector near to that spot. (The practical aspects of "near fields" becoming "far fields" approximately a wavelength from the source are described in Section 2.12 and summarized in Mackay, 1970.) Thus a complete electrocardiogram (ECG or EKG) usually involves the placement of six chest electrodes in the vicinity of the heart, which yields a sort of crude image to sense or locate such things as an infarct by the detected differences in activity at different places.

A vectorcardiogram is a sort of three-dimensional image that has become familiar. A difficult case is that of the fetus *in utero* where a somewhat insulating coating prevents uniform spread of potentials through the maternal abdomen, thus preventing observation of the VCG by surface electrodes on the mother. In that case placing a tiny telemetry transmitter within the fetus is a way of observing these potentials during normal activity and development (Mackay, 1970). Activity of the brain is often recorded simultaneously from a number of electrodes distributed over the scalp, and the map of the distribution of activity constitutes a low resolution map.

A review of bioelectric recording has been made (Geddes and Baker, 1975). It might be hoped that the magnetocardiogram would be less affected than the electrocardiogram by the torso boundary and internal inhomogeneities such as the lungs and blood mass in the heart, because their boundaries should have minimal effect on the perpendicular component of magnetic field (Geselowitz, 1979; Cuffin, 1981). The MCG may or may not have advantages over the ECG, but injury occurrence associated with heart damage probably involves dc currents best assessed magnetically. Other methods continue to evolve. For example, voltage-sensitive fluorescent and absorption dyes are regularly used to measure action potentials in a variety of excitable tissues. Using a fast laser scanning system it was possible to map the spread of excitation in amphibian and mammalian hearts exposed and stained with a fluorescent dye (Dillon and Mored, 1981).

When electric currents are applied to traverse and probe a body they may be low current dc but the frequency applied is usually in the range from a few dozen cycles per second up to about a hundred thousand cycles per second (a cycle per second often being called a hertz, abbreviated Hz). Voltage can be applied to a pair of electrodes and the resulting current measured, or a constant current can be forced to flow and the resulting voltage across the electrodes measured, or current can be introduced through one pair of electrodes and the resulting voltage measured across a suitably positioned second pair. Various combinations of electrodes attempt to control the region of sensitivity. With the help of a metal "guard ring" around a sensing electrode, and kept at the potential of the latter, recording can be confined to a restricted region in spite of overall spreading of lines of current flow into adjacent regions (e.g., Graham, 1965).

The electrical search for acupuncture points by resistance irregularities is another attempt to form a medical image suited to a particular application (Reichmanas et al., 1975). In this particular example small steady currents were employed in a bridge circuit in order to make resistance changes more noticeable as the exploring electrode was moved about. Measured electrical resistance varies with time perhaps because of blood redistribution; this is part of the basis of both the so called lie detector, and also an instrument used in one religion to determine what apparently matters to the subject. (For large scale engineering purposes, the low radio frequencies, and even lightning storms, can probe by induction into the earth for ore bodies which modify conductive properties and can thus be imaged.)

Measurement of volume change (plethysmography) seems possible with applied currents through surface electrodes. Certainly one can follow fluctuations ranging from those of respiration to those of the renal artery. Calculations suggest the possibility of observing changes in the head such as those produced by developing hydrocephalus (Murray, 1981). Using implanted electrodes it has been reported possible to compare blood flow changes in various parts of the brain in response to various stimuli in unrestrained cats by monitoring electrical impedance fluctuations (Birzis and Tachibana, 1964). Effectively they produced a low resolution image of the redistribution of blood flow in response to sleep, wakefulness, and arousal by various stimuli. The currents used in the measurements did not affect the brain while flowing directly through it.

2.3. RADIO

As we increase the frequency of our electromagnetic probe it becomes expedient to think of it as a radio signal. However, it should be emphasized that the distinction is rather vague. As examples of frequencies similar to the above, we have fed tiny radio transmitters acting at 50 kHz to unrestrained porpoises in order to do physiological monitoring upon them, and some special radio transmitters apparently have been tested by the government that run at frequencies down to about 30 cycles per second. For comparison, typical commercial AM radio stations transmit in the vicinity of 1 MHz and FM stations near 100 MHz.

At high frequencies the wavelength becomes small with respect to the dimensions of the human body. It then becomes practical to form radio waves into narrow beams with the help of small antennas or apertures, and the usual rules of geometrical optics essentially apply. Spreading due to diffraction need not render the beams useless.

The interaction of radio waves with tissue has been widely studied (Schwan 1965; Bottomly and Andrew, 1978). In part this was over concern for health factors, a major one being diathermy heating effects. The extent of penetration of a radio signal into the body will depend upon its frequency if for no other reason than the shielding effect of its partial electrical conductivity. Measurements made on small animals have often been extremely misleading because where the wavelength is comparable to the size of the subject, a small change in dimensions can make a large change in the interaction, with the frequency being kept fixed. It is also known that different tissue types will interact differently with any passing radio wave.

There is a large difference in microwave absorption between fatty and nonfatty substances, and since breast carcinomas tend to be nonfatty, though buried in fatty tissue, radio imaging should be useful. To obtain fine detail a high frequency would have to be used. Lenses to form images with radio waves are well known. Another approach is to scan the subject in many directions with a narrow beam, and from the collected attenuation measurements to then compute a cross sectional image, as will be discussed in the next chapter. Such an approach has been explored by Rao et al. (1980). Their phantom studies were done at 10.5 GHz where a wavelength of 1.4 cm existed in polyethyleneglycol which was used as a coupling medium between source, subject, and detector. Radio waves also are extremely sensitive to the presence of moisture which suggests their use to quantify small changes in lung water content, especially during the initial stages of pulmonary edema. In dog experiments, Iskander et al. (1981) explored this possibility at a frequency of 915 MHz. They measured the transmitted signal which they found to be 50 times more sensitive to changes in lung water content than was the reflected signal.

Microwave radio radiation emitted from structures in the body can be measured from outside with less spatial resolution but more penetration than when using infrared thermography. In the gigahertz frequency range one can search for breast cancer and measure subcutaneous temperature (Barrett and Myers, 1975; Porter and Miller, 1978). A nonbiological example also is instructive. One cannot look through a pile of sand but radio waves can penetrate several meters of sand. Thus radar can image subsurface features in a desert. As expected, the final image is not as fine as with visible light.

2.4. NUCLEAR MAGNETIC RESONANCES

Interspersed here in the frequency sequence between radio and infrared is mention of a special effect—nuclear magnetic resonance imaging—to which an entire later chapter is devoted. The *nuclei* of some atoms will absorb or emit radiation of a particular frequency if placed in a steady magnetic field. (The atoms which can be so studied have either an odd number of protons or of neutrons in their nucleus which allows them to act as small magnets; most biologically interesting elements have an isotope of this sort.) For example, in an applied field of 1000 gauss, protons will selectively respond to a frequency of about 4 MHz, being very slightly different if the hydrogen is in the form of water, alcohol, or fat. The effect can be considered to be like the Zeeman effect, but just as a mechanistic picture of electrons in continuing orbits around nuclei in atoms helps us to understand or remember the origin of optical effects, so a mechanical description will be given later that may help us with visualization. Different materials have different frequencies, all of which are in the usual radio range of frequencies and can be measured by ordinary radio techniques.

Nuclear magnetic resonance in condensed material (e.g., water or paraffin) was independently first observed in 1946 by Bloch and by Purcell, a feat for which they shared the 1952 Nobel prize. To form an image of a subject one must distinguish one region from another. If one could shape a magnetic field so that it is strong enough at just one point to produce resonance at a frequency being recorded then the subject could be moved around with respect to the magnet to record the response at different places and produce an image. This idea was known for decades but the first demonstration of imaging was by Lauterbur (1973) who instead sampled a line at a time by superimposing a small gradient on the overall magnetic field in order that different strips through a sample give detectably different frequencies. An image could then be computed from several such observations in different directions, and the difficulty of concentrating a magnetic field into a small volume could be avoided. Since that time image quality has improved steadily and commercial imaging equipment is becoming available. In Figure 2.2 is a sagittal image of a midline section of a living human head produced by Philips equipment having a superconducting magnet with a field strength of 6.02 kilogauss. Slice thickness is about 6 mm, allowing clear visualization of the spine joining the brain. Fat appears white, for example. The distribution of mobile protons and their relaxation times is what is seen. Gas or bone do not interfere, and there is little hazard. This view would be difficult to produce in other ways.

2.5. INFRARED

If matter is heated to increasingly high temperatures then the electromagnetic energy it radiates will increase in frequency upward from the radio range until it eventually is perceived to be "red hot" and eventually "white hot" as blue light emerges. The region of wavelengths somewhat longer than the human eye can perceive, beyond the red in Figure 2.1, is called the infrared. These frequencies correspond to various resonances in many organic molecules, thus making infrared absorption spectroscopy a valuable analytic tool. Combining an infrared and a nuclear magnetic resonance observation can sometimes determine details of organic structure. The distribution of frequencies radiated by a hot body is generally presumed to be broad and smooth, though this is not always the case.* The wavelength of maximum radiation for an object at the temperature of the human body is calculated to be about 9400 nm (Fig. 2.1).

A good absorber of radiation will also be a good radiator at the same wavelength, according to Kirchoff's law; one often speaks of a "black body" as being perfect in these aspects. Thus holes or black ink on an electric clothes iron appear dark, but if photographed with infrared film by the heat of the iron they appear bright as in Figure 2.3. Note also the changes in appearance of the plant, arm and shoulder, and background pigments when viewed by applied infrared rather than visible light. Ethnic differences in skin color of the human body do not detract from the possibility of making infrared examinations.

Infrared energy displays somewhat greater penetration than does visible light, and it also can be applied externally for visualization in certain clinical situations. In that case it should be realized that a "heat lamp" is not as copious a source of useful infrared as is a hotter "photoflood" lamp, though the latter will also have considerable visible light that will be filtered out before forming the image.

The longer wavelengths of "far" infrared can be detected by their heating action, for example, by scanning their pattern with a tiny blackened temperature-sensitive resistor called a thermistor. The "near" infrared can be focused by an

---

*Thus, at the time of this writing, sporting goods stores sell gas lanterns for camping that are incidentally rather radioactive because they contain thorium. Thorium has an electron structure rather like the rare earths, which allows radiation in narrow frequency bands from a solid. If radiation is restricted in the infrared, the temperature would go up, causing more radiation in the visible. Thus the fragile Welsbach mantles are used rather than heating a more sturdy structure; it is best not to accidentally inhale or swallow any of a crumbling gas mantle.

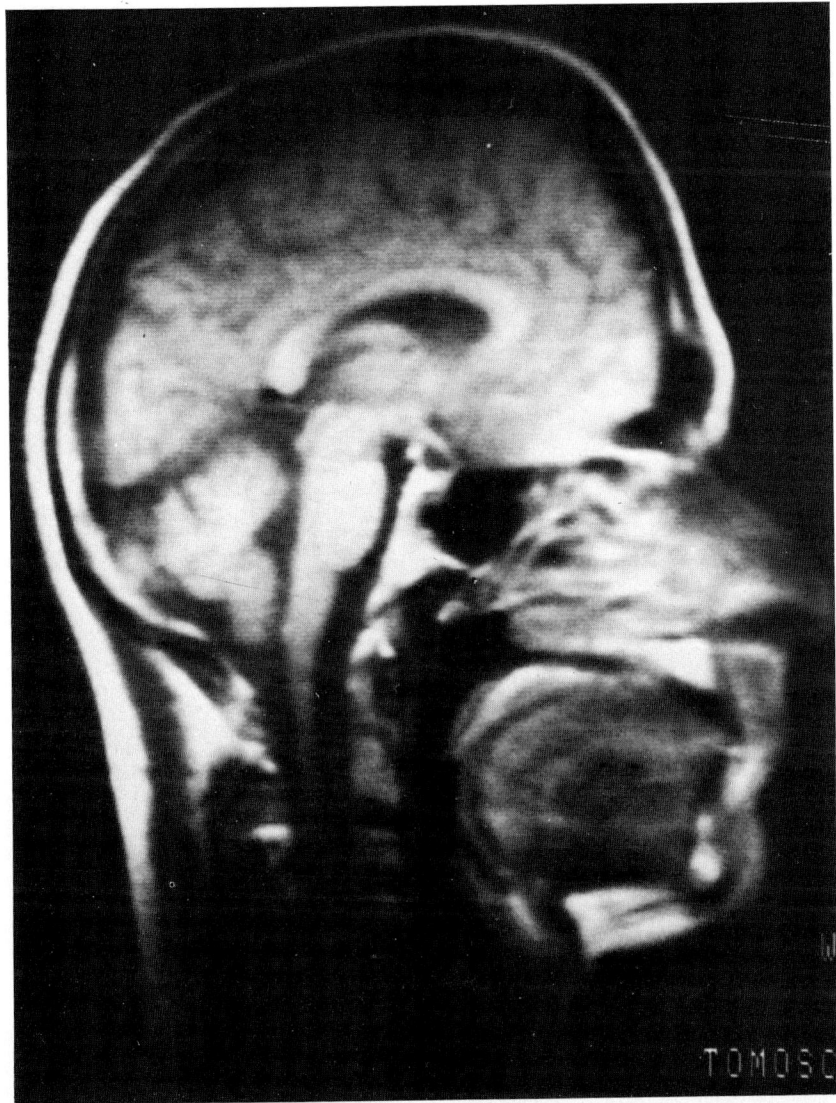

FIGURE 2.2. Nuclear magnetic resonance image along the midline of a human head and spine. The distribution of mobile protons is seen as emphasized by a relaxation time through the use of saturation recovery spin echo technique. Philips equipment having a magnetic field strength of 6.02 kilogauss was employed. Slice thickness is approximately 6 mm.

FIGURE 2.3. Upper left: The normal image of a cold iron illuminated by visible light. The words are painted on with black ink. Upper right: No illumination was provided, the image being recorded on infrared film by the heat of the iron. Good absorbers such as holes or black spots are good radiators and appear bright though temperature is the same as the surroundings. Lower left: a scene recorded by illumination with visible light. Lower right: The scene recorded with infrared illumination.

ordinary camera lens and photographed with an ordinary camera on special photographic film. During World War II some consideration was given to modifying the retinal pigment in the human eye to make it more like that of some fish through a special diet so that selected humans would be able to see otherwise invisible infrared searchlights. A more practical approach proved to be the construction of image converter tubes, sometimes called snooper scopes, in which an infrared image falling on an extended photocathode liberates a pattern of electrons which could be accelerated against a fluorescent screen to provide a visible image of an otherwise invisible scene. In some cases it is useful to observe fluorescence which itself is in the infrared range and is caused by illuminating an object with shorter-wavelength visible radiation (blue or green) while eliminating reflected infrared from the subject by placing over the source a blue infrared-absorbing filter.

In fluorescence one generally takes in a high energy quantum such as ultraviolet or blue and reradiates one of lesser energy such as yellow or red (Fig. 2.1). In rare cases the approximately simultaneous arrival of two infrared quanta may result in emission of a visible quantum. However, a reasonably practical glowing screen for the detection of infrared images can be produced if the screen is first activated by placing over it a suitable radioactive substance. Typical infrared image tubes are sensitive in the range of 700–1300 nm (nanometers or millimicrons), while typical infrared films respond out to 900 nm and in extreme cases out to about 1350 nm.

The most sensitive detectors for most wavelengths remain semiconductor devices which themselves display a lower background of noise if they are cooled by a thermoelectric cooler or, less conveniently, by a backing of liquid air or dry ice.

Infrared energy generally follows the usual rules of optics. Glass will not pass the far infrared (and hence the functioning of greenhouses for plants) though some plastics, semiconductors, and natural crystals such as rock salt (NaCl) pass a wider range of wavelengths. The focusing action of camera lenses is thus replaced by the effect of concave mirrors when extended response is required. There are many books on infrared, including those of Massopust (1952), Simon (1966), and Robinson (1973).

Observing a body by its own radiation is usually done with some sort of scanning system employing a sensitive detector and is called thermography. Regions of metabolic or circulatory irregularity would be expected to display a different temperature from the corresponding region on the other side of the subject. Such an observation is done in a cool room to emphasize the flow of heat toward the skin surface. One would expect such a procedure to be of value in evaluating peripheral vascular disease or to judge the effect of angioplasty. Many groups had hoped that such a technique would be of value in locating tumors in the female breast. In practice, screening tests apparently yield too many false negatives, and thus the method seems to have fallen into disfavor. A periodic reinvention for this latter purpose, which would hopefully inexpensively average temperature differences between the two sides over an extended period, is a brassiere fitted with thermistors; one would hope the subject was not a secretary immobilized before a warm window or drafty door.

There are applications where infrared images are recorded when the source of illumination is exclusively in the invisible infrared region. A filter that is opaque to visible light but transparent to infrared can be placed either before the source in a darkened room or before the camera lens. In the latter case it must be polished more carefully but the room can be bright. Typical lenses focus slightly differently for infrared and visible illumination, and focusing should be done on the near side of the region of interest. More specifically, if visible focusing is correct then the distance between lens and film should increase about one-quarter percent. Infrared shows maximum transmission in water at about 1100 nm. There are several rather well-known differences in the appearances of photographs taken in different parts of the spectrum.

An infrared photo will show the pattern of blood vessels below the skin surface (Fig. 2.3), making examination of the superficial venous system simpler. Thus the formation of a new varicose vein near the site of an old one can be followed and recorded. Infrared can be used to penetrate a scab in order to demonstrate the underlying condition. Infrared will penetrate a turbid or opaque cornea for observation of the iris; pupil size can actually be measured in an apparently opaque eye. In zoological studies it is well known that chitin is rather transparent to infrared. It is also known that the keratin of hair can be penetrated by infrared radiation. (The fur of some bears may give a greenhouse effect to warm the bearer, however.)

It is readily observed with a flashlight that visible light can shine through a finger or the female nipple; somewhat more detail can be seen using infrared. All such images can be combined to incorporate other information. The mouse in the frontispiece to this book was also photographed in the infrared through a war surplus image converter tube, but slight distortions in the electron optics made register with the X-ray image somewhat uncertain. But the principle is useful.

2.6. VISIBLE LIGHT

The visible range of wavelengths for humans is usually taken to be from 400 to 760 nm, and this spectrum of colors is shown as the dark band marked "visible" in Figure 2.1. A person who has had the lens of their eye removed in a cataract operation is typically able to see 350 nm which would be called the ultraviolet. With intense sources wavelengths out to 1050 nm are able to evoke sensation in the human eye (Griffin et al., 1947). In any case, medical images by visible light were the first and probably remain the most important.

In a sense the earliest form of recorded images were the anatomy drawings and paintings by such as Leonardo, Vesalius, Sabata, and more recently Netter. These drawings were so real looking that they took on a reality of their own, which in the early days was in a few cases incorrect; in those cases health care was sometimes conditioned by inaccuracies in the drawings or images and what was thought to be real. Modern day medical journals are filled with photographs that help to guide the visible inspections made by any physician.

Considerable thicknesses of tissue will pass light in a translucent way but will interfere with image formation. Thus a simple way to monitor heart rate is to shine a light through the fingertip and record fluctuations in transmitted intensity with a photocell placed on the other side to give a periodic pattern that sometimes includes a dicrotic notch.

Simple looking can be augmented, and a few examples follow. A subject illuminated from the side by sheets of light from parallel equally spaced slits will have traced upon the body a pattern of contour lines. A similar appearance of contours due to a Moiré pattern forms if viewing of a variable elevation surface is through the same periodic grating as the illumination. Such images have applications in following exercise or development. Images upon which are superimposed the momentary position of an observer's gaze or center of attention and interest can be instructive. Subject eye position can be determined by electrodes placed near the eyes (electrooculography) in order to measure direction of gaze. (See also Mackworth and Thomas, 1962.) It is interesting to follow the eye movements of an electrical engineer looking at a circuit diagram, or a person of one sex looking at a person of the opposite sex. We made some such observations on radiologists reading chest films to see what they actually did, but the results were not definitive at that time.

Color vision confers advantages, whether judging jaundice or more subtle changes. The figures in the frontispiece to this book were attempts to appeal to this sense, and they were able to demonstrate a shift in color associated with a shift in structure. More will be said of this in the final chapter. There have been suggestions that a fair range of colors can be seen when only two colors are in fact present; these ideas which apparently depend upon simultaneous color contrast are often associated with Edwin Land (1959). Indeed he suggested that, should much of the color range presently illuminating the earth be eliminated, we might still see objects in their true present colors (to which it should be added, "with eyes as they are presently evolved"). For a discussion of these interesting matters see Walls (1960). Two negatives of the frontispiece mouse were observed in such a color process, and this is the specific image of a "rat" to which he refers.

Medical images often contain detail that cannot be seen by the eye. Some further information on the visual process and properties of the eye will be given in the chapter on image processing, where one aspect will be to make the most of what information is available.

2.7. ULTRAVIOLET

Invisible frequencies beyond the blue end of the visible spectrum are called ultraviolet, and their spread and frequency are limited by absorption by the ozone in our upper atmosphere. Ultraviolet light causes tanning and sunburns, and also can cause skin cancer; the latter suggests that the energy per quantum is becoming comparable to chemical binding energies (see Fig. 2.1). Dermatologists are able to use ultraviolet light in some of their treatments. On the surface of the earth, sources of ultraviolet light include naked flames and electrical discharges. Such radiation is reasonably well passed by quartz but is largely absorbed by glass such as windows or camera lenses.* The ability of such radiation to produce visible fluorescence has application in fields ranging from entertainment to forensic medicine.

The human body is not hot enough to radiate much ultraviolet light, but such radiation can be shone on tissue for various purposes. Such wavelengths interact strongly with pigments. Thus they are used to demonstrate surface texture and pigmented skin lesions. Details of pigmented or depigmented areas almost invisible to the eye resolve crisply. Moles and freckles become more noticeable. A typical source of ultraviolet light is a mercury vapor lamp with a quartz tube.

In some cases the near ultraviolet returning to the camera is absorbed in order to record only the visible fluorescent light coming from the subject. Semen is somewhat fluorescent and ringworm is noticeable under ultraviolet light. Fluorescence photography is useful in bacteriology and various forms of microscopy. Certain surgical procedures are aided by the injection of fluorescein which then glows under ultraviolet light. Some microscopy samples can be observed without staining, not through fluorescent effects but because of selective absorptions in this range of frequencies.

Ultraviolet light photons have sufficient energy to break organic molecular bonds, which can cause damage in imaging but which allows laser surgery at wavelengths under 200 nm without major tissue heating or disturbance of adjacent tissue (Trokel, et al., 1983). At such wavelengths each photon has an energy of about 6.4 electron volts.

2.8. X-RAYS

The acceleration of a charge causes the release of electromagnetic radiation whose wavelength can be about the size of atomic dimensions (Fig. 2.1). One way of producing such radiation is to start electrons moving in an evacuated tube and then suddenly bring them to rest by slamming them against a metal target. In X-ray tubes this is typically done by accelerating the electrons through a voltage of a few thousand to a few hundred thousand volts. Of course, there are other ways to produce X-radiation. The acceleration of the curved path in many charged-particle accelerators causes the particles to radiate energy in a way considered wasteful in one sense but which provides a useful radiation source. The sun gives off X-rays which are absorbed by our atmosphere.

Unlike visible radiation with which we see and to which the majority of objects are usefully opaque, all materials are somewhat transparent to these Roentgen rays or X-rays. Higher energies or shorter wavelengths generally show greater

---

*At the time of this writing, the transparency of plastics to near ultraviolet (between the blue and the actinic or tanning wavelengths) creates a problem of potential retinal damage, not universally recognized, for those having a plastic lens replacement after a cataract operation. Similarly, some dark glasses open the pupil and pass more harmful rays.

penetrating power of X-rays, though there are some irregularities to this generality which allow for specific analyses of composition to be discussed in the chapter on spectral information.

Various thicknesses or densities of material traversed will differently attenuate the passing X-rays, thus producing an invisible pattern which can be converted into a visible pattern by placing a fluorescent screen or photographic film at a suitable place. Most people in our modern society are somewhat familar with X-ray pictures, but it seems appropriate at this point to include an example. The example is of a procedure for obtaining biological information from complex systems with minimum disturbance that is supplementary to many of these imaging techniques: it is the use of tiny radio transmitters to sense and telemeter information after suitable placement in the body (Mackay, 1970). In Figure 2.4 is seen a radiograph of a snake (boa) that has swallowed a mouse within which is a telemetry transmitter of pressure and temperature. This not only yields information about peristalsis in cold blooded animals, as well as the pressures of eating or being eaten, but it also gives information about attenuation of radio signals in traversing the body, which will be relevant in some of our later considerations of the limits of nuclear magnetic resonance imaging. It might be mentioned that this almost became a four-layer picture as a cheetah bounded into the room where we were making this radiograph. Alternative techniques for producing and improving X-ray images will, of course, be included in the remaining chapters, one of which is devoted exclusively to X-rays.

We close this brief section by noting two wavelength-related facts. Since the wavelength is very short, X-rays are appropriate for exploring molecular structure, it being especially convenient in the regularity of a crystal. A discussion of the possibility of using coherent X-rays to form holograms of minute biological objects for later three-dimensional scrutiny has been given (Solem and Baldwin, 1982). However, in most cases the fineness of detail found in an X-ray image is limited to a much poorer value than a wavelength by technical limitations of the machine producing the image, or else by random fluctuations in the number of photons traversing the subject itself. This effect cannot be safely reduced indefinitely by increasing the duration of observation or the intensity of beam because of another wavelength-related factor of safety. The shorter the wavelength the higher the frequency and energy associated with each quantum (Fig. 2.1). These energies are sufficient to modify molecules in the body, that is, this is ionizing radiation. Since dosage of ionizing radiation has literally become emotional to the extent of becoming a presidential campaign issue, some people unfortunately refuse to have necessary X-ray examinations made, sometimes preferring the vastly greater risk of a dangerous ailment.

2.9. RADIOACTIVITY AND NUCLEAR MEDICINE

Radioactivity involves charged particles (alpha and beta rays) which do not travel far through tissue, and gamma rays which are the same type of thing as X-rays.

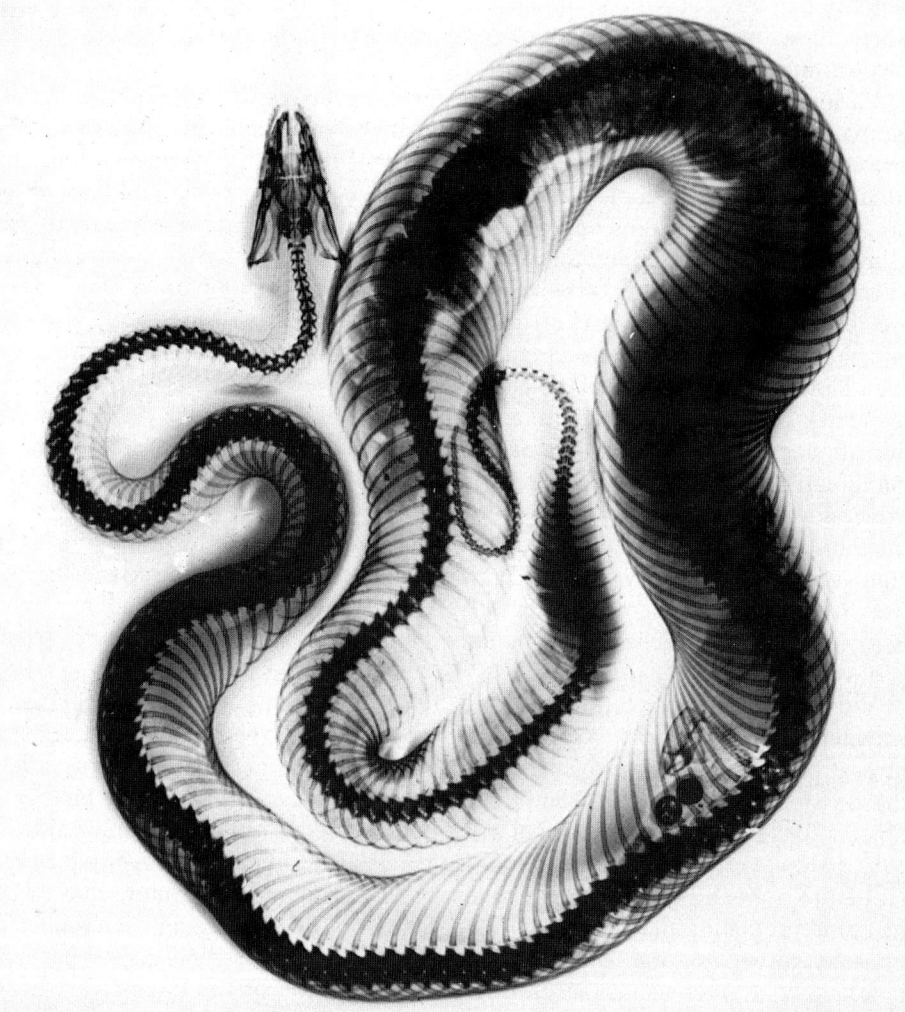

FIGURE 2.4. An X-ray image. The boa has swallowed a mouse containing a radio transmitter that senses and telemeters pressure and temperature, allowing the undisturbed monitoring of peristalsis in this cold blooded animal, as well as the pressure of eating or being eaten.

Gamma ray energies span about the same range as the emanation from an X-ray tube, and indeed a potent radioactive source is sometimes used as a substitute for an X-ray machine. Radioactive materials behave chemically about the same as their nonradioactive counterparts and disperse in the body in a similar way. Many sorts of tracer experiments (Hevesy, 1923; Chiervitz and Hevesy, 1935) became possible with the development of the cyclotron by Lawrence around 1936, which allowed the production of many elements in radioactive form. Autoradiographs of tissue sections can even be made with such fine detail by coating the section with photographic emulsion that it is appropriate to examine the result under an electron microscope (Rogers, 1973).

The production and handling of radioactive materials is a rather broad and special topic that has been extensively covered elsewhere, and so it is totally omitted from this volume. Nuclear medicine makes a definite contribution in most hospitals, though its imaging role apparently has diminished somewhat with the advent of X-ray CT scanners, with or without new tracers, which are to be discussed later. The activity remains especially vigorous in physiological observations of time sequences. We here mention gamma cameras and positron emission tomography, with the inclusion of a clinically important example.

The position of a radioactive element that has migrated to some site in the body as a result of a normal or abnormal physiological process can be determined in several ways. Assuming it to be a gamma emitter whose emanations can penetrate the body, one could place near the body a lead sheet containing a pinhole (or other collimating aperture) and scan the region beyond with a Geiger counter. At different places different counting rates would be observed and recorded but the process would be slow or the radiation dose high because while the counter is at one position, all other trajectories are being lost. A photographic film could record information from all positions at once, but greater sensitivity would be desired. Many gamma cameras use some modification of a scheme originated by Anger (1958, 1967). Essentially a large scintillator crystal that glows where struck by a gamma ray is placed in the region of space to be monitored, and flashes are recorded by sensitive photomultiplier tubes. Several detectors are placed against the crystal in different positions. A spot of light signifying the passage of a ray will be closer to some than to others, and the ratio of their impulse heights is sufficient to localize the spot within the crystal for electronic recording and display. Pictures made with a modern version of such a camera follow.

Tumors can metastasize to bones, and be observed by radioactive tracers, while it is impractical to search all of the bones in the body by X-ray. Perhaps the first images of radioactivity in bones were those of radium workers who licked the tips of their brushes while painting the luminous hands on watches. Suitable technetium-99m compounds will concentrate in any region of active or metabolically disturbed bone, where the gamma emissions will be noticeable in an image. In Figure 2.5 is seen the scan of the entire body, 2 hours after administering 12 millicuries to a woman whose breast cancer had metastasized to her spine, producing the intensified spot seen here. Scan time was 30 minutes to collect 300,000 counts. For finer detail, rather than using a scanning procedure, a pinhole or collimated image is projected on the camera without any movement, which produces an image such as seen in Figure 2.6. The bright spot on the spine is to be seen here if the photograph is not held too close. (More will be said in later chapters about grouping more points in a noisy picture into a small visual area by moving away in order to average adjacent areas in such a way as to better allow estimation of darkness differences.) For comparison, Figure 2.7 presents a standard X-ray picture in which no lesion is seen, even in retrospect, in the second lumbar vertebra where the problem was localized. The evidence does not appear

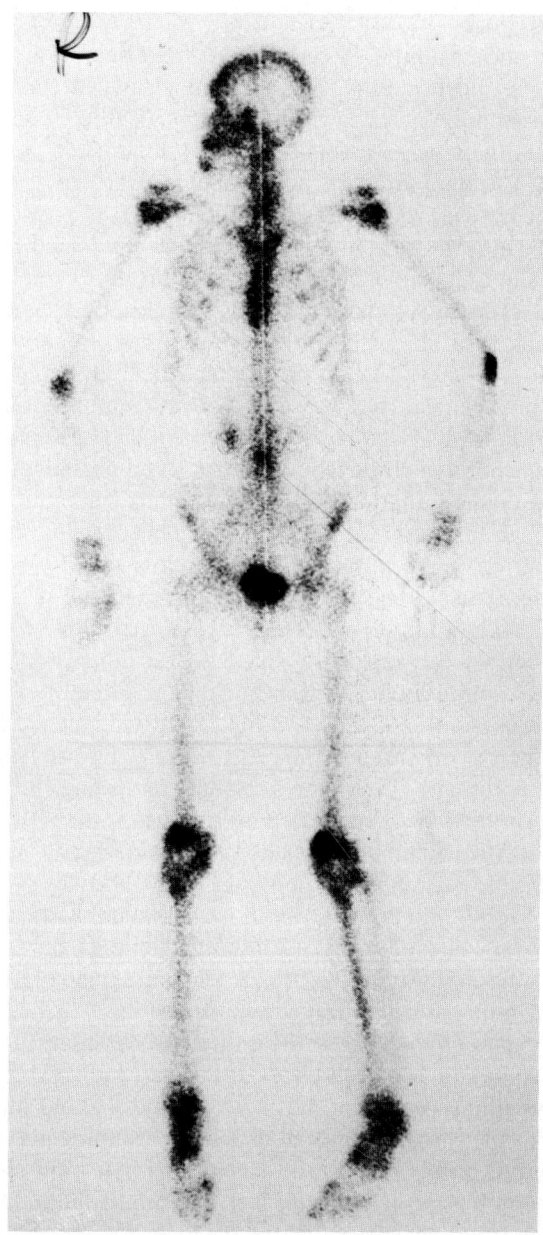

FIGURE 2.5. Scanning image of the radioactivity distribution in the bones of a woman. Slightly increased uptake is noted in the upper lumbar spine, approximately in L2. This image is from the front while the next figure is the same subject from behind.

RADIOACTIVITY AND NUCLEAR MEDICINE 21

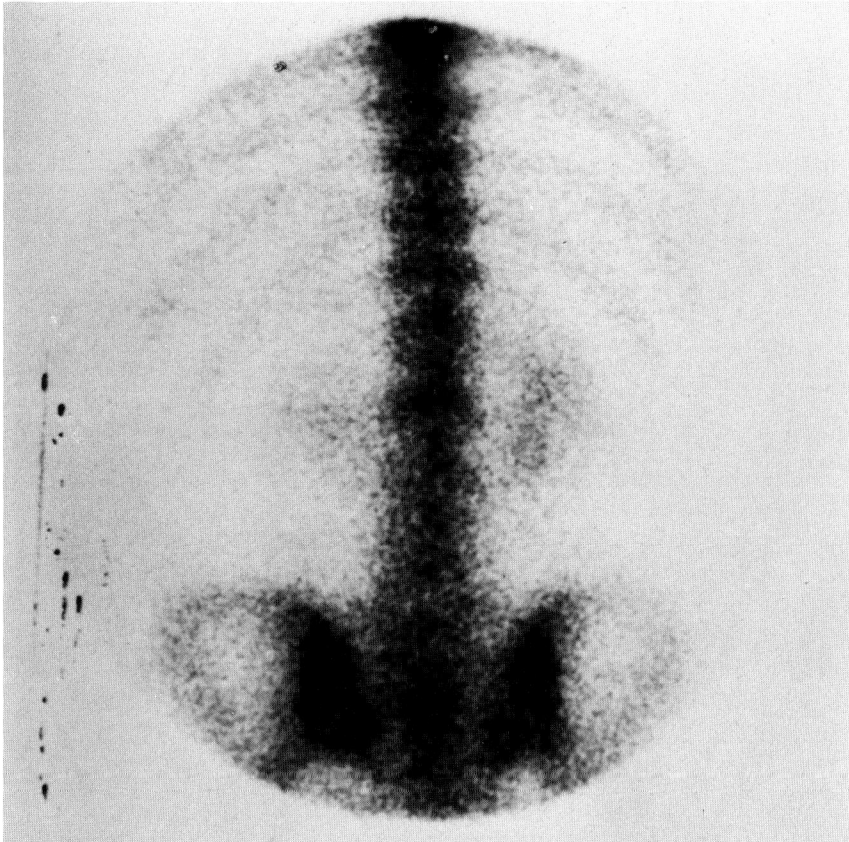

FIGURE 2.6. An image with stationary camera involving about 100,000 counts collected in 10 minutes.

in an ordinary X-ray picture for several weeks. However, at the time of the previous radiograph the lesion could be seen as a black spot in an X-ray CT scan as at the upper right in Figure 2.8. At the present time the medical literature does not clearly indicate if bone lesions are first detectable in the radioactive scans or in X-ray CT images, though a guess would be that they are first noticeable in a radioactive scan.

Some radioactive disintegrations release positrons. These do not travel far in the body before they join with an oppositely charged electron. The resulting annihilation releases two gamma rays that travel in almost exactly opposite directions approximately from the site of the original disintegration. If a ring of many detectors is placed around a subject then a pair of rays originating within the plane of the ring and traveling in that plane will activate a pair of counters at the same

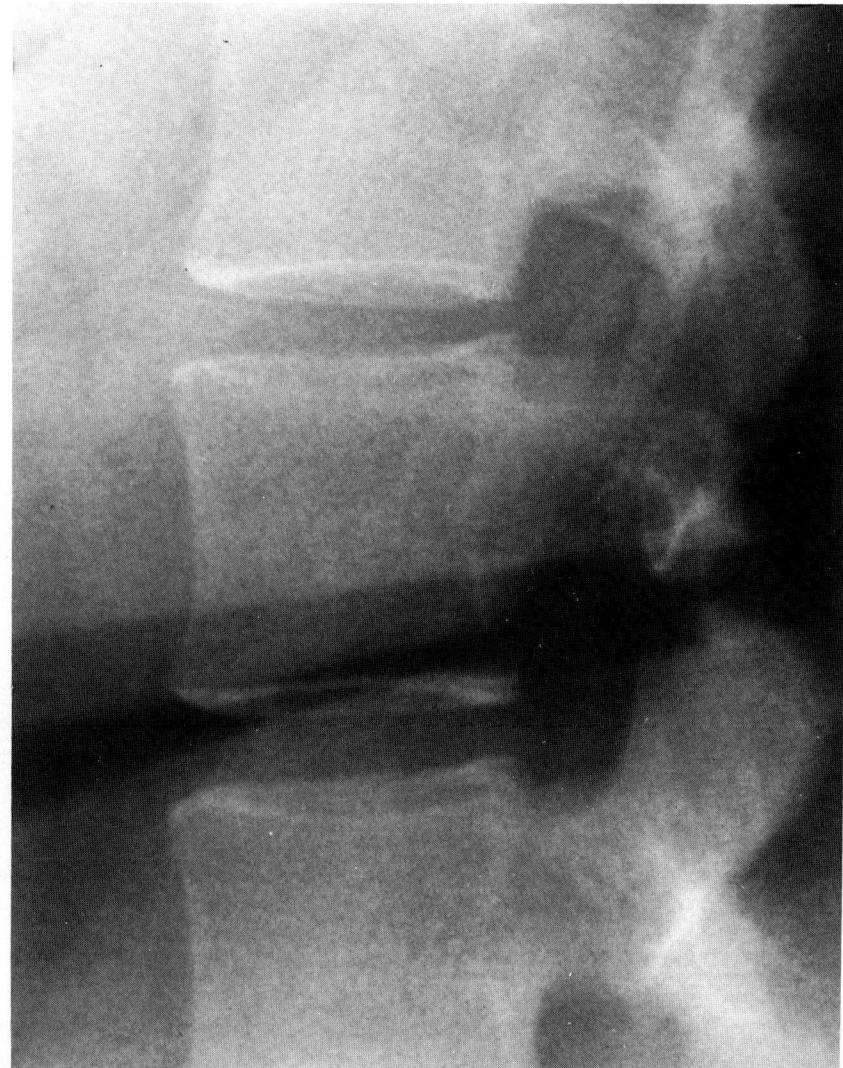

FIGURE 2.7. The second lumbar vertebra at the center shows no lesion in a standard radiograph, the subject being the same as in the previous two figures.

time. Electronic recording of this coincidence (Brownell and Sweet, 1953) determines one line through the source of that radioactive disintegration (Fig. 2.9). Events in the plane whose products go out of the plane are lost, but any recorded coincidence represents an event in the plane. Recording of all such pairs of events produces a map in the plane of the ring of all the relative intensities of the various positron emitter concentrations. Some modern circuits are fast enough to measure the small difference of arrival times at two detectors due to a difference in position

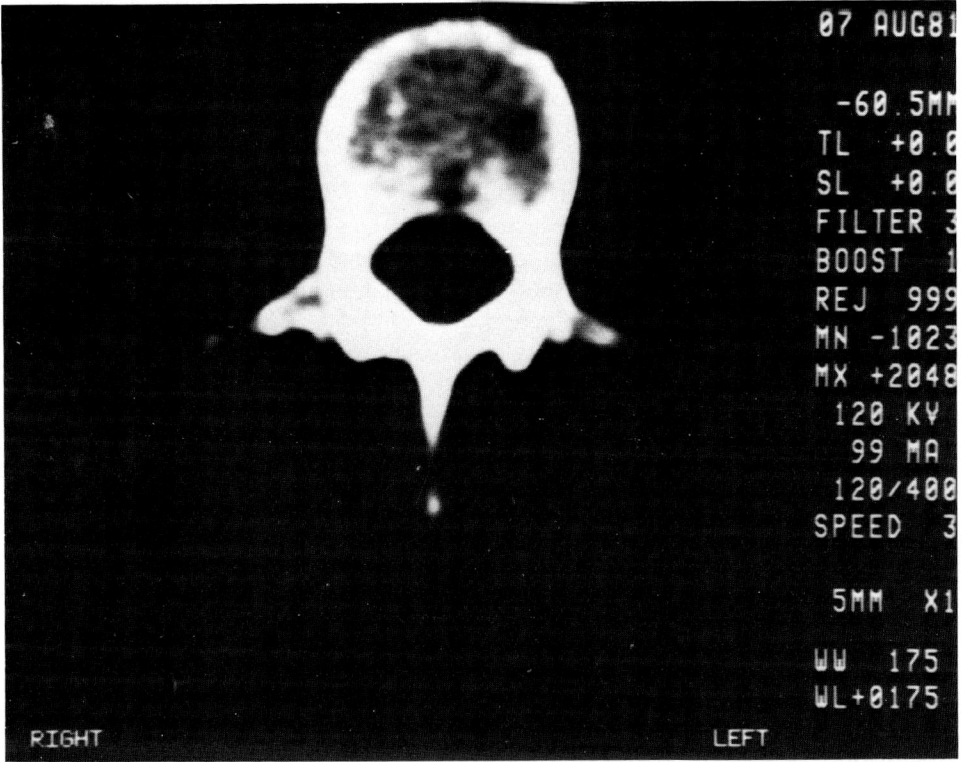

FIGURE 2.8. On a computed tomography image the black spot at the upper right in the spine shows a lesion not seen in Figure 2.7.

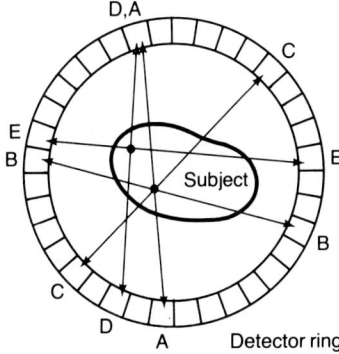

FIGURE 2.9. In positron emission tomography, the subject extends through a ring of detectors. A gamma ray pair from a positron emitter forms an approximately straight line connecting two detectors at one instant. Each such line is stored in a computer, with the intersections denoting regions of radioactive concentration. A cross section image in the plane of the ring is produced. If each line were merely displayed on an oscilloscope to a camera with open shutter, the pattern that would build up on the film would be of this sort, though it would be somewhat blurred and would require image sharpening in ways to be discussed later.

along the line connecting them, and later units will incorporate this information to further specify the disintegration position. The process is rapid and gives a cross sectional image or tomogram showing the distribution of the chemical into which the radioactivity was incorporated. Some useful isotopes for studying biological systems are oxygen-15, nitrogen-13, and carbon-11; with respective half-lives of 2, 10, and 20 minutes. Positron emission tomography (called PET) has had some interesting research applications in studying brain function and also has some clinical application, though an associated cyclotron can be required for isotope preparation. Resolution of a centimeter is obtained in under a minute. For a summary article on this method see Ter-Pogossian et al. (1980), and Schwartz et al. (1983) for an example of application.

It is possible to use radioactivity as a source of radiation for imaging in a different way called Compton radiography (Lale, 1959). Gamma rays from a monochromatic source such as ^{99}Tc are directed into the body, and at right angles to the incident line are observed the lesser energy photons that have scattered out. For example, a ring source around the head at 140 keV will result in an upward directed pattern of photons of 109 keV that can be directed by a collimator onto an Anger camera for viewing of an image that is a map of electron density in the brain in the irradiated plane. Various combinations of collimators can also be used in different scan orientations to build up two- or three-dimensional images with a density discrimination of a few percent, in all cases taking advantage of energy discrimination, which is less practical in transmission X-ray images involving the rapid arrival of many quanta having a spread of energies to start with.

2.10. COSMIC RAYS

Some of the highest energies are produced in nature rather than in laboratories, and in a sense cosmic rays are simply mentioned here to indicate the end of the frequency scale. Radiation incoming to the earth consists in part of various charged particles which, because of their energy, are very penetrating. Imaging can be done with them. For example, Alvarez et al. (1970) used the natural cosmic ray muon flux to search for the possible existence and position of an unknown chamber in the great pyramid of Chephren outside Cairo. They produced their "X-ray" image by placing a spark chamber in the already found Belzoni chamber to record the direction and intensity of the penetrating rays. They were able to detect the cap on the pyramid through 100 m of limestone and to visualize the corners, but they found no unknown chamber. This application goes beyond the largest animals and is the realm of what engineers doing similar things would term "nondestructive testing."

Again there is conversion of an invisible distribution into a visible display, though it has been reported that the human eye can detect the passage of the mu meson component of cosmic rays (D'Arcy and Porter, 1962).

2.11. ADDITIONAL MODALITIES

Sound and Ultrasound

Sound waves are a mechanical disturbance, not a combination of electric and magnetic fields such as were involved in the previous modalities. If a person is punched in front their back will shake, thus suggesting that sound waves penetrate tissue. One can listen to the beating of a heart through the wall of the chest, thus supplying more specific evidence. Sound does not readily pass in either direction between dense media and air, for example, between air and water or tissue, which has implications for the examination of the lungs and bowel, and for passage of sound into and out of the body. Perhaps the first attempt to couple sound from a dense medium into the ear was made by Leonardo da Vinci who extended a hollow tube into the water and through it heard distant ships. A stethoscope is also somewhat of an impedance matching device. Percussion can outline the lungs and heart to give a low frequency low detail map of the chest, as can auscultation or passive listening through a stethoscope. In the last three paragraphs of the first of his three classic papers on X-rays, Roentgen (1895) speculated about the possibility that his new rays were longitudinal vibrations, but he apparently did not have sound waves in mind. The sonar systems of World War II introduced the ideas of explorations of surroundings by high frequency sounds, and the associated technology development was helpful for the modern body-exploring methods to be discussed here.

The human ear can hear frequencies up to about 20 kHz while some porpoises can hear up to 150 kHz. Surprisingly, one can ''hear'' frequencies higher than the frequency limit of hearing. If sound at a frequency of 100 kHz is projected into the human lower jaw one will perceive a tone that is the highest frequency one hears normally, apparently by direct stimulation of the cochlea (Deatherage et al., 1954). Similarly, some seals can tell if a frequency of 100 kHz is present, and find the direction of the source, but cannot tell if there is a difference in two frequencies higher than 62 kHz (Møhl, 1967). The lower jaw of a porpoise is a good receiver of high frequency sounds (though for them the ear stands still and the body shakes in response to sound). Detailed maps of regions of sound sensitivity over the body have been made (McCormick et al., 1979). This is another sort of medical image, usually prepared for fish or whales, since entry of sound into their bodies from water is generally less localized.

The generation of intense sounds can be wearing and their application can be damaging. Loud audible-range sounds have some hazard; Quasimodo, the hunchback who allegedly rang the big bell in Notre Dame, had to be deaf. Higher frequencies, once they have gained entrance to the body, would be expected to interact with finer detail but not to penetrate as far. The velocity of sound in soft tissue is such that at a frequency of 1.0 MHz, the wavelength is about 1.5 mm. If a subject is scanned with a beam of some such frequency and the various echoes are collected and displayed then a cross sectional image of the body is produced. The

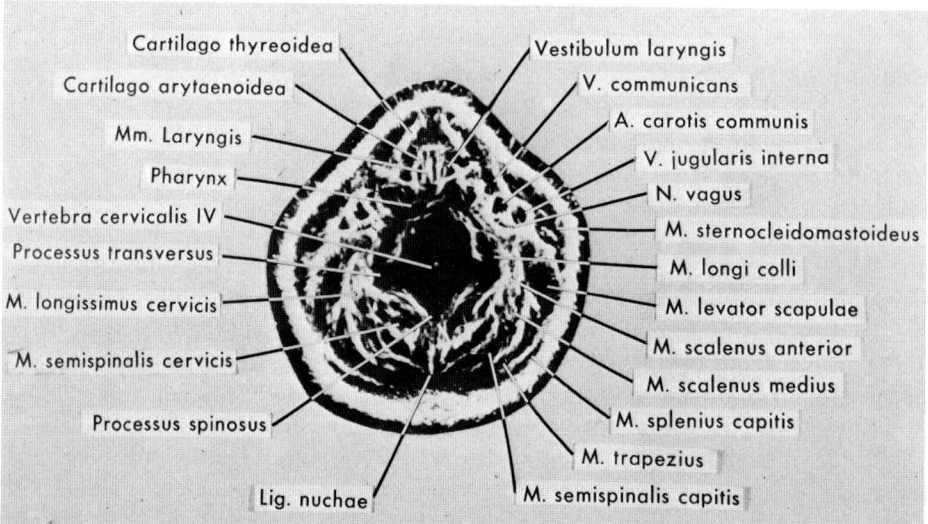

FIGURE 2.10. Pulsed ultrasonic systems can show the details of even densely packed structures such as the human neck. Soft structures are well visualized and movements can be followed. This is an early picture by D. Howry using a compound scan, the subject being immersed in a tank of saline. Sound enters the vocal folds through their outer edges rather than traversing air.

picture in Figure 2.10 was given to the author about 1960 by Douglas Howry and shows the detail that can be seen even in a densely packed structure such as the human neck. This image was made with Howry's compound scan which approximates the viewing of every point from every direction so as not to miss specular reflectors that may return the pulsed high frequency energy in some direction other than back to the receiving transducer. The early studies suggested that pulse reflection techniques (French et al., 1950; Wild and Reid, 1952; Howry and Bliss, 1952) gave better results than transmission techniques (Dussik, 1942). Howry did not write extensively, but he did compile an early atlas of ultrasonic results (1965). At useful intensities the high frequency sound seems free of harmful effects including ionization, though at higher intensity it can even produce brain surgery by its effects on tissue. This general topic is of sufficient importance so that an entire later chapter is devoted to it.

Corona Discharge Photography

There is a type of image that warrants mention because of its wide publicity in recent years. This is the so called Kirlian photography, sometimes referred to as the recording of auras.

In 1777 it was discovered that dust settling on a cake of resin formed a finely detailed starlike pattern if the resin had been sparked (Lichtenberg, 1777). (This would seem to be an interesting relative to the xerox process.) These Lichtenberg figures also can be developed on a photographic emulsion to which a surge of electricity (e.g., 10 kilovolts) is applied from a step up transformer such as an automobile ignition coil through whose primary is discharged a capacitor. The usual arrangement separates the emulsion from a metal plate by a sheet of glass and applies the voltage to an electrode contacting the other side of the emulsion. Useful applications of the effect had been made in surge recorders called klydonographs. Positive and negative voltages produce different appearances, and pattern changes near electrode pairs allowed recording time differences of about 10^{-8} second. In some experiments the exposure of the emulsion was found to be produced by light associated with the electric discharge rather than by direct charge bombardment (Merrill and von Hipple, 1939), though bending of the streamers in a magnetic field indicated the role of electrons. They also studied modification of gas composition, and it is clear that gases adsorbed on the surfaces could affect the patterns. (See also Fig. 10.12, upper right.)

Similar recordings can easily be made in the same way with a living object replacing the electrode, being certain that the subject is grounded and the surge is applied to the plate. Patterns are said to change with the health and even the psychic state of the subject. It has been shown that moisture is a principal determinant of the form and color of Kirlian photographs of human subjects (Pehek et al., 1976). Since it is common experience that things ranging from emotions to anticholinergic drugs can influence dryness of the mouth, it is not surprising that there would be changes around hands, for example, associated with a variety of conditions. Presently there has been little accepted medical application for these effects.

2.12. NEAR FIELD VERSUS FAR FIELD

There is another frequency dependent aspect of practical importance in manipulating both electromagnetic and sound or pressure fields. Near to a source of sound or of electromagnetic radiation the dominant conditions are different than far away, where "near" and "far" are with respect to a distance of a wavelength and thus depend on frequency. Shaping either the magnetic field or the low frequency radio field in a nuclear magnetic resonance imaging situation, for example, has limits that can be considered in these terms. Another specific example is given in the chapter on ultrasound where there is a relevant consideration in the use of transducers for generation or detection. A few general rather "loose" comments are given here.

For some people the terms "near field" and "far field" mean which mathematical terms are included in the description of diffraction (Fresnel vs. Fraunhofer diffraction), but examples of physical representations can be given. A simple case is the vibration back and forth of a small sphere in a fluid. Two effects are produced: (1) fluid flows from in front of the advancing sphere around to the side, backward, and in behind with a speed and direction that depends upon the path direction and speed of the sphere; (2) a pressure wave propagates outward with the speed of sound. Close to the source the most noticeable motion is the flow, while far away the compression wave particle displacements are the more noticeable. The two effects become equal at a distance of $1/2\pi$ times the wavelength λ. Inside this distance, in the near field, displacement diminishes with the cube of the distance and the direction of motion observed at a point depends upon the orientation of the vibrator. In the far field one has pressure diminishing with the first power of distance, and the wave progressing radially outward in most directions though more strongly along the vibration axis than perpendicular to it, with none radiating just perpendicular. By contrast, the near field has a component everywhere.

The same considerations apply to the electromagnetic case. As an example consider an alternating current applied to a small coil. Near the coil will be magnetic lines of force in the familiar pattern seen where iron filings are sprinkled around a permanent magnet or coil. The field starts strong but drops relatively rapidly with the cube of the distance. Beyond $\lambda/2\pi$ one instead notices that lines of force loop around on themselves independent of the source, and one has radio waves propagating outward at the speed of light. They travel until they eventually reach some obstacle or detector. Here field strength in the wave has become noticeable since it decreases more slowly than the other contribution, that is, with 1/distance, as with sound above; the energy in this spreading wave drops with the familiar square of the distance, being proportional to the square of the field strength. Observation of equipotential lines with a directional antenna indicates direction to the source in this far field, no matter how the source reorients. Associated with a changing magnetic field will always be an electric field, and a changing electric field has a magnetic field, so these waves always have two components acting together.*

Aspects of detection and control of pattern shape are different within and beyond the transition distance, which is not really abrupt. Practical differences in the electromagnetic case have been tabulated elsewhere (Mackay, 1970). Steady magnetic fields or electric current paths have near field forms that resemble the patterns of iron filings around permanent magnets, and can be modified or controlled to the same extent. Some simple cases are well known. Thus the field around a small permanent magnet decreases with the cube of the distance and is twice as large on the axis as to the side the same distance (just as the earth's magnetic field is about twice as large at the poles as at the equator). Modeling for visualization of more complex cases can be done by observing the flow paths of water between "sources" and "sinks" (Moore, 1949).

The vibrating sphere above was assumed to be very small with respect to a wavelength, and a wave was radiated in many directions. If the frequency became very high so the wavelength became tiny with respect to the sphere then the sphere might better be considered as an array of many points (smaller spheres)

---

*In a partially conducting medium such as ocean water or tissue, both attenuation and velocity of electrical disturbances depend on frequency, with attenuation being roughly 50 dB per wavelength (Mackay, 1970). Comparing sound and electromagnetic waves in the ocean indicates that in these respects a 1 Hz electromagnetic signal is like a 100 kHz acoustic signal.

moving and radiating in synchronism. The sum of contributions can be largely in one direction just as some complex radio antenna combinations can concentrate transmission or reception in one direction. If the source is an array of individual small sources then the pattern of current flow and interference between regions can be perceived at a greater distance than can the pattern around (close to) any one component dipole of the above sort. One can visualize currents flowing between and extending outward from the group of radiators to a greater distance than from one alone so the transition distance from near to far field can depend upon source size as well as on wavelength. A specific example is given for the ultrasound case as in Figure 6.1. Roughly speaking, the transition from near to far field is where one ceases to be able to recognize the outline of a source on a diffusing screen, whether one is considering the electrocardiogram pattern at different positions on the chest, the pattern on a frosted glass from a light, or the sound distribution in front of a loudspeaker.

2.13. EMANATION VERSUS TRAVERSAL AND SUMMARY

Some emphasis has been given to the fact that in some cases a given modality is applied to the body to explore it and emerges either after traversing the body or after transmission to some depth followed by reflection back from some internal structure, while in other cases the same modality may simply emerge from an internal structure either in the normal course of its functioning or because of selective uptake by some marker in the course of functioning. Thus one can tap out the margins of the heart, lungs, or liver while listening or, at least with the heart, one can simply passively listen at different positions in order to determine abnormalities. One can apply exploring electric currents or one can observe the currents that evolve spontaneously during functioning; one can apply infrared energy or observe the distribution of heat energy being released. Transmission of X-rays through the body to form an image has been compared with ingesting a radioactive material which will go to a selected type of site and mark it by radiation similar to X-rays. Nuclear magnetic resonance imaging is like employing a radioactive tracer that can be turned on at will, with a half-life that depends upon the surroundings, and which has a remotely adjustable frequency or energy.

Generalities are difficult but it can be said that radiation applied from outside can be controlled in momentary position and direction while internal emanation tends to be in all directions. Lenses and mirrors are suitable for dealing with the latter situation if they exist. The various mirrors, lenses, or slits in imaging systems must usually be larger than a wavelength, and their size will often limit the blur or resolving power of the entire system.

Even when the entering radiation is under control, scattering can degrade an image as when attempting to see through fog or the ocean. In the classic case of X-rays, an opaque metal grid projecting from in front of the detector has all its elements directed along lines converging to the source so that any photon going to the side and being scattered back to the detector from an abnormal direction will be obstructed; such a process works equally well in a scanning system with slits or a parallel system where the image is formed overall at once. Even the position of material causing blurring can affect the outcome as is indicated by analogy with the common observation that when looking at a picture, it is seen clearly if a piece of frosted or nonglare glass is placed in contact with the picture but is obscured if the same piece of frosted glass is placed against the face.

The motion of electrons produces photons, that is, electromagnetic radiation ranging from radio waves to X-rays. When the charge is accelerated there results an electric field component called the radiation field that decreases with only the first power of distance. Photons are also generated when electrons in atoms undergo transitions. Much of other interaction with matter of such electromagnetic radiation is with electrons, though there are exceptions. In the X-ray region of the spectrum interaction is with the innermost electrons of atoms, in shells of different energy. Because of their location these electrons are shielded from disturbance by other atoms, making sharp regions of absorption possible even in solids. The far ultraviolet shows a broad absorption in dielectrics due to the outer unshielded electrons in atoms of solids or liquids. Infrared absorption bands represent natural frequencies of atoms or even molecules as a whole, which are heavy and thus have lower frequencies. In gases the frequency of rotation of whole molecules may be observed. All substances possess absorption bands somewhere within the electromagnetic frequency spectrum. Near these, velocity changes somewhat with frequency.

Photons in the visible range are energetic enough to produce effects on a distinctly individual basis, and in weak beams granularity becomes evident. Photon energies in the ultraviolet are of the order of magnitude of many chemical reactions which thus can be triggered by them. Production of ionization is possible at energies upward of a few electron volts for each quantum, thus making X-rays different from radio in potential for damage, though at high intensity the latter can cause damage by heating and perhaps other effects. At X-ray frequencies the photon energies are high enough to make their interactions with matter distinctly granular.

Low frequency fields go from one coil to the next, just as electric currents go from plus to minus, or flow near a vibrator goes from high to low pressure. When an emanation extends from a source to a sink it cannot arbitrarily be shaped or guided in between by geometry changes at the ends, that is, shaped from outside a region of interest being probed. In the far field or radiation field where emanations act detached from their source with no necessary destination, narrow rays can be formed if the source size is greater than a wavelength, or focusing and images are possible in the usual sense.

There are also some significant similarities, including thermal effects, among the many wavelengths. Extensive exposure to any modality from radio to x-rays can lead to cataract formation in the eye, though that effect probably has differing mechanisms.

The next chapter will discuss some aspects of imaging that apply to all modalities.

3
IMAGE FORMATION

3.1. SCAN VERSUS PARALLEL PROCESSING

Sequential measurements at different positions are often not as fast as those made simultaneously, nor do they allow as accurate comparisons during overall changes. For example, a pair of electrodes can be moved or scanned from place to place on the body to build up an impression of the distribution of electrical activity by using a single recording channel, but it is clearly faster and better to have many electrodes and recorders acting at once (in parallel) for simultaneous recording from separated regions. Slowness does not imply lack of detail, and an array of electrodes can actually cancel or reject bioelectric signals from a specified depth to augment lateral information in this example. Computed tomography gave increased X-ray accuracy, making it possible to compare adjacent regions more subtly, but at the cost initially of monitoring one ray at a time with a mechanical scan. Modalities of relatively high frequency often are handled with many channels or rays of information at a time, an example being the formation of an optical image with a concave mirror.

The image-forming methods of Figure 3.1 are each applicable to a number of modalities. The imaging properties of a pinhole are probably familiar, having been known for centuries in such things as the camera obscura. Light from any one point on the subject can illuminate only one point on the image screen, other light simply being blocked. Though there is a somewhat best distance for a given pinhole in a pinhole camera, an image forms over a large range of object and film distances. Application is not limited to light. If, for example, the pinhole is in a sheet of lead then an image of radioactivity or of an X-ray distribution can be formed.

32 IMAGE FORMATION

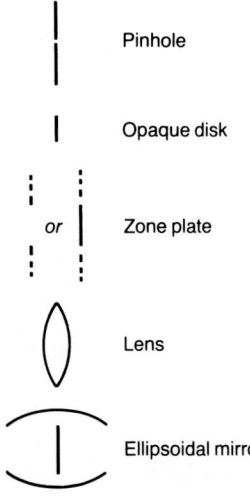

FIGURE 3.1. Systems in which radiation from a point on a subject to the left is restricted to a point on an image that forms to the right. Under proper conditions these will function with ultrasound or X-rays and other modalities, as well as with visible light. Multiple or intersecting rays can be focused at one time allowing the image to form overall simultaneously.

The second drawing of Figure 3.1 recalls that at the center of the shadow of a circular obstacle can be seen a bright spot, sometimes called Arago's bright spot.* An obstacle placed before a sound detector has been used to locate the direction of a sound source in the ocean, using an analogous technique. An extended source of many points will give an extended image having corresponding variations in brightness. The first two methods block out or reject much of the incoming radiation to give a dim image, but a concentric pattern of obstacles and openings can give a much increased intensity (Jenkins and White, 1937).

The bottom three lines of Figure 3.1 depict systems that collect radiation over an appreciable solid angle and thus give bright or intense images. The first of these consists of an alternating series of opaque and transparent annuli with radii proportional to the square roots of whole numbers to form what is called a zone plate. Some parts of a wave front approaching a point on the right would interfere or cancel the effect of other parts of the same wave. A zone plate blocks out those parts of an extended wave that would tend to cancel the other parts, and the center can be either transparent or opaque to determine which phase is accepted and which is rejected. A zone plate has many of the properties of a lens, including an effective focal length for rays coming from far away, and radiation from a given distance will come to focus at a particular place or places.† (Holograms, to be mentioned later, can be considered as the superposition on a photographic film of many zone plates, each being formed by one point on the subject so that in viewing the hologram the many zone plates give many points of light corre-

---

*Such a bright spot can be demonstrated in the shadow of a penny held a few meters from an automobile taillight bulb or an arc light if viewing is with a magnifying glass.
†Fresnel lenses, which in thin plastic form have become common household items in recent years, should not be confused with zone plates, though they have a pattern of tilted fine grooves in concentric circles.

sponding to the many points on the original subject; no lenses need be involved in the formation of the diffraction pattern called a hologram, whatever the modality employed.)

A converging lens for light is convex since the velocity of light in glass is less than that in air, and such a curvature thus brings all rays diverging from a point back in phase at a corresponding point in the image. Lenses for high frequency sound are often made of plastic and submerged in water or other tissue-like material. Since the velocity of sound in plastic is greater than that in water, such a converging lens for sound will be concave. (An example of this is to be seen in Chapter 6.) Beams of charged particles also can be focused by curved electric or magnetic fields. X-rays are essentially undeviated by passage through matter and thus it is not practical to make lenses for them.

Concave mirrors also can collect and focus radiation to a point whether the radiation be radio, sound, or invisible light. X-rays hitting a surface approximately perpendicular will not be reflected back and so the usual sorts of searchlight mirror do not exist for X-rays. However, mirrors for X-rays can be made in a different way. Consider an elliptical "football" silvered inside: a light inside near one end will be focused near the other end by reflecting from all points, and the two ends can be removed to leave the center shell, which is the surface shown in longitudinal section at the bottom of Figure 3.1. Both focii of the ellipse are outside the included region, and a disk at the center of the tube can block any direct or unfocused rays. Modern microinch lathe methods using a diamond tool can fabricate such a barrel-shaped mirror. Several coaxial reflectors can be placed inside each other to accept a greater expanse of rays, all at grazing incidence. This last configuration actually can function as a converging mirror for X-rays as well as other radiation. X-rays display a refractive index very close to unity and real, but careful refraction measurements show it to be slightly less than unity.* According to classical theory this is because the refracting medium contains electrons that have their own natural frequencies of vibration and, unlike the situation with visible light, the frequency of X-rays is much higher than that for most of these electrons. This unfamiliar circumstance implies that when X-rays pass from air into any material the beam will be bent away from the normal to the surface, as was observed by Siegbahn. Thus at grazing incidence an X-ray beam in air will be totally internally reflected upon contacting a solid, as was demonstrated by A. H. Compton; with "soft" radiation (relatively low frequencies) the effect is less distinct because of absorption effects (Nahring, 1930 vs. Valouch, 1930). In any case, X-rays can be focused by grazing reflection (Underwood, 1978), though not

---

*This means a velocity greater than the velocity of light. Classic formalism defends this by saying the limiting velocity of relativity theory is the velocity of signals or energy which is group velocity and is always less than "c" even if the wave or phase velocity in a medium is greater. Phase velocity may be greater than free space velocity c, and the product of phase and group velocity is c^2 under some conditions where velocity changes with wavelength. Dimensions of a standing wave pattern give phase velocity while timing transmission of a pulse gives group velocity (as does introducing modulation in the prior case).

well with lenses, and it is the method used in the Einstein Observatory satellite to record the X-radiation from hot astronomical events.

Related to a point aperture or pinhole is a point source of radiation placed on the opposite side of the subject from the detector. This shadow casting of objects by a point source of radiation is the traditional way of producing a radiograph. Looking at the uniform sky through a pinhole will allow one to see debris floating in the back chamber of one's own eye (entopic perception) by this same mechanism (a sudden increase in "floaters" perhaps indicating partial retinal detachment); things here will appear inverted by the action of the visual system since the lens of the eye will not contribute its usual inversion. Just as a pinhole can be enlarged into a patterned aperture to give a blurred image requiring reconstruction for proper viewing, so an extended source casts an image in which each point on a subject is represented by the pattern of the source; manipulation needed for viewing will be mentioned in Section 3.3. In all these cases, many rays can simultaneously arrive on an array of detectors from the multiple points in the subject.

The human visual system is able to function rapidly because its many slow components all act in parallel. All the light rays entering the eye from an appreciable distance are focused on the retina where separate receptors simultaneously monitor different regions of the visual field. It would be a much slower process if a single receptor were scanned from place to place at the focus of the lens in order to gradually build up a pattern of relative intensities. Occasionally an alternative form of scanning slows human vision as when a person walks into a darkened room and gradually builds up an image of the far wall by scanning its surface region by region with the small beam of a flashlight. In later sections we will see examples where different modalities are used sometimes with a scanned source, sometimes with a scanned detector, and sometimes with a broad source and a detector that records the overall pattern all at once as on a photographic film. It might be mentioned here that the situation with modern television camera tubes is that information is everywhere stored continuously in the image and its sudden removal is by a rapid scanning process which does not slow the acquisition of information by interfering with the storage that is taking place without interruption.

Ultrasonic systems often scan to form an image, some by mechanical movement and some not. Nuclear resonance imaging also involves scanning. Modern integrated circuit techniques make it considerably more practical to simultaneously process the signals from a distribution of small detectors all at once. Though X-ray examinations have traditionally used the simultaneous recording of all rays through the body at once, scanners have been considered to implement the extraction of spectral information (Jacobson and Mackay, 1958; and dose reduction without detail reduction Mackay, 1960); the recent revolution in some aspects of diagnostic radiology by images computed from projections depended upon making available mechanical scanners that recorded one or a few rays through the body at a time (Hounsfield, 1980).

Discussion of the speed of formation of an image can be deceptive if one does not state clearly what is assumed to be held constant. Because an electron has a very high ratio of charge to mass, it can be forced to accelerate or move rapidly. We expect a fast response from electronic circuits, and find it faster to scan an electron beam over the target of an X-ray tube to shift an emerging X-ray beam than it is to mechanically move the entire tube. But mechanical scanning can be reasonably fast for reasons crudely indicated by a simple analogy.

Laying down an X-ray image is in some respects like watering a lawn overall to a given depth. One can wave a hose back and forth to eventually put a certain amount of water at each spot, or one can have simultaneously active a number of unmoving parallel sprinklers doing the same thing. In the X-ray case one must have a certain number of quanta falling in the same region in order to know if that region is lighter or darker than the adjacent regions, and so there is a crude analogy between drops of water and quanta in the present situation. It would seem that waving the hose would not only be inconvenient but would take a much greater time, just as scanning an X-ray tube fronted with a small aperture over a person would require a long time to radiate every spot. But this is not strictly true in its implications. First of all, the mechanical scan could be done with minimum stopping and reversing by spinning in front of the tube a disk perforated with a spiral of holes which would successively trace out raster lines (a Nipkow disk from the early days of television). More important to the consideration of the overall time is the diameter of the hose compared with the diameter of the pipes feeding the sprinklers. Very roughly speaking, if there were five sprinklers each fed by a pipe the size of the hose, then they should be able to deliver water at five times the rate, while if the hose had a capacity equal to that of all five pipes then it should be able to deliver water as fast as all the sprinklers together, and if it could be scanned over the lawn appropriately rapidly it would be able to complete the watering at the *same instant* that the sprinklers would be switched off having completed their unmoving delivery. Thus an X-ray tube forming a beam and running at high current can be moved as rapidly as mechanical considerations will allow and still form an image, assuming data from each of the positions can be separately accumulated and processed sufficiently rapidly also. Wheels can spin rapidly to distribute radiation information, though *if* a scanning process proves to be the limiting factor in acquisition of an image then a nonmechanical scan can speed it up.

Parallel collectors or detectors minimize exposure of the subject to radiation dose relative to the situation where one might scan a single detector while steadily irradiating the entire subject rather than also scanning the source. (The storage properties of television camera tubes however make it perfectly reasonable to aim one through a light microscope without increasing thermal damage to the specimen, even though a scanning process is involved in taking out the information; more will be said of this in the next section.)

Positron emission tomography as shown in Fig. 2.9 is the epitome of parallel processing, with its many detector pair combinations all sensitive at once instead of scanning with one pair, and thus it is good for following dynamic events such as brain functioning in response to various stimuli. One could move a pair of detectors on opposite sides of a subject around in some plane containing both to gradually pick up the degree of activity along a number of lines, but at each position enough time would have to be spent to collect a number of events. Translation plus rotation would be used to build up many lines not all through the center of the subject and not all in one direction, rendering the process slow Incorporation of distance information from difference in arrival time at the two detectors would allow formation of an image by one translation motion or by rotation, just as it does in the ultrasonic scans to be mentioned later, but the result is still slower than the use of many detectors. If the two detectors were replaced by an x-ray source and one detector, then absorption along a line could be measured rather than emission along a line, and the pattern of absorption rather than emission determined; this is computed tomography by x-rays, which is relatively slow for the above reasons and in which time-of-flight during transmission does not yield the position of dense regions.

A final comparison between series and parallel transmission of information can reinforce these ideas from familiar experience. In a television transmission the light intensity at a point is converted to a proportionate electrical signal which is transmitted, after which the electrical signal appropriate to the next point is transmitted in sequence. The information is then received in sequence and appropriately spread out as a pattern of light and dark. By contrast one has the images produced by a fiber optics bundle. It has long been known that light shining in the end of a glass rod or even into the source of a stream of water will come out the far end because total internal reflection will prevent its escaping from the sides of the rod even in spite of bends. If a group of rods is placed side by side and a pattern of light shone into one end then the same pattern will come out the far end because each bright region will have a corresponding bright region at the output, and similarly for intermediate or dark regions. Such a bundle can be several feet long, or with modern techniques for controlling fine fibers, miles long.* Such bundles are used in endoscopes and also to transfer an image directly from an oscilloscope to light-sensitive paper without the need for lenses to compensate for curvature of any of the surfaces which abut the ends of the bundle. A parallel channel system can be used to take in a dim image and give out a brighter one. If the input has a photocathode to emit electrons and the output a fluorescent covering, and if each fiber is instead a short tiny tube of, say, 15 micrometers diameter, then one has a microchannel intensifier giving an outgoing brightness increased by perhaps 10,000 times over a wide spectral range, because at each electron reflection along the tube wall several electrons are released by secondary emission (Lampton, 1981).

Intermediate cases using some scanning with several channels or rays simultaneously active are sometimes useful. In Figure 3.2 is shown a point source of X-rays falling upon a row of detectors which record one line across a subject. If there were many adjacent rows of such detectors then a single pulse from the X-ray tube

---

*There is a naturally occurring fibrous mineral called ulexite (a calcium-sodium hydrated borate) that is not flexible but normally shows this optical property if the two ends of a piece are merely flattened and polished.

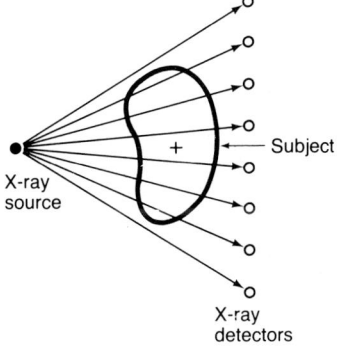

FIGURE 3.2. An example of an imaging system in which data are simultaneously collected along a line while subject motion perpendicular to this line provides information about the other dimension in order to yield a two-dimensional image. The image is a projection in which planes at different distances are superimposed. An example of such an image produced with equipment normally employed to yield a cross sectional image is seen in the next figure.

would record an entire image from the photons transmitted through the subject in the same way that a photographic film placed in the position of the detectors records an overall image all at once. With a single row of detectors the subject reclines on a couch which is pushed through the assembly while parallel lines of information are electronically periodically recorded, for example, every centimeter. The parallel recording gives a line rather than a point of information at any instant, and the subject motion is the scan to form the entire image. This procedure allows the making of a classical-type radiograph of a subject in a CT scanner (discussed in Chapter 4) in which one would usually have the subject fixed while the tube and detectors rotated around in order to build up an image of a cross section or a slice. Moving the subject closer to the tube increases the magnification, though a smaller part is then included in the image. An example of the result is shown in Figure 3.3. The subject is a man with a knife in his spine, and he will be seen in cross section in Chapter 4. This "scanogram" contains quantitative information and is equivalent to an X-ray photograph on film made under carefully controlled conditions so that any line could be scanned on a densitometer to yield a density profile. It is also the line-by-line equivalent of Figure 10.5, though somewhat better controlled. Small differences can be emphasized in the final viewing to make the display more effective, as will be discussed in Chapter 10.

3.2. ABSOLUTE LIMITS

In some sense a form of noise limits all our observations. For example, if our hearing were made more sensitive we would begin to perceive the thermally induced Brownian motion of the air molecules bumping our ear drum, and thus quieter sounds would still not be heard. Roughly speaking, one might then try to speak more slowly to allow averaging the signal over the newly found irregular background noise, but then the frequency response of the whole system would be correspondingly reduced. Similarly, it is known that a properly dark-adapted rod cell can detect a single photon, but this has little to do with our ability to see detailed images in dim light. In this case the limit is quantum fluctuations since a

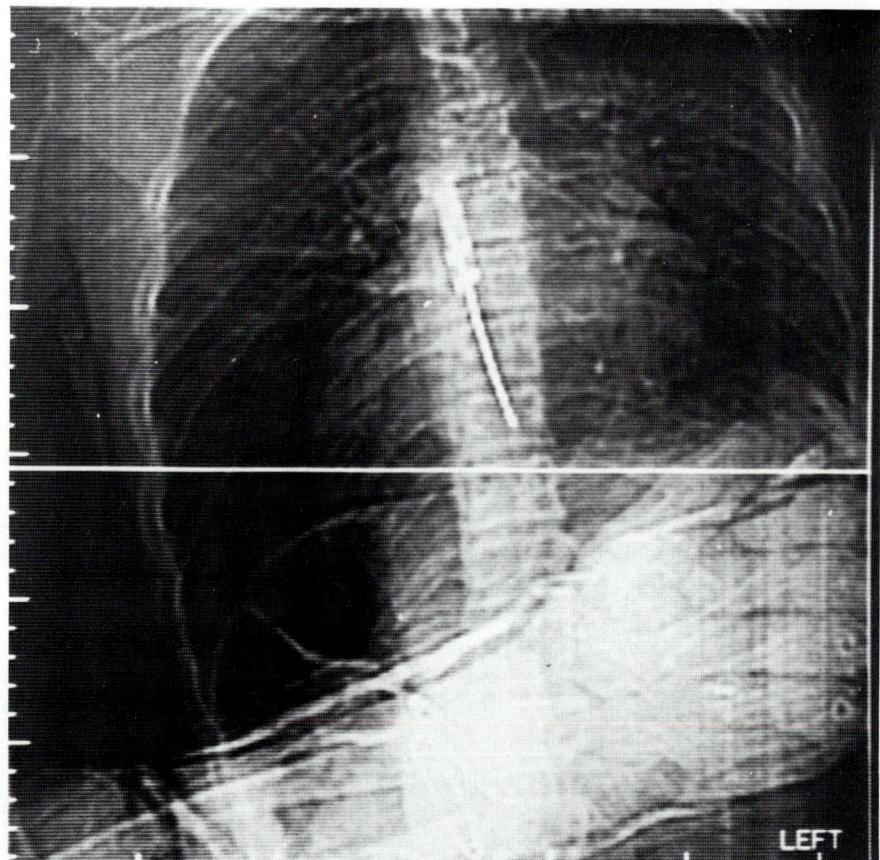

FIGURE 3.3 An image of human subject formed one line at a time by pushing the subject through computed tomography equipment (which was not rotating as it normally would be). The information was stored in a computer in the form of numbers representing the brightness of each point. A knife is seen in the spine of this man. A cross sectional image or tomogram of this is to be seen in Figure 4.9.

dim light involves few photons, and fluctuations in their number make it ill-defined or unclear which of two adjacent regions is the brighter or if they differ at all. In a radiographic image made with minimum dose to the patient there will result an image whose different regions each have a few spots; if it is difficult to judge where the density is high, one can average over the spots in a larger area thus trading geometrical discrimination for greater perception of shades of brightness. This is why some radiographs are best observed by standing back a bit, as was also suggested for Figure 2.6.

Any signal passed through an amplifier will have added to it some amount of noise, if for no other reason than there are thermal fluctuations in the resistors and irregularities in the transistor currents. These all can and should be made small with respect to fluctuations intrinsically associated with the original image emerging from the subject, even when some television scanning process is introducing further noise.

Fluctuations in an image limit things ranging from the effectiveness of image intensifiers to the possibility of contrast enhancement, and we shall see that it places the ultimate limit to which radiation dose can be reduced for a given subject. Thus it is useful to consider a simplified version of the origin of fluctuations in an image.

The view to be adopted in this section of the formation of a light or X-ray image (or other) is that the radiation comes not as waves but as photons or quanta, the variation in whose number introduces fluctuations in the image. The formation of the image might be likened to standing a person before a white wall and throwing at them many small mud balls; when the person steps away the outline will not be perceived unless enough balls were thrown.

The human eye can perceive differences in brightness between two adjacent areas if they differ by about 1%. (Under some conditions some can do better than this, and if the boundary is not sharp, discrimination will often be poorer, but this is a convenient figure.) Thus to distinguish two adjacent regions as being distinct, at least 1% more photons must emerge from one than the other. This suggests discrimination of these two regions would be possible if 100 quanta came from one, and 1 more or 1 less came from the other. But there will be statistical fluctuations in n events of about $\sqrt{n}$, which means a percentage or fractional fluctuation of $\sqrt{n}/n$ which is $1/\sqrt{n}$. This means that at times, rather than having 100 quanta emerge from an area, there might be as few as 90 or as many as 110, which would certainly mask an actual difference of 1 between two such areas. If, however, the scene was brighter so that 1,000,000 quanta emerged from one pile and 1% more, or 1,010,000, from the other, then expected fluctuations would be limited to the square root of a million, which is 1000, and the 1% difference would be persistently noticeable. In a dim image fine discriminations of brightness are impossible, not because of the properties of the eye but because of quantum fluctuations in the imaging medium. The final result is presented to the eye but detail can be no finer than the limit imposed by fluctuations at the place in the overall process where the density of quanta is lowest. In the case of an X-ray system this will generally be where the X-rays emerge from the subject (or, more precisely, the fraction of these absorbed at the detector is the number involved in the image). *No subsequent process can improve this detail.* Thus when one emerging X-ray quantum is trapped in a fluorescent system placed against a film (and called an intensifying screen), some 1000 light quanta may be produced and 10 grains of emulsion made developable in the film. One can then readily see the position of the original quantum, but it is clear that the amount of information or detail cannot be greater in any place than it was when the X-rays left the subject. These arguments have all been given in more detail elsewhere (Jacobson and Mackay, 1958), and studies have been made on various images to determine the statistical certainty required by the eye to separate two noisy image elements (Rose, 1957). It is often found that signal-to-noise ratios in the range of 1–5 are required, but this varies with the observer and task.

One can sometimes trade resolution of geometrical detail for fineness of brightness discrimination. Thus when quanta are emerging everywhere over a subject, one can choose to compare two larger regions with regard to the number of quanta emerging in a given time (the exposure time, in a continuing process, or the eye's summation time) in order to decide if regions are distinct and separate. One has more quanta from each larger region over which to average, at a given dose rate, and thus over these two regions noise will average out better, but the fineness of geometrical detail being studied is worse. Stated conversely, as the area over which you average decreases, the more roughness is seen, and it is necessary to increase the number of photons to make it look smooth or fill in the holes; a simple demonstration is the difficulty in seeing hair through a microscope viewing a printed picture. This is why standing back from a radiograph will sometimes allow one to see patterns that would not be noticeable up closer. It is also an aspect of the utility of image smoothing to be discussed in Chapter 10. If one decides it is necessary to see finer detail that is half as wide in each transverse direction, then the same number of quanta must be passed through each little unit of area that is one-quarter as large as in the prior case; delivering four times the number of quanta overall will deliver four times the dose to the subject. Increasing transverse detail fineness by a factor r requires increasing the dose by r^2.

This rule involving the square of detail size is the case of trying to see finer detail in a fixed object of unchanging contrast. If one wished to detect a smaller object or void (i.e., structure differing in density from surroundings), the same change in length along the beam also must be considered, since merely making an object thinner will make that region less different from the background. If a little cube buried in slightly different surroundings is just noticeable, and is to remain noticeable after each side is made half as long, then dose must be increased by *four* factors of two. As above, two factors of two are still required to pass the same number of photons through the area across the observer's view. This either can be considered as defining structure across rays to the detector, or definition can be referred to the detector itself which must be twice as fine in both directions to see twice as fine detail, and yet needs as many photons as before. (In observing anything whether it involves hearing, sight, touch of a surface, etc., there is an implied smoothing across the aperture of a detector to average out fluctuations.) Final difference in transmission between two regions is also proportional to the distance along the beam over which the absorption difference between adjacent columns persisted. The little cube is now half as thick, which makes its differences from the background only half as much, and also a given error in judging thickness now matters twice as much percentage wise; to restore the difference requires twice as many photons, and to restore the ratio of the signal to the noise also requires twice as many. Dose depends on the number of photons traversing the subject, and in this case where resolution is maintained in all three dimensions as size is possibly changed in all three, dose varies with the inverse fourth power of the resolution distance. Stated in reverse, the fourth root of exposure is a slow improvement in detail. In computed tomography we will find either a cube or fourth power variation of dose with detail, depending upon whether slice thickness is maintained fixed (see Chapter 4).

In the radioactive emission case, dose is less governed by resolution since the radioactivity in the body will all run its course, though the required duration of the observation can be reduced with increased activity and dose if time is a consideration. Evaluation of an image improves in a complicated way as noise and distraction is reduced, whether one is considering locating a tumor or a hostile submarine. This aspect of psychophysics has often been studied and has been reviewed (see Biberman, 1973).

There are various devices which take a dim image and make it brighter in order to be seen by the human eye, one having been mentioned in Section 3.1. In the case of X-ray systems, these image intensifiers often take the form of a fluorescent screen which glows in response to a pattern of X-rays, against which is placed a layer of material that emits electrons wherever hit by light. This combination is placed in an evacuated tube so that the electrons can be accelerated by a steady voltage in order to hit a fluorescent screen, thus converting the pattern back into a visible one. By reducing the width of the cloud of emitted electrons through suitable electron optics, one increases their density and so increases the brightness of the pattern at the final fluorescent screen. Also, each electron picks up considerable energy from the applied electric field, and in crashing into the final fluorescent screen emits correspondingly many photons of visible light. Both effects brighten the visible image, which can be enlarged with a suitable magnifying glass without loss of apparent brightness.

A specific example would be the generation of 8500 light photons for each X-ray absorbed. If a 25 cm circle of electrons is reduced to a 5 cm circle, the fivefold decrease in diameter gives a 25 times increase in concentration of electrons. Remarkably, the optical image can then be magnified five times without loss of brightness if, in optics terminology, the exit pupil of the magnifying system is kept larger than the pupil of the eye. The acceleration of the electrons brings overall amplification up to about 100. This is enough to bring the light level of gastrointestinal fluoroscopy up to where an image orthicon television signal is larger than its beam current noise, so the detector does not interfere, if one is doing the final viewing via a television system.

Some such image intensifiers absorb approximately 1 quantum in every 20 that strikes their front surface, and then deliver a satisfactorily bright image. In a sense, such an intensifier can only become 20 times as good, no matter what technological changes are made. Noise generated or added in an intensifier may be small. A figure of merit for detectors is "detective quantum efficiency," which includes noise introduced as well as fraction of input that is effective; this is described in the next chapter.

In a scanning or television system the information received at various points is sequentially converted into a voltage signal having different values at successive times. This variable voltage is converted into a two-dimensional brightness pattern at the output monitor. Some noise is added to the signal in the camera tube and voltage amplifiers, and this can and must be made less than the noise fluctuations associated with the X-ray (or other modality) pattern emerging from the subject if the overall process is to function with minimum irradiation and dose to the subject. Even the transfer of information from a bright television input onto a motion picture film can have pitfalls that require increased dose to the subject; in part it was this that led us to introduce the use of magnetic tape recorders into radiology (Mackay, 1961).

The most basic requirement in radiology is that of sufficient quanta to carry statistically significant information about each image element. With modalities other than X-rays, other sources of noise play the role of photon variations. In any simple scanning system the limitation of statistical fluctuations will demand that a certain time interval be spent on each image element. If there is a given emerging beam intensity this will then limit the speed with which a line or frame can be covered or reproduced. However, if the detector can store or integrate incident flux at every image element in the whole image until called upon to suddenly deliver the stored signal which has slowly built up, then either a whole line or a frame can be done in the time otherwise required by each image element. With b image elements per line, and b lines per frame, this results in a time saving of b and b^2, respectively. In most television tubes (e.g., Vidicon or Plumbicon, Orthicon, Iconiscope), the effect of the photons is stored as electric charge in the capacity of a film or grid structure of some sort. The effect of no quantum capable of interacting is lost due to attention being elsewhere in the scanning process at the time of arrival. Limited storage can also take place in the persistent glow of many phosphors when subject to radiation.

A recent type of converter from a light image to a television signal is the charge-coupled device (Kristian and Blouke, 1982). Electrons from photons collect at each point on a silicon chip. There can be approximately 500 × 500 such points in a centimeter square. An array of surface electrodes shifts the charges in an orderly way from each point to the output in succession, so the whole camera action, including scanning, is on the chip. Quantum efficiency (the fraction that is effective) is up to 70% (vs. a few percent for the eye or photographic film), the noise is low, and there is a high linear dynamic range from light to dark. Applications for these will increase.

One can actually see quantum fluctuations. Thus Figure 3.4 may help emphasize that these are extremely practical considerations rather than being merely a convenient theoretical fantasy. The pictures are X-ray cross sections of a living subject computed by CT methods to be discussed in the next section. CT images were chosen for this example since they can be modified in photon density by two means: reduction of tube current (or time) or reduction of slice thickness. Such images are usually shown as if the subject were seen from the direction of the feet. In each of the four present cases the accelerating voltage in the X-ray tube was 120 kV and the scan time 3 seconds. In the upper left picture the tube current was 99 mA and the slits limiting the slice thickness 2 mm wide, while in the upper right picture the current was 20 mA and the slits 10 mm wide, thus involving essentially

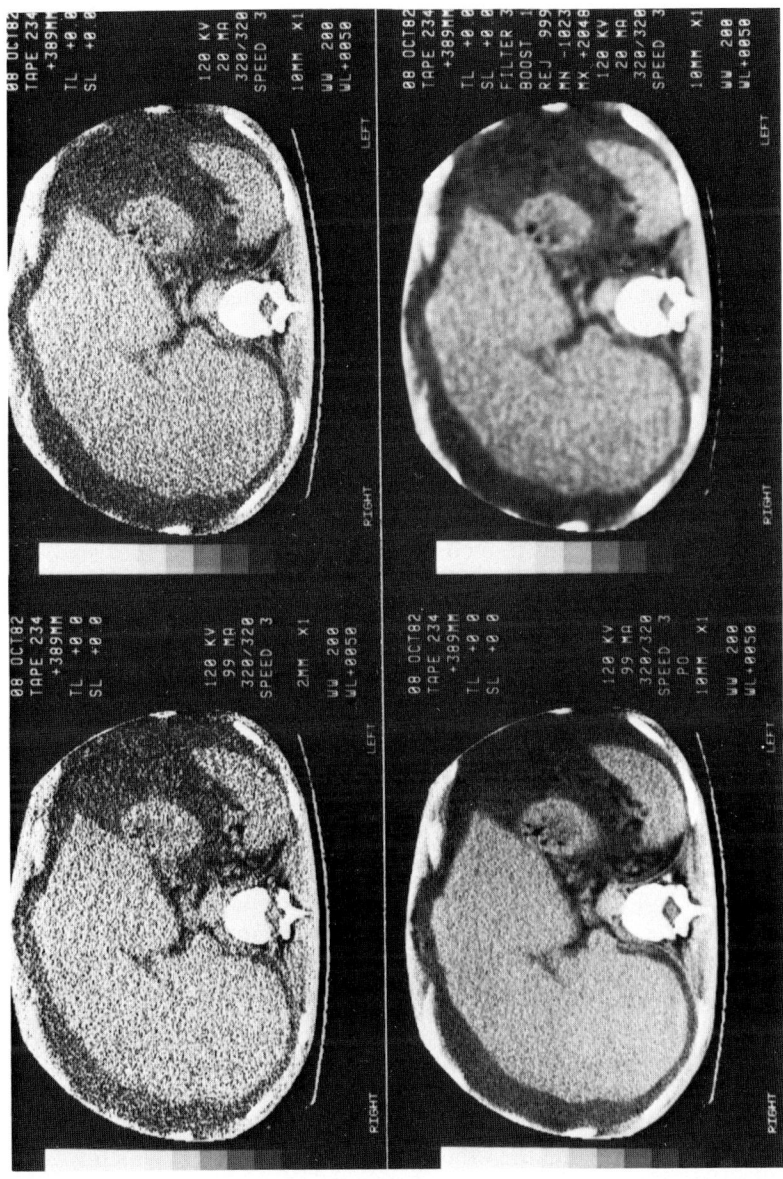

FIGURE 3.4. X-rays images in which quantum fluctuations can be seen, and which obliterate one entire organ (the right adrenal gland) unless dose is increased as at the lower left. In this case, even with computer processing of the image to smooth it (lower right) this structure can not be seen. Relative to the picture (not the subject), to the right of the spine is seen the spleen, above which is the stomach. Above the spine, the round structure is the aorta, at the left of which is the vena cava, below which is the adrenal gland. The large structure above is the cirrhotic liver.

44   IMAGE FORMATION

the same number of photons.* The subject is seen to have ascitic fluid around a lumpy cirrhotic liver with an enlarged left lobe. Considerable quantum mottle obscures the finer details. At the lower left is an image with 25 times as many photons since the tube current was restored as well as the slit width. Finer detail can be seen. The stripe contacting the left side of the spine (the right side of the subject seen to the left) is the crus of the diagram, and the narrow stripe contacting its left side and touching the liver is the adrenal gland which has disappeared in the noise of the upper two figures.

Computer processing of images to bring out unnoticeable detail will be discussed in Chapter 10, and one of these procedures was applied to the upper right image to produce the image at the lower right in Figure 3.4. Smoothing by averaging over 5×5 groups of image points removed much of the distracting granularity but in this case did not really restore to noticeability the small adrenal gland. Images can contain detail that is not noticed by the eye; brightness changes of less than 1% can go unrecognized even though they are readily recorded, and slowly changing intensities denoting the boundary between two structures may also go unnoticed. Thus contrast enhancement and edge emphasis are two useful processes. The Xerox process happens to incorporate some of each and this is why a Xerox copy sometimes appears better than the original (though it also can appear worse). An interesting example of the existence of unnoticed information on photographic plates is in astronomy where the estimated size of a nebula is often twice as big using a microphotometer record as when the photograph is viewed by the eye directly. In this section it is appropriate to mention when contrast enhancement may be appropriate.

It has been mentioned that a contrast of 1% can be seen. If the brightness of a scene is changed either by increasing the illumination or by placing before the eye a neutral filter, then all intensities are multiplied by the same factor and contrast is not changed. Contrast can be altered by adding or subtracting an overall intensity from an image rather than by multiplying, as will be discussed in more detail in Chapter 10. To return to our early example of 1,000,000 quanta coming from one region and 1,010,000 from the adjacent region, there is a definite difference between the two since the variation in each is limited to about 1000. The difference of 1% would be at the limit of noticeability under reasonably favorable circumstances. But if one subtracts the same number of quanta emerging from each region then the difference becomes very noticeable. If one were to subtract 1,000,000 quanta from each (subtract the smaller from the larger), then 10,000 would come from one region while the other would be totally dark, which would be most noticeable. Contrast enhancement by differencing can be extremely useful but it is profitable only if there is statistically significant information in the first place. Contrast enhancement tends to emphasize high frequency noise such as

---

*In a computed tomography system, the number of quanta along a ray depends upon the width of the ray, which is fixed, and its height, which is the thickness of the slice. For a given tube current or exposure time, percentage fluctuations thus increase as a thinner section is viewed since there are fewer photons over which to average.

ABSOLUTE LIMITS 45

FIGURE 3.5. A familiar type of image degraded by limiting either the brightness detail or the geometric detail. At the lower left the best image has 16 possible steps of brightness, and geometric resolution finer than most features of the face. Above it are pictures with 2, 4, and 8 possible brightness levels, and beside it two stages of reduced geometric resolution.

quantum fluctuations. In a subject with only fine details there will nowhere be an extended region where a statistically significant estimate of the average number of quanta can be obtained, and quantum noise is a limitation in images with many fine details. It is not useful to pass such a picture through a contrast-enhancing machine. Conversely, a picture containing only a few large relatively uniform areas in which there are smooth gradients can profitably be processed by a contrast-enhancing method even though this radiograph involves no greater dose to the subject. In the latter case information is possibly lost by being unnoticeable if contrast enhancement is not used before viewing by the eye.

We have already seen a nuclear medicine image made with roughly 100,000 photons in Figure 2.5 and can somewhat compare it with a 35 mm slide made with 10 billion photons. In closing this section it may be helpful to show pictures of a more familiar type of subject under various degrees of geometrical resolution and discernible steps of brightness. (These images are produced in the digital processes to be discussed in Chapter 11.) At the upper left of Figure 3.5 is an image of a face where only two levels of brightness can be discriminated. The two images at the upper right allow for 4 and 8 brightness levels, while the lower left image includes 16 levels. To the right of this best image are seen 2 images displaying increasing geometrical degradation, each with the possibility of 16 brightness steps. Loss of either brightness detail or geometrical detail can interfere with analysis of the image. The lower right image is more face-like if viewed from a distance, and even the author's glasses are then more noticeable.

Especially in radiology, quantum ideas can directly indicate optimum conditions. Thus when we calculated the dose per emerging quantum for different size subjects both with and without contrast agents, in order to get optimum contrast per unit dose, we found predicted values of preferable voltage that often agreed with those dictated by decades of experience (Jacobson and Mackay, 1958).

3.3. COMPUTED IMAGES

An X-ray picture is a projection of a three-dimensional object onto a two-dimensional surface. From a group of many such projections one can reconstruct the complete form of the original object. Similarly, a two-dimensional object can be projected as a pattern along a line, and from a group of such lines the original flat pattern can be reconstructed. The projections in many directions through the object can be stored either as variations in darkening of a photographic film or as a series of corresponding numbers in a computer. The computation of the original pattern from its projections can then be done either optically or by computer. Optical or hybrid methods can have some advantage in speed due to the parallel processing of many points at a time, while a digital computer tends toward more precision and flexibility. The impact of some of these methods upon radiology and medicine in general has been immense. One application is to make X-ray images of a slice (a tomogram) through the body, with no confusing shadows from regions in front of or behind this section, and with a radiation dose to only the observed regions; doing this from a set of line projections will be covered in more detail in Chapter 4. Examples of such images were seen in Figure 3.4. The most important application to date has been the X-ray case, but other modalities can be handled similarly.

Originally the equations for reconstruction were given by Radon (1917) who was interested in gravity. Much later the first application of mathematical image reconstruction was to a short-wave radio case in astronomy. The pattern of emission over the sun was desired, but the microwave antennas could not focus on a point, only on a strip. The problem of determining local activity from a set of strips was solved by Bracewell (1956) and applied to fan beams as well as to parallel rays. Other applications with other modalities included electron microscopy (De Rosier and Klug, 1968; Gordon et al., 1970) and optics (Rowley, 1969). In the meantime a nonmathematical method had been used to determine the distribution of radioactivity in a slice (Kuhl and Edwards, 1963) by what is termed "back projection."

In 1963 Cormack did an analytic reconstruction of an X-ray image; that is, exact formulas were employed. Hounsfield (1973), working for EMI Ltd., evolved a complete system (patent application August 1968) not involving those analytic methods but rather employing iterative ones. It became available and widely used. This last generated increased wide interest in all the previous. Ultimately analytic methods have proven faster for computers than iterative ones. For their contributions, Cormack and Hounsfield shared the 1979 Nobel Prize in Physiology and

Medicine (see Cormack, 1980; Hounsfield, 1980). Applications are not just medical. Overall detail is such that, for example, tree rings can be seen in fresh lumber. When compared with conventional radiography, the methods show poorer spatial resolution but better contrast resolution.

Quantitative values of attenuation emerge in producing these digitally computed tomography (CT) images, perhaps making it possible to characterize different tissue types by numbers. In the X-ray case, two Hounsfield units are taken as one older EMI unit, based on a Hounsfield unit being 1000 times the fractional difference from water in the linear attenuation coefficient of the material (see Chapter 5). As an example, the aorta has a density of about 50 H, but in anemic patients the lumen is seen less dense than the wall. Warming water from room to body temperature causes density changes that produce a change of about 5 H.

The methods are not restricted to biological materials having a density about that of water. For example, rocky materials can be examined, including the inclusions in meteorites, using commercial medical equipment (Arnold et al., 1983). The only necessary modifications are to the software constants in the computer, which are user-accessible in some units. For example, water which is usually normalized to zero in Hounsfield units is given a value of about -750. It is then possible to distinguish regions of high and low atomic number on a scale of millimeter detail.

Computed tomography, which will be explained in simplified detail in the next chapter, is not the only example of imaging processes requiring the use of computation to form a final image from some intermediate pattern or collection of patterns. Thus an array of receivers in one hole recording the signals produced as a diverging source moves in another hole can be used to image the material in between, whether it be earth, ocean, or tissue being studied with microwaves, X-rays, or acoustic sources. In some procedures a single recorded intermediate pattern may have to be "decoded" to become recognizable.

A pinhole image is useful for many purposes. In an emergency a small opening can replace most broken eyeglasses, for example. But a pinhole accepts only a very small fraction of the incident radiation, giving a weak image requiring a long exposure. A group of pinholes accepts correspondingly more input but requires sorting out the multiple images that result. In Figure 3.6 is seen an imaging process in which a considerably larger aperture of some particular shape (here "coded" as an annulus) gives this shape for each source point; some form of computer is then essential for reconstructing the image, where each point has become a particular pattern (here a ring). Here the template acts as a "matched filter" that is to be correlated with the detected pattern at every spot. As shown, it can be done using ordinary white light in an incoherent correlator, the film is the second step being against a diffusing sheet or light box giving rays in many forward directions from each point. All points at an original distance a are displayed; that is, one layer can be displayed while others are somewhat more blurred; the action is not perfect and other aperture shapes can be more effective. Some light goes in the directions of partial overlap of the intermediate film and template, resulting in final illumination falling off from the center of the point in the image inversely with the radius;

48   IMAGE FORMATION

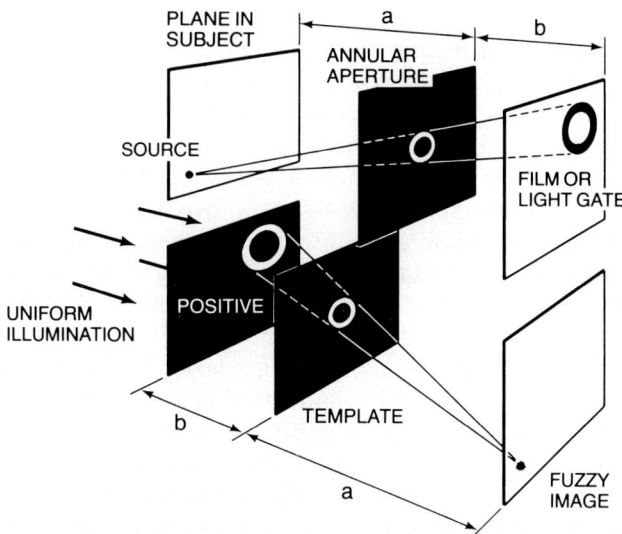

FIGURE 3.6. Imaging being done by a coded aperture, in this case an annulus. What is recorded must be reconstructed into an image by some form of computation; in the present case this is shown below being done optically rather than by computer.

this can be corrected as will be seen in Figure 4.5 at the expense of emphasis of noise. As in Figure 4.5, film can be replaced with a "light valve" which converts a pattern of light to a pattern of opacity to allow the final image formation without delay.

In Figure 3.6 was depicted the imaging of a source of radiation using a patterned or coded aperture (Simpson and Barrett, 1980). If an object was to be imaged instead by transmission, perhaps of X-rays, a preset pattern of sources would deliver a corresponding pattern on the film for each point in the subject. Thus a ring source would put a ring pattern at the detector due to all rays through any one point in the subject, but this can be clarified (Stoner et al., 1976). The pattern of overlapping rings gives images blurred somewhat in the same way as the X-ray images of Frank produced as in Figure 4.2, and similar compensation is required (Fig. 4.5). With compensation a sharp image can result from a broad source, which has some practical applications.

Rather than an annular opening, one can use a checkerboard pattern with random squares made opaque, or a group of pinholes can be used. An aperture that is a zone plate gives a pattern of zone plates (by casting shadows not by diffraction), the size of each depending upon the subject–source distance. This result can act as a hologram (Mertz and Young, 1961; Barrett et al., 1974). Usual zone plates of lead do work with X-rays but are too small when designed for medically useful wavelengths. However, the above hologram is feasible for viewing by visible light if the density of the image does not become too high, and each of the overlapping zone plates contributes a bright spot to the final image. The method is appropriate for dilute objects such as X-ray stars but further subtlety is required for nuclear medicine.

The usual action of a lens itself can be regarded as a computation, which view is useful for some purposes. Thus a pattern of light falls on the surface of a lens, and the lens can convert this pattern into an image of the source of the light. This can be considered as a form of optical computation, which alternative view of imaging will be relevant later: the function of a lens can be considered as the producing of a Fourier transform. Just as an audible sound pattern can be broken up into a group of frequencies, or a group of frequencies of suitable amplitude and phases can be combined to yield a sequence of pressures (or sound), so a pattern of light and dark in space can be considered as the superposition of cyclic patterns of brightness of many frequencies. Fourier analysis of a periodic signal is in terms of frequencies that are multiples of the original repeating signal frequency, but an irregular pattern can be analyzed into a group or spectrum of all frequencies, with the relative amount (amplitude) and phase of each specified by a graph. This conversion back and forth between a pattern in time or space and its frequency components is called a Fourier transform. Light from a distant scene falling on the surface of a lens is distributed in a diffraction pattern since different parts interfere with each other. This distribution is sorted out by the lens to give an image of the source that is providing this smeared pattern of illumination, and the action is a Fourier transformation; a person's eye "knows how to make" a Fourier transformation. In a later example of use, a negative placed near a lens and illuminated with parallel light will show a spatial frequency pattern at the focus, and any modification of the pattern there by a mask will affect all points of a final image reformed by a second transformation with a second lens (Figs. 4.5 and 10.9).

3.4. EQUIPMENT EVALUATION

The performance of an amplifier is often specified by the range of frequencies that it will pass. Thus a good phonograph system might be expected to pass a range of frequencies from 60 Hz to 20 kHz (to 20,000 cycles per second). Similar concepts can be used to specify the performance of an optical or visual system. Thus the number of equally spaced lines per centimeter that can be resolved can be specified. In the acoustic case a loud signal (large amplitude) will pass somewhat even at a high frequency, and similarly in the optics case, the difference in brightness between the alternating light and dark stripes must be specified (the contrast). Even the human eye itself has been studied in this fashion. Thus it is known that the retina cannot distinguish a grating, even with 100% contrast, when its individual bars subtend less than about 0.6 minutes of arc at the eye (Westheimer, 1963), and that at low contrast a sinusoidal grating of 3 cycles per degree is the most noticeable spatial frequency for humans—the value being 0.3 for cats (Campbell and Maffei, 1974). The idea of a spatial frequency will arise in a different context when the modification of an image to emphasize or suppress certain periodicities in an image arises in image processing (Chapter 10), the effect being like the use of electrical filters to emphasize or suppress selected frequencies in the output of an amplifier.

A related consideration is that of a point spread function or a line spread function which specifies the way the output from a tiny source of radiation is spread or blurred in going from stage to stage in the receiver or detection process. The spreading of a point of light or other radiation can be measured directly or it can be calculated from frequency response measurements of the above sort. For the human eye, for example, there is good agreement between the two methods of analyzing the quality of the image in the eye (Westheimer, 1963). As will be discussed later, if the way in which each point of an image is blurred is known, then to some extent that blurred image can be rendered sharp by subsequent processing of the image in such a way as to "unspread" the intensity distribution around every point (Chapter 10). Thus, for example, one would expect to be able to render more sharp an image produced with an X-ray machine having a relatively broad focal spot without disobeying the laws of information theory or entropy (Mackay, 1962).

The performance at each step in an imaging process, up to and including the eye, can be specified in some fashion, with the overall image degradation often depending upon some one limiting factor. In the X-ray case it is desirable from dose considerations that the major limitation be the random quantum fluctuations which, being variable from instant to instant, cannot be compensated for by later image processing. However, a greater X-ray intensity is often used for maximum performance with a given piece of equipment so that parameters of the equipment itself (e.g., slit widths) limit resolution. In the latter case some sort of post-exposure processing may sharpen the image. In any case, it is desirable to be able to specify the performance of a particular apparatus, both for the comparison with other pieces of equipment and also to determine if a given piece is functioning as usual.

In general, specification involves observing what comes out of a piece of equipment that is observing a known test object. Any one institution may have some object for noting daily performance that is not much more than a piece of wood within which is a pair of nails that either just can or cannot be resolved by a piece of apparatus. It is, however, difficult to communicate what this means. Various groups still attempt to agree upon some "phantom" that can be considered as a standard and reproduced in all laboratories. One temptation is to use liquids of various dilution in order to determine the smallest noticeable change between adjacent regions of specified size, but care must be exercised in that case so that the wall of the container separating the solutions does not intrude a noticeable line of demarcation where otherwise would be an invisible boundary.

A useful form for the presentation of performance data for equipment evaluation and comparison is shown in Figure 3.7. These data are for a particular computed tomography scanner. There are seen to be two general regimes, performance to the right being limited by quantum effects, and that to the left by machine parameters. Among the latter are the slit widths or aperture transfer function and focal spot size, the ray spacing or sample frequency, the pixel size in

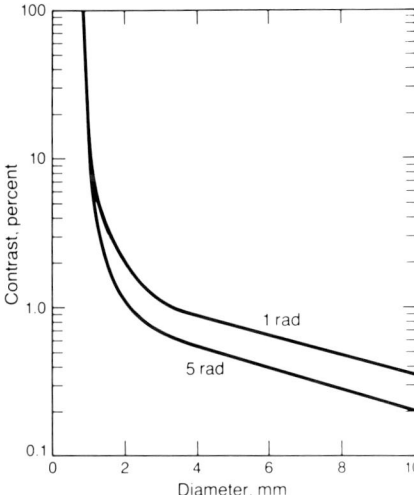

FIGURE 3.7. Curves depicting performance of a specific instrument observing objects of a specified size and contrast. At low radiation levels performance is limited by fluctuations in the number of photons, while at higher radiation levels the properties of the machine limit ultimate performance.

the reconstruction matrix, and the form of the reconstruction algorithm. To the left and above the bend the mechanics of the machine are independent of the X-ray dose. Thus a description of a particular piece of equipment could be given by specifying a number at high contrast (say 12%) and at low contrast (at say 1%), but it is probably best to give the complete curve. (See also Hospital Physicists, 1981.)

In general one analyzes an optical system with a sinusoidal test pattern (where the lines do not have abrupt edges) of 100% contrast as an object, and one measures the contrast in the image. The ratio of image-to-object contrast is plotted against spatial frequency or number of cycles over unit length. The resulting curve is known as a contrast transfer function. When resolution is limited at each of several successive steps, the overall modulation transfer function is equal to the product of those of all the intermediate steps. One often speaks of a "modulation transfer function" as having a value between 0 and 1 at any particular frequency, or it can be given in percent. It tells how well an imaging system transfers different frequencies into the final image, a large value extending to high frequency meaning good spatial resolution. In scanning, an object spacing larger than the aperture width gives a value of about 1, and about 0 for objects with spacing less than the detector width; the precise value does depend on the abruptness of the edges of the objects. The figure was produced from an image of rows of cylindrical holes of different diameters in polystyrene containing water and using volume averaging to obtain low contrast differences. As test materials various dilutions of glycerin or sucrose have been used in various laboratories, but the colloidal polystyrene balls used in electron microscope calibration seem very interesting.

As expected, the limiting resolution is rather independent of dose, but low contrast visualization depends mainly upon dose (fluctuations). Contrast times diameter is one figure of merit, and another is contrast times the square root of dose for the lower values; but in general it is helpful to specify the entire curve and some details of the object from which it was derived.

Since contrast in a display can be modified, it seems preferable that it be contrast in the object that is specified, and this has been commented upon (Hasegawa et al., 1982). High frequency emphasis can somewhat restore a blurred image, for example a CT image made with wide detector aperture, but this also emphasizes noise, while smoothing an image helps improve the noticeability of large structures. These possibilities (to be discussed in Chapter 10) may be operative and adjustable almost without realization in some modern commercial systems where one turns knobs until appearance is good, and when active, they obviously can modify one's interpretation of the detail in a given test object. Of course, a classical radiograph can subsequently be processed in an optical or television system so as to make more subtle detail perceptible, and in that case the processor can simply be construed as part of the overall system, and its effect rightly included in the performance curve.

4

COMPUTED TOMOGRAPHY

A tomogram is a picture of a slice. The traditional images of internal body structures made by the transmission of radiation into one side of the body and out the other leave the pattern of one structure superimposed upon that at a different level in the body, sometimes causing confusion. Thus methods were developed which essentially would image a single plane at one time. Tomographic concepts are not new, yet in recent form these methods have proven sufficiently successful to change some aspects of what is considered preferred medical practice. A brief overview of some particular aspects was given in Section 3.3. It is instructive to understand the methods, their diversity of applicability, and what suddenly made them important.

4.1. INTRODUCTION

The classical form of tomography may have originated in 1921 when the Dutch radiologist Ziedses des Plants moved the X-ray tube and film during exposure to blur all planes but one of interest. In Figure 4.1 is seen the source and film pivoting in unison at opposite ends of a rigid connector. Motion can be in a straight line or in an arc extending out of the plane of the figure. The latter result can also be achieved by rotating both subject and film about horizontal axes in the plane of the figure. Structures in planes other than the plane of interest will have their image fall at different places on the film during the motion while those near the pivot will

54 COMPUTED TOMOGRAPHY

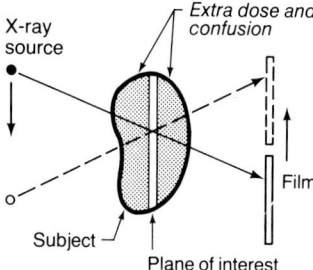

FIGURE 4.1. Moving the film and source oppositely during the exposure is the classical way for blurring regions other than the one of interest in order to produce an image of a section or tomogram. Structures in overlying planes still introduce some confusion and are subjected to radiation dose.

stay fixed and relatively sharp in the developed image. The overlying structures on each side of the plane of interest do contribute a somewhat confusing variation in background density, and these also receive needless X-ray dose. Thus it is preferable to irradiate the plane of interest from its edge by a sheet of X-rays, or other radiation, and then somehow compute the distribution of absorbers within the slice from an observation of the pattern of emerging radiation in the plane of the slice.

A clever method for accomplishing this was described by Gabriel Frank in a German patent issued in 1940 (while a major war was in progress). The implications of his drawings (Fig. 4.2) will be mentioned since, starting from modern computing methods, people sometimes reinvent these simplifications, and they are also conceptually constructive. At the top, slice 18 across a vertical object is projected through slit 21 as a line along the cylinder of unexposed film 22. At the bottom, subject holder 42 and film 41 are both rotated together by motor 43 in order to record a sequence of projections of the slice through the subject as viewed from different directions edge-on. At the center is seen the reconstruction of the image of the cross section by the point of light 25 shining outward through each successive line along the developed cylinder of film. Each projection is successively imaged by the cylindrical lens pair onto the flat film sheet and is recorded as that film rotates at the same rate as had the subject.

The previous sentence is misleading because the system will not quite work as shown. Shading in the center section of the figure indicates Frank understood his method but showed the lens wrongly oriented.* Instead in the figure lenses 26 and 27 should be oriented vertical so that each spot on each original projection would be smeared into a correspondingly bright vertical streak of light on the final film 28 as it rotates. (In practice one would follow a spherical lens with a cylindrical one to control the length of a streak.)

It is instructive to consider a radio-opaque spot away from center in the slice of

---

*A possible source of confusion is that the actual image from a cylindrical lens is perpendicular to what one sees when looking through that lens, for example, when looking through the side of a water-filled drinking glass. Peering through a glass rod held near the eye at a spot of light one sees a bright line perpendicular to the axis of the rod (the Maddox rod of ophthalmologists) while the actual line image is parallel to the rod. However, light not already diverging, such as a laser beam, can be spread vertically alone by a horizontal glass rod if one is beyond the focus.

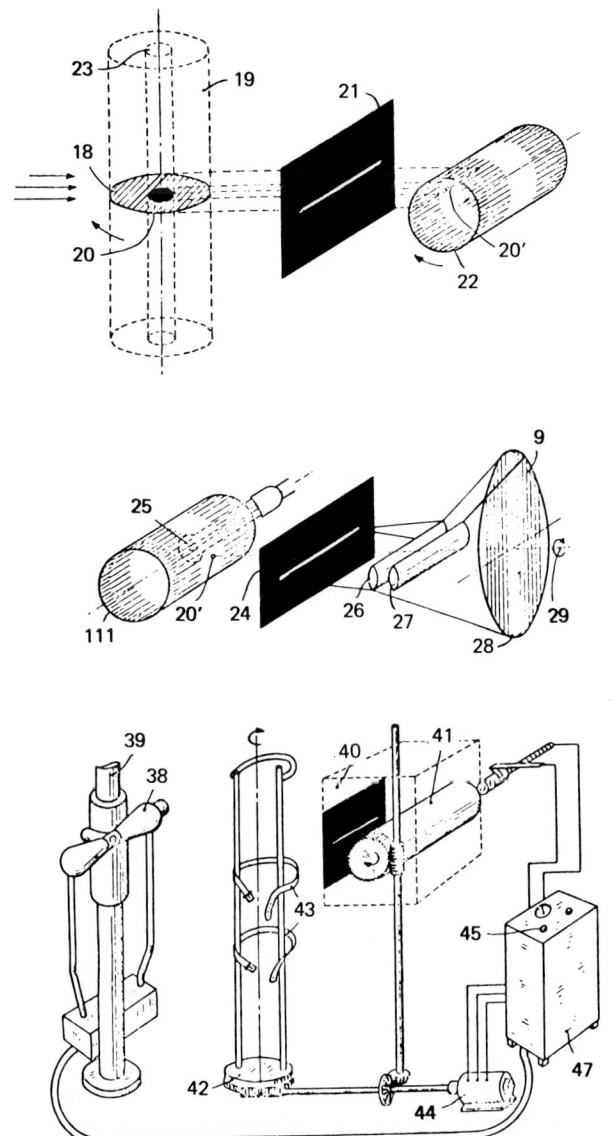

FIGURE 4.2. Three of the eight figures in Frank's 1940 patent (not counting an introductory swastika pattern), showing computed tomography. A set of transverse projections stored on a photographic cylinder is reprojected back on to a sheet of film that rotates as had the original subject in the sheet of X-rays. The method is equivalent to what is now called back projection and requires some image processing to remove the effect of interaction among points in the plane of interest.

56   COMPUTED TOMOGRAPHY

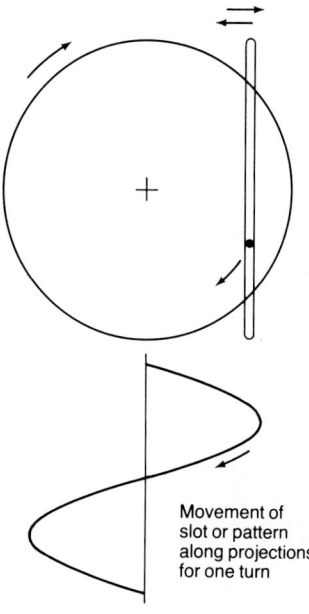

FIGURE 4.3. Just as a rotating disk with projecting pin can drive a slot from side-to-side, so a line of light moving from side-to-side will continuously be over a single point on a rotating disk of film. Some blackening will take place at spots near this point and will overlap adjacent points.

the subject. For one revolution of subject and film cylinder, the pattern around the cylinder of aligned projections is one cycle of a sine wave from side to side for any one point in the subject slice. (If the X-ray source is distant from the subject so the rays do not much diverge in their sheet then the trace around the cylinder will be sinusoidal, while in a fan beam it will be slightly modified.) A vertical line of light of about uniform brightness will move from side to side over the film sheet as it turns, corresponding to the motion of the bright spot from side to side as the cylinder turns. The effect of this can be understood from the mechanical analog of Figure 4.3. If this depicts a rotating wooden disk with a nail projecting outward through a slot in a sheet of metal, then the metal will be pushed from side to side in a sinusoidal fashion. If instead the metal is externally driven sinusoidally from side to side as the disk is turned then only the one point where the nail had been will always be under the slot for a complete turn; light shining through the slot will steadily blacken film at that point if a photographic film sheet is attached to the disk. The slot corresponds to the strip of light from the cylindrical lens. One point will steadily darken during the cycle, thus producing a black spot corresponding to the one opaque spot in the subject. A distribution of many spots in the subject will yield many simultaneous streaks during the adding of the effects of the stored projections, thus producing a final image with many corresponding spots.

Frank's patent specified essentially parallel rays, which is not what most sources produce. If nonparallel X-rays were used in a fan beam that diverged as it crossed the subject, this could be compensated by slightly tilting the final recording film so that the vertical streaks from the points on each projection similarly diverged as they crossed the turning film. In the drawing of Figure 4.3 the motion of the slit or band of light would then be like a windshield wiper (driven by a turning crank in a slot pivoted at one end) rather than side-to-side parallel to itself. After consideration of the next figure this point should become clearer.

In 1957 Takahashi evolved considerations similar to those of Frank. The determination of the distribution of radioactivity in a slice by Kuhl and Edwards (1963) also used similar considerations. These processes are all equivalent to what is now often called "back projection." Images can be produced this way but they are not as crisp as desired unless some form of contrast enhancement or edge emphasis is used upon them (Jacobson and Mackay, 1958).

In Figure 4.3 all of the final sheet of film would have received some exposure as the slit scanned back and forth over its surface, the motion past the slit being slowest for points "nearest the nail." Thus other regions are darkened to a lesser extent depending upon how close they are to the steadily exposed point. This process does eliminate unwanted planes but at the expense of blurring the desired plane by other points within it. The result is that a dark spot on the subject becomes a dark spot on the optically computed tomogram but it is spread out somewhat or blurred into a "point spread function" which can be compensated for by procedures which will be clarified in Chapter 10, but which amount to modifying either the projection or the final recording by supplying some extra brightening near any spot displaying an increasing brightness, and taking a little away just beyond. The appropriate form of edge emphasis both for these projections and for positron emission tomography (Fig. 2.9) is the same as for the radio astronomy case mentioned in Section 3.3 and is predictable (Bracewell and Riddle, 1967).

It was suggested by Frank that the projections (which he called "partial pictures" or Teilbilder) be stored on a sheet or cylinder of film from which the final cross sectional image could only be seen by reconstruction, but the intermediate storage can be bypassed. For further reduction to essentials, one can contemplate a strange form of photography that would yield a negative of a cross sectional image from projections but without storing the projections as an intermediate step (Fig. 4.4). A sheet of X-rays is shown passing through a "slice" of the subject to be imaged and going on to enter the edge of a sheet of film rotating in synchronism with the subject (or entering the edge of a fluorescent disk on whose surface the film is placed). Between the object and the film the pattern of X-ray intensities at any instant is one projection. Each structure in the object at any instant produces a corresponding band across the film, and rotation causes the superposition of all such bands to form an image. Though the projections are not stored and used later, a fluorescent screen in between will show the momentary projection, which can be modified optically or otherwise to emphasize changes along its width.

The absorption of X-rays across the width of a film after entering an edge is not even conceptually practical since they either will be absorbed mostly in the momentary left part of the revolving film or else most of them would pass through the film and out the far side; this problem of forming a uniform stripe for each ray is shown in the side view at the center of the figure as solved by slightly tilting the axis of film rotation so the left hand edge is slightly low, and assumes uniformity

58 COMPUTED TOMOGRAPHY

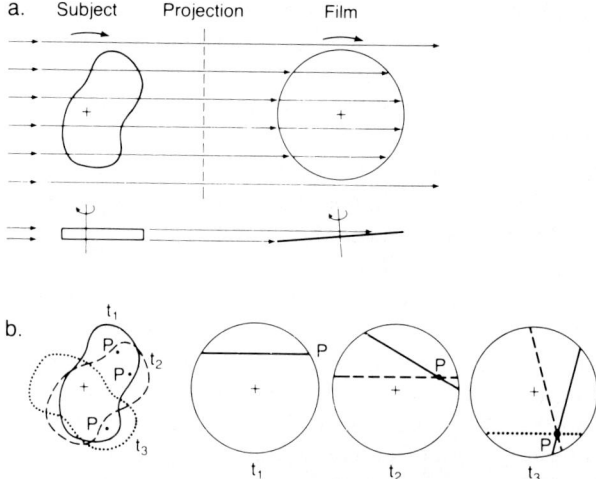

FIGURE 4.4. (a) Rotating a subject and film together in a sheet of X-rays in principle produces a cross sectional image, without actually storing projections for a two-step process. The film could be slightly tilted as shown below to spread the blackening across it. (b) Successive positions of the subject at times t_1, t_2, and t_3 with lines of radiation and brightness through the sample point P, all such lines steadily adding their effect at one point on the rotating film at the right.

through the subject slice thickness. If one actually wished to test the formation of such an image, one could use a combination of fluorescent screens and fiber optics or cylindrical mirrors to spread the effect of the quanta, without requiring too many. The building up of a pattern is shown at the bottom of the figure. Opaque regions in the object would appear light on the film, that is, a negative would be produced. Again, since each stripe somewhat darkens the film beyond the appropriate point, to produce a sharp image would require some form of edge emphasis or deblurring such as discussed in Chapter 10.

If in the subject being considered the point being recorded were instead radioactive, and a collimator such as a thick lead plate with many small parallel holes were placed in the position designated as "projection," then a radioactivity tomogram would result; if the collimator holes diverged from the subject, an enlarged image would result. This simple example helps emphasize the comparison between X-ray tomography and positron emission tomography (Fig. 2.9), and for dose or exposure time reasons, each event would be converted into a glowing line for any practical recording.

A point that darkens every line through it must come out darker than other points when viewed from many directions, no matter how these directions are sequenced. The bands across the film in Figure 4.4 need not be parallel. Note that the method shown at the top of Figure 4.4 will still work in a sharply diverging fan of X-rays (with a larger film) *unless* a fluorescent screen is placed at the position marked "projection" in order to make the projections visible for manipulation by reflection or refraction. Conversion to visible light (or storage) obviously takes

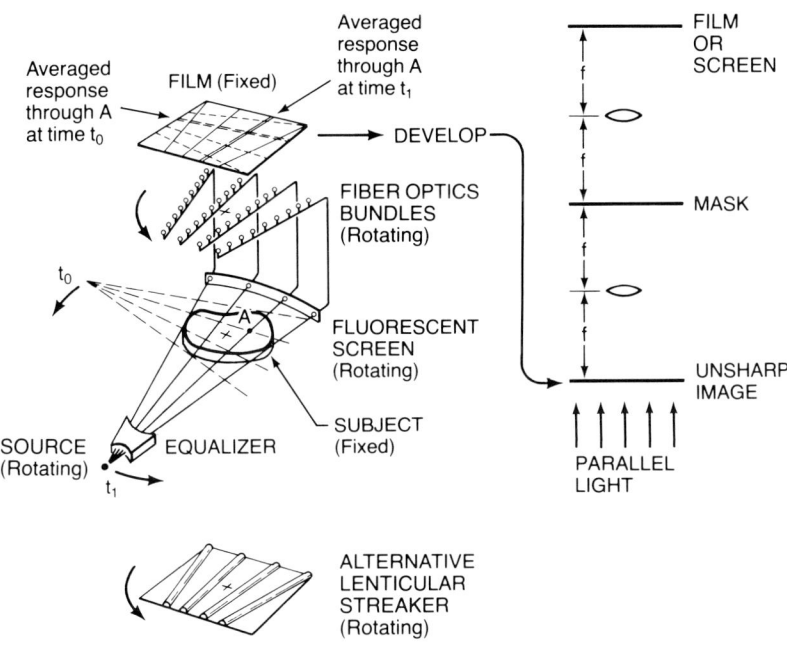

FIGURE 4.5. Rapid production and addition of projections by a fan beam from a rotating source, with optical processing of the intrinsically somewhat fuzzy image. Such a unit has not been built, and it is suggested to clarify the principles. If used for CT motion pictures, image size reduction and a light valve rather than film would be preferred. The cycling of two rays through point A is shown with the corresponding buildup of darkening on the film. Below: rotating radial cylindrical lenses replacing fiber optics bundles to spread light coming from a fluorescent ring and mirror around the subject, only four being shown for clarity. Rotating half a cup-core transformer could supply power.

away the direction the X-ray beam had at each spot, though the overall process could still be made to work by adjusting the magnification onto the final film from the near to far side (cause the streaks to diverge appropriately). The uniform intensity streaks over the film must each mimic the path of the X-ray beam through the slice of body, which emerging beam then gives an average of the darkening along that path contributed by all body points along it, just as the streak then deposits an appropriate average darkening along the corresponding path in the final film. The same processes effectively take place when projections are stored in a computer for recombination and display.

An example with nonparallel rays that summarizes the various aspects is given in Figure 4.5. An X-ray source spins around the subject, the "bow tie" filter of plastic equalizing the intensity of rays at different sideward distances from the center of the "round" subject. (One could instead assure detector effectiveness by placing around the subject a bag of water with suitable outer shape.) Along the fluorescent screen is the momentary projection. At many points the glow is picked up by narrow fiber optics bundles whose upper ends are spread into lines above the subject along the corresponding X-ray paths. These lines all spin together and the output light is added (recorded) on film or on the target of a television camera. Size can be reduced for easy spin or film advance if lines on the film remain parallel to lines through the subject. Fiber optics make the action obvious but are not essential. Below is shown the form of a plastic disk with raised almost parallel cylindrical lenses that could spread light from an upward directed ring-reflector outside a belt of fluorescent material around the subject, a negative photographic mask equalizing any illumination nonuniformities.* The film can be advanced after each half-turn-plus-fan-angle of the tube and optics since a complete tomogram is then recorded.

In Figure 4.3 the image point turns while the stripe of light moves to stay over it. This is the same as revolving the stripe over any one point, which spreads the light (all of which preferably would have gone *only* to the point) over circles of linearly increasing circumference around the point, causing illumination to decrease slowly away from the point, being proportional to $1/r$. This smearing effect is shown in Figure 4.5 being corrected by an optical system which emphasizes higher spatial frequencies (Mackay, 1962 or Chapter 10 here).

The mask of Figure 4.5 is a patterned absorber. One alternative is to place there a sample of what one is searching for, to yield a dot where found. For those familiar with the concept of a "matched filter," one can be made here by forming in place a hologram comprising a film made from a subject consisting of a single spot; the mask will not be axially symmetric, will require simple laser light, and will give out a single spot of light at suitable position and brightness in a later complex picture for each point-smear of the form being produced. However, it is not necessarily the best mask for removing blur. To undo the effect of $1/r$ spreading at each point, an axially symmetric mask with transmissivity proportional to r is correct, though the logarithmic response of the film, or lack of it, can change this somewhat. (Compare Fig. 10.4.) One can often use ordinary noncoherent light for a tolerant system with such a simple filter (Rogers, 1977) if it is rather monochromatic light.

Since the entire pattern is in one place at one time, laser light can be shone through the total image at once for coherent processing, even though the light of the projections was not coherent; then intensity at one image region can drop it at an adjacent region. Photographic film is convenient in the laboratory, but in an actual application one would wish not to develop film in the observing process. Projection can be onto the target of a television camera if some electronic processing can be done during the output. One can instead purchase light modulator cells

---

*There is an alternative way to form streaks or lines of light from points of light. A spot of light viewed by reflection in a tilted piece of metal that has been brushed horizontally with a wire brush to produce minute ridges and grooves will appear as a vertical line of light; curving the surface can increase uniformity of illumination into the eye or camera lens. Similarly, an automobile windshield scratched by the wiper can show a radial streak from a distant street light at night. In the above application one could look through a phonograph record-like pattern of concentric ridges centered above the X-ray tube but rotating about the axis in the subject, rather than constructing radial cylindrical lenses, though lenses waste less light. Peering through a star sapphire or at a reflection from the mineral tiger's-eye approximates the action of these unusual optical systems.

in which there is a sheet of light-sensitive cadmium sulfide against one of liquid crystals, so where light falls there is immediate proportionate darkening (Bleha et al., 1978). Adding or storage followed by abrupt removal is possible, so this can replace the film for immediate or real-time coherent optical processing. A change in bias produces a negative, and so optical subtraction of images in real time also is possible. Such a "light valve" has some amplification involving its own power source unlike some semiconductor sheets that can do similar things with different colors for "read" and "write" but which require considerable input energy.

Instead of working upon the final image to eliminate the blur, one could instead modify the individual projections with a light valve during their recombination. Since the extra $1/r$ contributions are to be subtracted there must somehow be a negative component on both sides of the projected-back streak into which a point on the projection is spread (e.g., on both sides of the spread fiber optics bundle). This would be termed "filtered back projection." It is known from the studies mentioned in Section 3.3 that to process the individual projections before their recombination as streaks into a deblurred final image requires that each point on a projection be modified into a bright center with a negative intensity on both sides diminishing as the square of the distance along the projection from that point. If one had a storage tube of a type where charge could be added or subtracted from each region (like numbers in an array in a computer) one might rapidly alternate during the rotation between positive storage of the momentary back projection streaked across the image plane and a negative storage of the same projection streaked or spread by a cylindrical lens diaphragmed or filtered for side-to-side diminution according to the square of distance. A second light valve of the sort mentioned previously might comprise a correction channel storing negative values for subtraction. In a sense there is no "negative light," and in this application one must somewhere take away something already laid down during a previous projection, which, for example, phase reversal alone really would not accomplish. The final light valve thus must display the difference between the positive and negative patterns or channels to reveal an image that, in principle, would be sharp.

To increase speed requires considering the exposure method or pattern. Exposing the entire slice from all directions at once can cause problems with scatter. One can spin an X-ray tube around the subject, or have several fixed tubes activated in sequence, or ring the subject with an extended anode along which the electron beam travels either by deflection or by cathode motion. Powering a moving tube through slip rings is not ruinous because of electrical noise if the signal comes out optically. A transformer secondary can rotate around the primary, but typical power for a single tube is 10 kilowatts so some equipment mass and heating problems are involved and an increased power frequency desirable. Several adjacent slices could be formed during one turn if several tiers of film were available with suitable light guides. Presumably one can produce X-ray CT motion pictures in the fashion of this example in order either to obtain a dynamic view or, for example, after injection of contrast material to follow the vascular versus the parenchymal phase when the medium has diffused.

Though appropriate image processing has been demonstrated, the device of

62   **COMPUTED TOMOGRAPHY**

Figure 4.5 has not been constructed. It is described here to give a physical understanding of what is necessarily involved in forming a computed tomographic image, even when that is done by mathematical manipulations buried in a computer. A digital computer is able to store the value of brightness at each point in an image as a number, and this number can be increased or decreased as the result of successive projections falling across the particular location. Thus with considerable accuracy a computer can store and modify an image.

4.2. DIGITAL COMPUTED TOMOGRAPHY

In 1917 the Austrian mathematician J. Radon proved that a two- or three-dimensional object can be reconstructed uniquely from the infinite set of all of its projections (Sec. 3.3). Much interest in this fact was generated by the development by Hounsfield (1973, 1980) of a practical X-ray system with functional apparatus that gave excellent performance. For those not familiar with this, the approach can be made plausible by the simplification of Figure 4.6.

This figure shows an object penetrated by a series of transverse pencil beams of radiation, for example, X-rays or high frequency radio or ultrasound. The ratio of input to output intensity for beam 1 is a measure of the amount of material along the beam, and one can write an equation assuming all the absorption takes place at the four indicated points (or in the region around each) along the beam. In the X-ray case, for a monochromatic beam, the equation merely states that the logarithm of the ratio of beam intensity before and after the attenuation of passage is the sum of the four attenuation coefficients along the beam times the path length in each little region. Similar equations can be written for the next three beams. One produces a graph of the amount of material in each column as one goes across the bases of the columns in a scan, in this case for four columns yielding four equations. For an image one also needs to know how the amount of material changes

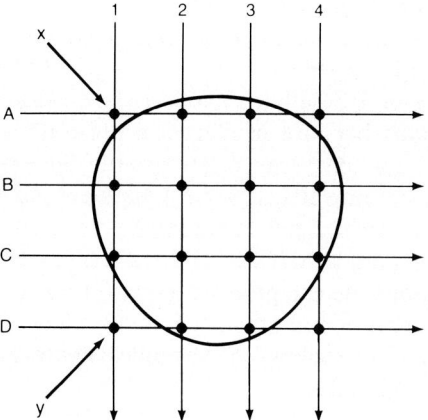

FIGURE 4.6. Forming a cross sectional image of a neck or torso extending out of the page by digital analysis of the attenuation of a sequence of rays in the plane of the page where the structures to be imaged lie. Attenuation is assumed all localized at the positions shown, and these then constitute the 16 subdivisions or picture elements in the crude image.

up and down the columns. From the four horizontal beams one obtains the amount of material in different layers without regard to how the material is distributed from side to side, and these four more equations are applicable with the original four to determine the absorption at the sixteen points in common, which constitutes a crude image. Since there are sixteen unknown absorptions and only eight equations, one must reshuffle the unknowns among eight more equations by projections in the directions labeled X and Y. The computing process is actually carried out in a different way, but it is clear that one can form an image of a cross section by projecting many beams at many angles into the edge of a desired "slice"; the more and finer the beams or rays the finer the detail. Examples such as are given in this book employ a total of 20,000 up to about 250,000 rays,* from equipment with a few hundred to a few thousand detectors.

What is stored is the projection toward the edge of a slice of the amount of material across the slice perpendicular to the line along which the recording is made. These data are usually acquired in a mechanical scanner in which the source and perhaps the detector both move about the subject. In the X-ray case, to which we will give particular attention as an example, this means rapidly moving an X-ray tube fronted by a small aperture to limit the beam or beams, and in some cases synchronously moving one or more detectors as an alternative to a fixed array of many detectors functioning simultaneously. It is both practically and conceptually useful to note that this one-dimensional projection of a two-dimensional slice is also available as any single line in a television X-ray system in which a television camera "watches" a screen exposed to the cone of X-rays traversing a subject during normal fluoroscopy; accuracy is not high but rotation of the subject does give the full set of projections at any one line on the screen, while the other lines fill in information on the adjacent sections for the full three dimensions. From these projections must be formed an image. The necessity for this extra step has yielded the name of CAT scan for computerized axial tomography or computer aided tomography, which has been reduced to CT scan meaning computed tomography.

There is no special requirement of parallel rays as shown in Figure 4.6. It is merely necessary that each point be viewed from many directions in order to realign the nearby points among the various equations. The requirement of viewing each point from a variety of directions also appears in ultrasonic imaging where a compound scan allows collecting echos from specular reflectors at all angles so none go unseen: thus it is not surprising that the first X-ray CT units used the same scanning mechanism as Howry earlier had used to produce Figure 2.10. In these early units a single pencil of rays moved back and forth parallel to itself while the entire assembly slowly circled the subject. Scattered radiation in the X-ray case was largely rejected because apertures before source and detector

---

*To define detail to the extent of n subdivisions across the width of the subject is here seen to involve n rays in each of n projections to completely specify n^2 points. With a circular image and n divisions across the diameter the corresponding number of projections is $\pi n/4$. Reconstruction from less data uses assumptions about the missing data.

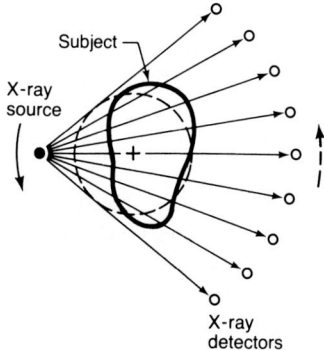

FIGURE 4.7. A diverging source of X-rays falling on a ring of electronic detectors provides a fan beam in a sheet through the subject. The source and detectors can rotate together, or the source alone can rotate within a complete ring of stationary detectors. All points within the circle around the axis of rotation are traversed by a complete set of rays. Reducing the distance from source to axis increases magnification. Apertures before the detectors provide one way of rejecting scattered radiation arriving from inappropriate directions.

minimized acceptance of photons arriving along other paths. Overall speed was increased by use of several rays and detectors simultaneously.

A single standard X-ray tube can supply several beams or diverging rays. The apparatus is contained in a large shield shaped like a doughnut through which the subject is positioned. The arrangement is depicted in Figure 4.7. Around the subject spins the source of the X-rays, in this case many rays diverging in a fan pattern. Each ray is shown terminating upon a detector to measure the emerging intensity. For the sake of rapidity these detectors do not count individual photons but rather record momentary total flux. They may be some form of solid state detector, or they can be ionization chambers. The detectors can be mounted at one end of a rigid support or gantry to whose other end is attached the source so that they can rotate together around the subject. This is the so-called "third generation" scanner. Alternatively, in "fourth generation" scanners a thousand or more detectors totally encircle the subject and stand still while only the source moves. Fourth generation scanners do not necessarily produce better images. All regions within the dotted circle are traversed by enough rays to yield a complete set of projections; only partial information is available elsewhere. If the source is shifted toward the pivot, with axis shown as a cross within the subject, magnification increases while the amount of the subject that can be viewed decreases. With a fan beam, spacing is not quite constant across the subject. A full image requires rotation of 180° plus the fan angle.

Consideration of this figure will reveal that, within the circle, any or *every point of the subject "sees" a rotating ray, or sequence of rays, through itself* as the source rotates. After a scan each point has seen rays converging on it and diverging from it. In fourth generation scanners each fixed detector sees a sequence of rays converging on it from a changing direction as the source rotates, giving the impression of a "reverse fan beam." In that last case a small source motion can be interpreted as producing a whole new set of rays closely spaced to the previous, at each of the widely spaced detectors shown. Data can thus be grouped either to give many rays at each angle or many angles (projections) with broader rays for one turn; visualization of fine detail is enhanced by the former as will be mentioned with Figure 4.8. Scatter rejection is difficult in fourth generation scanners because of rays entering any one detector at angles changing over 60° in some cases.

One detector monitors the direct X-ray beam, and all others are compared to it, thus compensating for changes in X-ray output. Since signals are periodic steps up from the condition of no X-rays, zero drift in the electronics can be compensated. The detectors can use some form of solid state device such as a semiconductor *p-n* junction or a scintillator such as bismuth germanate with a photomultiplier or photodiode, or they can be ionization chambers. For the latter, xenon which has an atomic number of 54 is employed. This has spectral absorption characteristics similar to cesium iodide (iodine being 53 and cesium 55) which is the phosphor most often used in fluoroscopy with an image intensifier. The gas is used at a pressure of 5–25 atmospheres to place more atoms in the beam. A sealed ion chamber tends to be independent of temperature and drift but is generally too direction sensitive for fourth generation scanners.

When defining detail with a noisy medium, output signal size is not the ultimate criterion for detector excellence. Thus it is better for the ratio of signal-to-noise (S/N) that twice as many quanta be absorbed even in a detector having half the amplification (due either to less gain or some energy loss) even though this gives the same average output energy in the signal. The "quantum efficiency" of a detector is the fraction of incident photons that contribute to the output, that is, the probability that any one incident photon will be detected. A fraction approaching 1 is desirable. It is also better if the detector does not add noise of its own since, if it does, it can give the same average output as a "quiet" detector but with greater fluctuations. Detector noise requires counting more photons (or a greater fraction of these incident) to obtain the S/N that a noise-free detector would find with a lesser number. The "detective quantum efficiency" is the fraction of incident photons that would have to be detected without additional noise by the detector being considered in order to yield the signal-to-noise ratio being observed (Jones, 1968). The comparison is between input and output signal-to-noise ratios rather than input and output counts. It is calculated by taking the square of the ratio of the actual output S/N ratio, measured following any essential amplifiers, to the input S/N ratio which is calculated* from the average photon arrival rate and observation duration to give $\sqrt{n}$. Quantum efficiency will affect the necessary subsequent amplification of a signal, but the generally smaller number called detective quantum efficiency is the characteristic to maximize in comparing detectors since it determines the limit of detectability.

Photographic recording of each projection gives greater geometric resolution than does use of an array of discrete detectors, but the detective quantum efficiency of film is low. It can be measured by taking the ratio of signals out and in (a density difference between two small regions resulting from the exposure differences) compared with noise out and in (standard deviation in density of a small area versus the square root of the number of photons on the observed area). The

---

*If a signal of n photons has fluctuations or noise of $\sqrt{n}$, as in Section 3.2, then $n/\sqrt{n}$ or $\sqrt{n}$ is the ratio.

result is typically in the range of a percent. Better film could be made at the expense of decreased shelf life or refrigerated storage.

One method for producing the image is to use a computer to gradually form a pattern that would produce the observed projections. In the computer is stored a set of numbers that correspond to the different densities of the subject at different spots; it is convenient to think that somewhere in the computer there is actually a row of columns of numbers arranged to correspond to the 16 or more spots of Figure 4.6, plus rows of numbers corresponding to the projections through these spots. The spots or picture elements are often each referred to as a "pixel." To commence the estimation one assumes some distribution of densities among the points, for example, an array of zeros as in a blank picture. One picks one row (or column), adds up the absorptions at the various points along its length, subtracts this from the actually observed value at the end of that ray, and distributes this difference equally among all the points along the ray. That line then matches reality. This is then repeated for each of the other lines that give rise to the projections, and each computation partially undoes the matching of the previous computations. However, if the corrections are repeated many times the picture should (but may not always) approach one of the object yielding the actually observed projections. One can improve performance by imposing constraints such as no value under zero for any pixel, or a known value for bone absorption when a value in that general range is encountered.

This iterative method of successive approximations, in which corrections are successively applied to an arbitrary starting image to bring it into better agreement with measured projections, is related to techniques long used by mathematicians for solving massive sets of equations (Jacobi, 1846) and was used in the first EMI scanners. If one starts with a blank image, the first iteration is equivalent to "back projection" of the projections as suggested by Frank to form an image (Fig. 4.2). Corrections can be made either point-by-point for all rays passing through that point, ray-by-ray by correcting all points contributing to that ray, or all-at-once by calculating all projections in advance.

Analytical reconstruction using exact formulas was introduced by Cormack (1963) and has been modified extensively since (see also Sec. 3.3). It can take several forms. In "Fourier reconstruction" the frequency components calculated as comprising each projection are taken as the frequency components in the object or desired image in the corresponding direction; the computer adds these frequency components from various directions (sine and cosine waves of appropriate amplitude) at each point to give the original object. In "filtered back projection" the profiles are modified in a particular way before the computer then adds them in Frank-like fashion. The form of the modification (Bracewell and Riddle, 1967) is known as a convolution integral and so this is also called the "convolution method." Radon's method was also in this general form and, by properly selecting the sampling interval and detector aperture to remove necessity for smoothing, it itself is not difficult (Galvin and Bjarngard, 1975). Negative values of the modified projection undo smear deposited by other projections but limit some modes of optical processing because one cannot readily remove signal laid down earlier

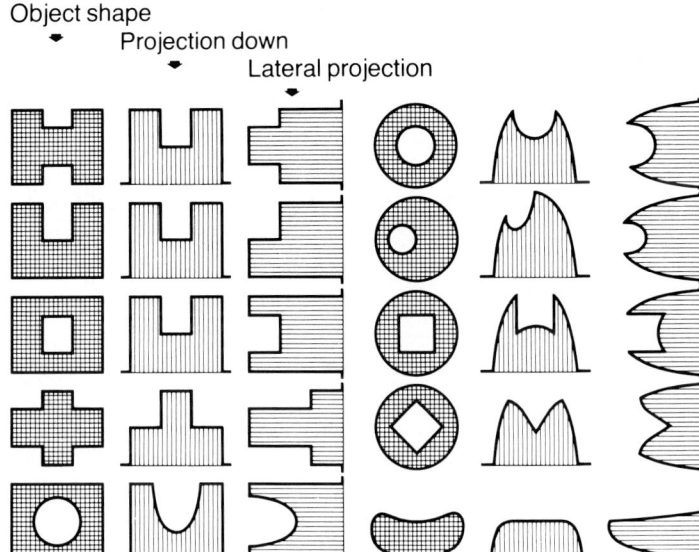

FIGURE 4.8. From the patterns in a vertical and horizontal projection one can guess the shape of an object in these simple cases. Internal structure is represented, though more projections than two would be required for more detail.

except in a digital storage array of numbers. In Chapter 10 further aspects of the modification of an image or its projections will be discussed.

Analytical methods are faster than iterative methods since a complete analytical reconstruction can be performed in the time of a single iteration, and, with filtered back projection, the projection can be modified or processed as it is collected for immediate image display after the scan. Processing times are a matter of seconds. Interpolation is sometimes used to fill in missing data or else the image is smoothed. Noise affects different reconstruction methods similarly when the data are complete. Among the many writings and reference summaries are Cho et al. (1975), Lewitt et al. (1978), and Brooks and DiChiro (1975).

Lest the formation of an image from a set of projections seem an incomprehensible piece of computer magic involving unrelated shapes, the simple exercise of Figure 4.8 may be helpful. Ten simple cross sectional shapes are shown, and to the right of each is shown the up-and-down and side-to-side projections or graphs of the amount of material in each column and in each row. Given just the two projections one should be able to infer the shape of the object that produced them, at least for these simple symmetrical objects. It should be noted that object numbers 8 and 9 are the same but somewhat rotated, thus emphasizing the need for projections taken at a multitude of angles. It should be clear, for example, that the recorded shape of the projection curves does indicate the presence and shape of internal items such as holes, and it should be plausible that great detail could be inferred if many projections were available.

68 COMPUTED TOMOGRAPHY

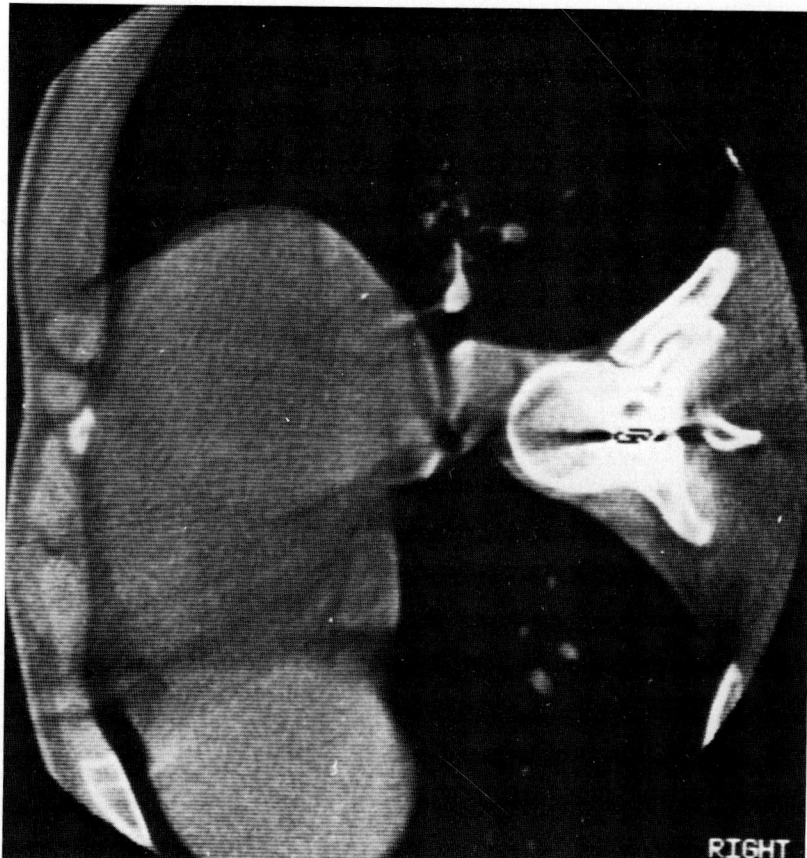

FIGURE 4.9. A computed tomogram of a man with a knife blade in his spine. Two sorts of artifact are produced by this opaque object.

Consideration of this figure also should make it plausible that decreasing the number of projections or angles of view does not much decrease spatial resolution, but decreasing the number of rays does decrease spatial resolution, that is, if limited by dose or computation time, more fine lines per view than directions of view are desirable (see also Shepp and Stein, 1977; Schulz et al., 1977). Generally no detail finer than about the width of a ray will be seen. In Figure 4.8 there is considerable detail as if from many narrow rays even though only two views or projections are given for each object.

In Figure 4.9 is shown an actual X-ray CT image. This is the same subject as in Figure 3.3, but instead of stepping the subject past the ring of photodetectors on a CT scanner to produce a somewhat traditional X-ray image sometimes called a "scanogram," the subject was fixed and the X-ray source spun to produce the data for a computed tomogram. (Figure 3.2 showed the detectors in a line to emphasize comparison with flat X-ray film but the correct drawing is Fig. 4.7.) The knife can be seen in the subject's spine; he was very lucky since only a few cross fibers were cut and no paralysis resulted. This figure is oriented in customary manner as if the subject were being viewed from the feet so that his right side is displayed to the left, or in this case, down.

Two sorts of artifact are demonstrated in this figure. Radiating upward from the knife are seen dark stripes that have nothing to do with body structures. Such are often seen in CT images in the vicinity of some relatively opaque body. For example, they are often seen around teeth with metallic fillings in images of the head, and in the chest, such a streak across the aorta can alter diagnosis of a dissecting aorta. By reference to Figure 4.6 it should be plausible that an opaque object at one point obliterates each ray through that point and thus eliminates information about points beyond; one can imagine the loss of necessary equations. Parts of each projection can either be unobservable or unreliable because they exceed the useful range of the detector or other components of the overall system including amplifiers and perhaps the computer program itself. This type of artifact extends its influence to parts of the image that are relatively remote. Smoothing techniques can be used in some cases to approximate missing information and somewhat restore a troublesome image (Lewitt et al., 1978).

Depending somewhat on the details of computation used to form the image there can be other distortions of brightness right at any sudden extreme change in opacity. For example, the knife blade here appears somewhat outlined. In the present case this seems due to a mathematical limitation called the Gibbs phenomenon (Guillemin, 1949). Various waveforms are often represented by sums of the many frequency components that are present in what is termed a Fourier series. Thus a flat-topped square wave can be considered as made up of a sine wave of frequency equal to the repetition rate of the square wave plus some sine wave of three times the frequency, plus some of five times the frequency, etc. It is often assumed that the sum of such a combination of frequencies will yield (converge to) a perfect flat-topped wave but this is not true. At any sudden step or discontinuity there is an 18% overshoot, followed by a lesser overshoot no matter how many frequencies are included; the duration of the error will become shorter as the frequency response is increased to include higher frequencies, but it will still be present at any edge. (Another aspect of this is shown in Fig. 5.4.) The problem is simply that when a scene is represented by a group of sinusoidal waves of different frequencies, they cannot add up to two different values depending on whether a sudden step is approached from the right or the left.

It should be emphasized that these effects can take place when processing of an image is being done and need not occur in ordinary images. For a comparison, in Figure 4.10 is seen an ordinary radiograph in which different sized pieces of metal also appear. Following surgery this patient complained of having difficulty in bending over. A large retractor had unfortunately been left inside him. It is seen that there are no disturbances of the image either from this large opaque body or with the smaller surgical clips. In an ultrasonic image of this same subject there could be noticeable diffraction effects and possible distortions that are produced at an interface where longitudinal waves can become partially converted to transverse waves in a solid. In a nuclear magnetic resonance image the nonmagnetic surgical clips would be no problem since they do not constitute closed current

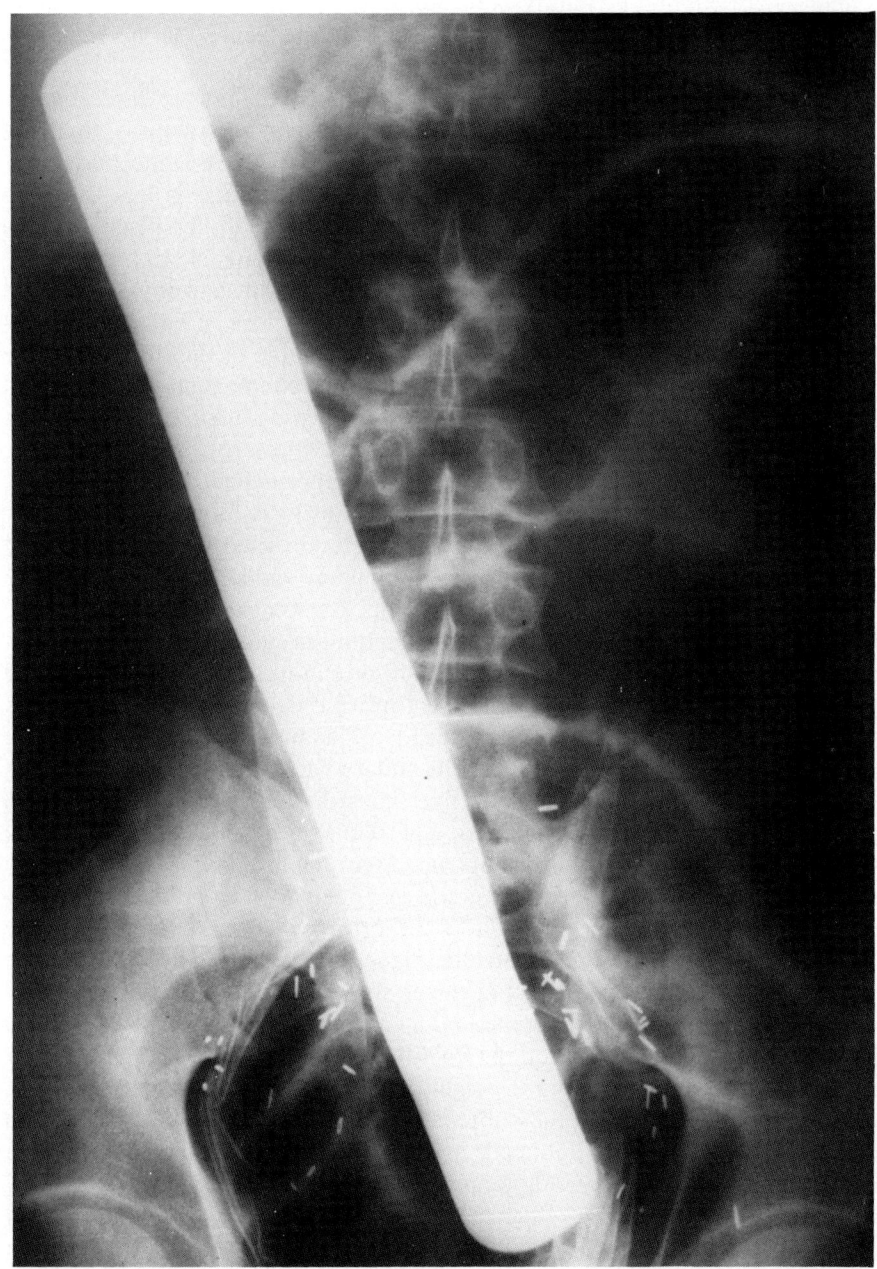

FIGURE 4.10. An ordinary projection image is not disturbed by large and small pieces of metal (surgical clips and a retractor left in by accident). With other modalities than X-rays the large mass of metal could create problems.

paths, but circulating and shielding currents could be induced into the large mass of metal by the changing magnetic fields to be discussed in Chapter 7.

There are several other sources of artifacts in CT X-ray images. The granularity associated with quantum noise (Fig. 3.4) can be considered as an artifact of random or irregular form. If all projections are not of the same object, that is, if there is subject movement, then streaking artifacts result. There are also systematic errors that repeat but can be reduced with effort. If the detectors are not all well matched then artifacts can result in a form of a ring that changes with time, intensity, and spectral input or voltage; fourth generation scanners eliminated this by making each detector contribute to each image element. If the number of intervals in a scanning system is too small compared with the desired spatial resolving power then a star pattern can result, though this usually is not a problem with present commercial scanners. Since absorption varies with X-ray energy, and most beams contain a spread in energy, the actual composition and the effect at each region will change progressively as the beam traverses a body; the effect of this "beam hardening artifact" on the computations is to introduce a central depression in the appearance of a disk-shaped object and make small extended contrast gradients difficult to observe. In early scanners this last problem was reduced by placing the patient's body in a compensating body of water so that the path length of each ray did not much depend upon the size and shape of the patient. A third generation scanner may use 500 samples per channel per scan and so a 1.5 mm detector window may move 5 mm between views. Thus a grid-controlled tube to turn on the X-ray in pulses, of say, 30 msec has been employed to reduce blur in each projection in some commercial equipment. In principle, the detector–source line of sight should be stepped in increments of half the detector width or less to avoid "aliasing" artifacts (Brooks et al., 1979) and increased noise, though in practice a sample space of twice this from adjacent detectors may not much degrade performance, probably because of averaging the object over the slice and the finite source effect.

In any hospital environment there are mundane practical considerations that can dominate performance and must be considered. Thus injection of an iodine contrast medium causes nausea in a small fraction of patients, probably due to osmotic effects in the brain. Vomiting into the detectors causes severe artifacts, and they should be protected against this as well as against spilled saline and so on.

Fineness of observable detail varies with dose, and thick patients appear grainy in practice because fewer quanta emerge. How detail varies with dose depends upon what is assumed to remain constant, and upon the sources of noise. Hopefully the noise is arranged to be limited by the number of X-ray photons interacting with the detector. In a signal of n photons there will be random variations or noise of about $\sqrt{n}$ according to the Poisson distribution, and thus if the signal-to-noise ratio is to be increased, the photon flux and dose must be increased proportional to the square of that ratio. Noise is actually increased slightly beyond $\sqrt{n}$ by variation in detector efficiency with photon energy variations in the beam from an actual X-ray tube, but the above generally applies. Relating dose to the photon flux accurately is difficult because of the scattering of electrons and photons out of the slice (and perhaps back into it) but a good approximation is possible toward the center of the slice. As the size of a subject increases, the dose to maintain constant both geometrical resolution and density discrimination (ratio of signal to noise in both) increases exponentially with radius.

If X-ray energy is fixed but finer detail is to be observed, then there are several possibilities. In Section 3.2 we noted that to see detail half as wide (not oriented in any particular direction) on a radiograph required passing the same number of photons through an area half as long in both transverse-to-the-beam directions, thus giving four times the dose. That is, for a fixed object of fixed contrast or unchanging dimensions along the beam, the number of photons per unit area and dose varied with the inverse square of a resolution distance. (One rad of dose corresponds to 100 ergs of absorbed energy per gram of tissue.) If instead one wished to observe an object half as large in all three dimensions, then in order also to improve the signal steadiness defining thickness it is necessary to increase dose with the inverse fourth power of a resolution distance. Similar considerations apply to a CT scanner. Suppose a structure passing through the slice is still to be just seen after reduction to half the size. The necessary increase in dose would be: one factor of two in order to restore the *difference* in number of photons through the thinner structure compared with the surroundings now that the absorption difference along the beam is half for half the thickness; this restores the differences but they must become twice as large in order to resolve the smaller meaningful differences toward and away from the detector in a thinner object, that is, a second factor of two to restore signal-to-noise ratio; a third factor of two is required to give increased resolution along the slice edge (along the projection) in order to cram twice as many quanta through the group of rays comprising one view. Dose then varies with the inverse *cube* of the resolution distance. If an obliquely oriented small structure is not to be blurred in averaging over the slice thickness, one prefers a slice thickness that equals the resolution distance, or is at least proportional to it. This last requires that all the quanta limited by the other three factors be passed through half as thick a slice with half the mass for twice the dose, in which case dose varies with the inverse *fourth power* of detail.

In real machines the slice thickness is often 5–10 times the smallest resolvable object dimension in order to reduce total dose, dose per slice, and number of scans to cover a given volume. Machines typically allow slice thickness adjustment from 2 to 10 mm; and for scanners that promise detail resolution slightly better than 1 mm, it clearly becomes meaningless if structure is then averaged over one of the thicker slices. A typical dose is 1 rad for 2 mm detail with a signal/noise of 200, and because of the fourth power law, to increase this to an image as sharp as a typical radiograph would require a dose orders of magnitude too large. In Figure 3.4 was seen the effect of decreasing slice thickness which proportionately decreased photon number while tube output was unchanged. The measurements of Figure 3.7 included the effects of contrast as well as size and dose.

If one requires many adjacent sections to be imaged, for example, to visualize part of the brain, then classical tomography (Fig. 4.1) will dose all sections in making each. Dose is a matter of deposition of energy in a region, perhaps causing damage, and ten adjacent CT sections do not require added dose over a single section (though they extend the region exposed to dose) while with classic methods, ten times the dose would be required. To be a bit more precise, small spillover from adjacent sections slightly increases dose, and dose for a series of abutting slices may be 20% greater than the single slice dose, depending upon distance between slices and the sharpness with which X-ray intensity drops with distance from the slice center. This is less damaging than in the classical case where dose is cumulative and increases proportional to the number of sections scanned. If one still wishes to perform classical tomography for small high-contrast structures in detail, one can use CT for localization, the CT not necessarily being oriented the same as the classical scan.

As an approximation, dose to a single slice of tissue is obtained by multiplying the "technique" in milliampere-seconds (number of views or projections times number of milliseconds for each set of rays times tube current) by a factor determined by calibration; the factor may be about 1. Maximum dose from a single scan is a useful parameter that indicates what a particular machine source is able to deposit on a surface, and which can be measured at the surface of a specified object such as a large plastic cylinder. Critical tissues can then be positioned away from places of maximum energy. With a 360° scan all points in the patient's skin will receive about the same dose, and at the patient's center about a third of that. For a 180° scanner the isodose curves in a slice are horseshoe shaped, with a skin variation in dose of perhaps 10:1 because fewer quanta emerge than enter, thus giving less dose to the far side. At greater kilovoltage and beam penetration, the depth–dose will increase to be closer to that at the skin. A single value of maximum dose resulting from one scan does not fully serve as a measure of total energy absorbed, but remains a useful parameter.

The speed of production of an image is always of concern, and becomes critical if one considers trying to image a beating heart, for example. Mechanical scanning is somewhat slow but modern X-ray units are able to store successive image information, roughly speaking, every few seconds if the data are stored rather than being processed at the time into successive completed images; indeed, the X-ray tube current in some cases must be reduced to prevent overloading the tube. Scanning without mechanical motion is possible for all modalities of interest, including nuclear magnetic resonance and ultrasonics, as will be discussed in later chapters. Scanning X-ray tubes in which an electron beam is deflected over the target in order to deflect the outgoing X-ray beam have long been known (Moon, 1950; Towe and Jacobs, 1981) and these can be improved. Companies in America and Japan have under development systems in which an anode around the patient (or a crescent half around) is scanned by an electron beam to produce a section image in 50 msec for real-time heart studies.

One can instead activate in sequence a series of separate sources in order to achieve the effect of a moving source. In the case of a ring of X-ray tubes one would prefer not to pulse them all on simultaneously because scattered radiation from each would then degrade all the other projections. For rapidity in the X-ray case, the detectors are generally not used as quantum counters, but rather an average current is recorded from an ionization chamber or the total glow recorded from a small scintillator.

Ultimately a limitation on speed is the requirement that enough quanta must emerge with each ray to define its intensity to a statistically significant degree. The rays must be closely enough spaced so as not to leave big spaces between. The fineness of detail to be observed cannot be smaller than the width of the rays. A large expanse can be covered either by making the rays broad or by having many of them, only the latter providing fine geometrical detail. Thus to cover a large subject in fine detail requires the computation of many points, each from many rays having many quanta. This can demand a relatively high dose to the subject and may require considerable time though the latter can be decreased with an extremely intense source (or sources) plus improvements in the mechanics of the system. The same considerations apply to computed imaging using other modalities where noise fluctuations also limit the certainty of measurement of a ray intensity, and where fluctuations can be reduced only by increasing intensity, or by averaging over a larger area (at the expense of geometrical resolution) or a longer time; sacrifice of resolution in brightness, whose distinguishable steps are comparable in magnitude to the random fluctuations in intensity, results when too little energy is spread over too much area.

To the extent that a bodily process is really cyclic, one can attempt to store successive projections from successive moments of the same geometry or phase in the cycle (or better, record all phases along with a time indicator such as an ECG, and select later). The heart can be imaged in this fashion to some extent (Robb et al., 1974) though there are beat-to-beat variations in shape which may be more extreme in sick subjects of special interest. In a standard television X-ray system each line is a projection. Thus a television image of a rotating subject yields the complete set of projections for many slices through the subject. If one attempts to exploit such a concept it is helpful to select an individual television tube which has as uniform a response as possible over its surface, though subsequent processing for nonuniformities is possible if there are no changes in relative sensitivity with time. Recordings of the heart in end-systole and end-diastole have been made by Lantz et al. (1975) with a fast television system having a frame rate of 250 per second and a linear horizontal scan time of 32 microseconds. The recording of many lines from each frame allows the production of a number of image slices, which permits the three-dimensional reconstruction of the organ. Such a process for image formation also has certain mathematical speed advantages over alternatives (Ramachandran & Lakshminarayanan, 1971), and these have applicability to NMR and other images as well. Associated with any attempt to speed a television system should be an observation of the ability of the camera tube to respond rapidly, for example, by recording a rotating disk.

Again referring to the X-ray case, from one set of projections around the subject one can compute the absorption for X-rays at "each" point in that plane across the subject. The subject then can be moved a small distance through the apparatus and each point calculated for a new plane parallel to the first one. This process can be continued to build up many cross-sectional images, each of which

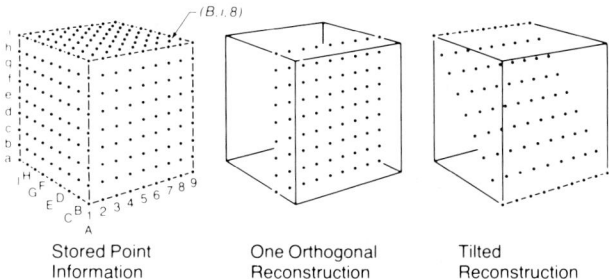

FIGURE 4.11. The computed tomograph image stores in a computer the brightness at each point in a plane as one number. Successive scans parallel to the first store information about the points in successive planes. Any sequence of points can then be displayed from within the entire three-dimensional volume in order to show the appearance of planes through the subject other than the original ones.

can be displayed upon a cathode ray tube or television screen. However, once one has information stored about all points throughout a volume of a body,* then these points can be called up in different sequences for the display of different sections through the body as in Figure 4.11.

Suppose in that figure that the first image computed and stored was the left hand face of the cube, followed by the plane parallel to it with the number 2, after which the third set of projections would be collected to form the parallel plane containing the number 3, and so on. All the points in the cube would gradually be accumulated and stored in the computer, and it is convenient to think of the position of the storage of a number as corresponding to the position of a point in the cube. Any of the original planes could be displayed by "calling up" the numbers representing the points in that plane and producing a proportional brightness on the oscilloscope screen at the corresponding position. On the screen instead could be displayed the dots shown at the center section of Figure 4.11. On the lower left of the screen would be displayed the point designated (1,B,a) which came from the first section, for example. Next to it would be displayed a point from the bottom of the second section and in one increment from the right side, and so on. By sequentially calling up all the points located inward one increment from the right face of the cube one can display a section through the subject perpendicular to all the planes that were originally scanned. In the display one can instead move to the left one increment for each successive step upward to a new line on the television screen; this results in the display of a plane at an angle through the body as shown in the right hand part of the figure. It should be clear that from this array of computed data one can display any section through the volume. Often the sections across a subject are more widely spaced than are the lines across the section in making the original projections, in which case the resolution in a tilted display will not be the same up and down as across the image.

---

*As a picture element is called a pixel, so a little element of volume such as one of these dots is sometimes called a voxel or boxel.

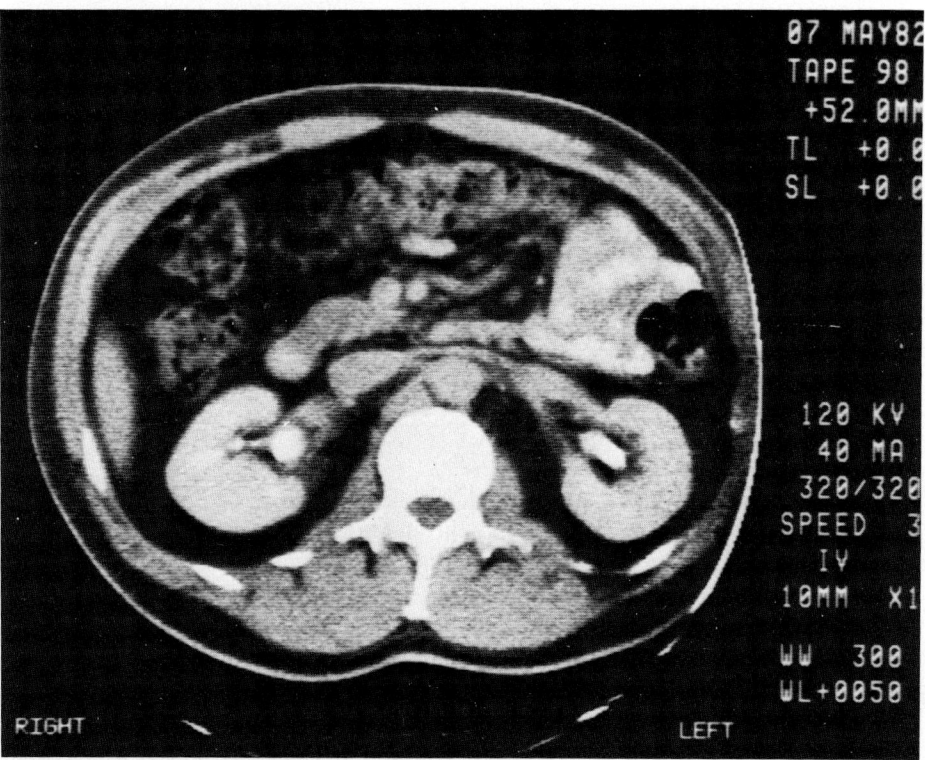

FIGURE 4.12. Computed tomogram of kidneys, seen on both sides of the spine, with the bowel at the top of the image containing gas, feces, and oral contrast agent. The kidney looks lighter than the muscle adjacent to the spine because of filtration of injected iodine contrast agent. The circle anterior to the spine is the aorta and the ellipse beside it, the vena cava; the renal veins from both sides are seen going to the latter. The bottom tip of the liver is at the left side of the image, which is the right side of the patient. Tube voltage was 120 kV. This is one of the initially formed series of transverse sections from which the next two figures are derived.

Examples of different display orientations from a set of cross-sectional images are seen in Figures 4.12, 4.13, and 4.14. The basic type of image is a transverse section through both kidneys as in Figure 4.12. The dense white at the kidney centers is iodine opacified urine, a consequence of an intravenous contrast agent. In Figure 4.13 is a coronal reconstruction (as if the front of the subject were removed) from several transverse images. (This is rather like the familiar intravenous pyelogram.) Since resolution in Figure 4.12 is about a millimeter and slice thickness a centimeter, the vertical and horizontal resolutions are now different. In Figure 4.14 is a sagittal reconstruction of the left kidney (as if the subject were viewed from the side) from the same data, again with differing resolutions in the two directions. For a comparison, an ultrasonic image of the same kidney is given in Chapter 6.

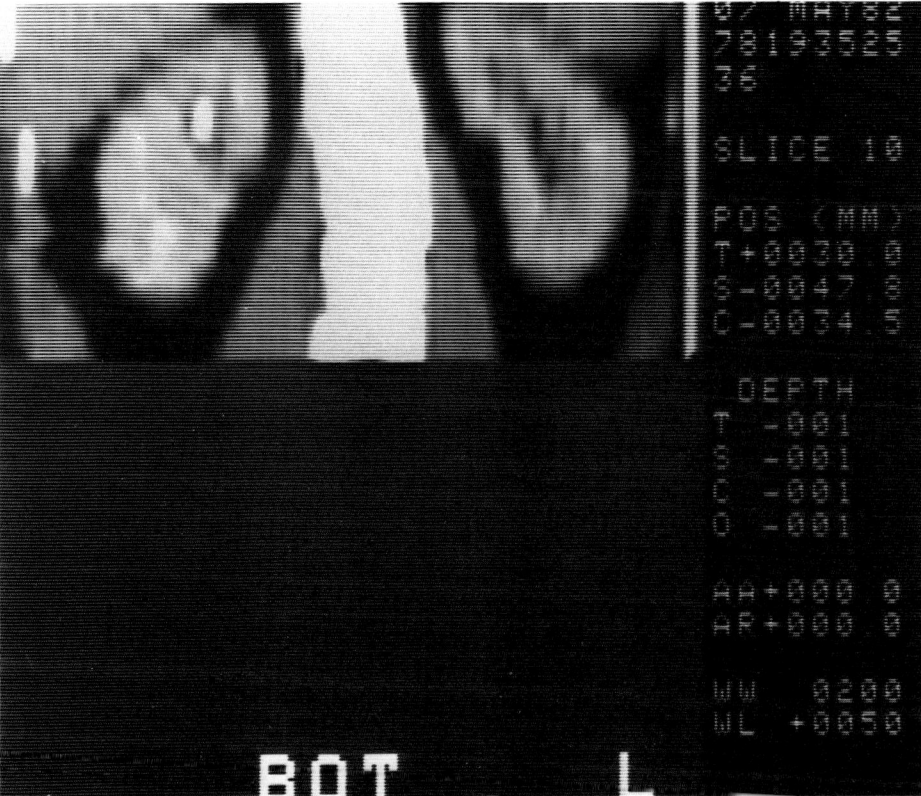

FIGURE 4.13. Coronal reconstruction of both kidneys from data of a series of transverse sections. Psoas muscles are seen on the sides of the spine in the center, with the liver at the upper left of the figure and the spleen at the upper right. White in the kidney is intravenous contrast agent. The "step" appearance is due to the slice thickness and advance being greater than the resolution in the individual original sections.

Computing procedures have become sufficiently rapid that one can change the orientation of the section being observed while watching it in "real time," once the individual points have been computed and stored. Where about 5 sections per second can be alternated, one can achieve an excellent three-dimensional impression of the inside of a subject by "running back and forth" in any direction. Interslice smoothing makes viewing less jumpy and more satisfactory, the amount being adjusted to suit. One can also do such things as observe a plane through the subject containing the optic nerve as the subject is apparently rotated around the line of the nerve. The effects are more impressive than can be described in print with static photographs.

Each point in an image corresponds to a number stored in the computer and these numbers actually represent in a quantitative fashion the X-ray absorption in

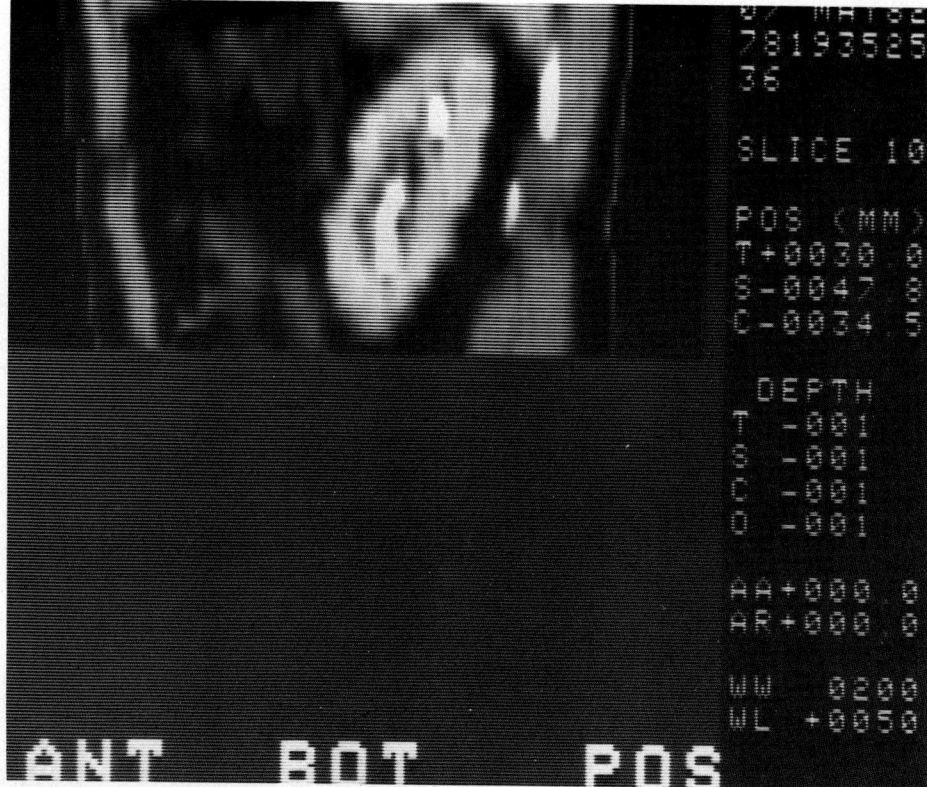

FIGURE 4.14. Saggital reconstruction of left kidney from data of a series of transverse images, again showing less resolution up and down. The anterior abdominal wall is to the left, and two white ribs are seen at the right.

the corresponding small volume of the subject.* Thus from a set of CT images one can plan radiation therapy. In principle one can automatically add the areas of all regions with the same CT number to get the size of a structure. However, the latter probably cannot automatically sort out the area of metastasis and its response to chemotherapy, and possibilities for quantitative evaluation of the size of structures are given in Section 10.6. People have attempted to characterize tissues by their CT numbers but it is always necessary to document that machine-related variations are less than the differences thought to be significant between different tissue types (Levi et al., 1982). If nothing else, the previously mentioned beam-hardening artifact will somewhat alter the absolute values of these numbers, depending upon what is adjacent and the scan conditions. Also, small structures can be misjudged because of "partial volume effects": just as a small bubble at the

---

*One hounsfield (H) represents 0.1% of the attenuation coefficient of water and the scale zero is water. Then air has H of -1000 and bone about $+1000$.

edge of an ultrasonic slice (partially in it) will be judged too small from its echo, so a small structure in an X-ray CT image will have its CT number modified if it is only partially in the slice.

Other modalities can be employed to form images computed from transmitted rays in the plane of the image. High frequency radio can be collimated into rays small with respect to body dimensions, and its use in image formation has been mentioned (Iskander et al., 1981; Rao et al., 1980). In a nonuniform magnetic field nuclear magnetic resonances can be observed a row at a time, and such image formation will be discussed in Chapter 7. With ultrasound one can form an image either of the distribution of absorptivity or of refractive index (velocity profile). Sound rays curve slightly in the body in response to structures where velocity changes, and this results in "forbidden regions," where direct rays never penetrate, surrounding any point where the index of refraction exhibits a localized maximum (McKinnon and Bates, 1980). In spite of small curvature, they conclude that it is seldom feasible to obtain better images than those reconstructed on the assumption that ultrasonic rays travel along straight lines.

Editorials in radiology journals have confused the role of William Oldendorf. He suggested a method (1961) for monitoring the X-ray opacity at an internal point by spinning a ray source and detector around a subject. The point of the subject in the plane of spin on the axis of spin gives unchanging absorption during each rotation (zero frequency or steady signal) while all other regions move in and out of the beam during rotation to give an ac signal that can be rejected. That center point can gradually be shifted around within the subject by translating subject or apparatus to build up a tomogram, though considerable time and dose would seem to be required relative to the projection methods that constitute what is normally considered computed tomography. No computer is required and the method could be useful for observing a single region. There is an interesting analogy here to the compound scans of ultrasonics and to a sensitive point method to be mentioned for nuclear magnetic resonance where the external magnetic field spins, allowing unchanging signals from only one point. The concept appears different from the usual CT though scrutiny reveals some possible correspondences that are instructive. In principle a line through each point in each direction around a half circle would be enough to first define every point, and then translation could build up an image; this is the same requirement as for one complete CT scan with the same amount of fluctuation in the beam, but when one alters the concept to one of rejecting all spin frequencies in the final signal other than zero frequency, then there is an implication of several turns for each point. However a spin of n turns about a point can be done with a ray $1/n$ as bright in terms of photons per unit area, if one averages over the large number of turns; this would take correspondingly more time unless angulation was faster. Thus with more careful computation and a less simple conceptual system one has not a spin to give point information but point information coming from one turn (or one half). The scheme of imaging requires then about as many turns as there are points (or half as many). One could try to do many points at once with several sources. Dose might actually be made similar to that in CT, but the mechanical requirements of rapid rotation during a mechanical raster scan to fit this different concept would tend to make it slow.

Computed tomography with X-rays was an extremely significant development, and it is interesting to speculate on what factors led to its sudden recent success. There had been earlier suggestions for imaging a slice by radiation along the slice only, the advantage being as in Figure 4.1 where points in the image are affected by corresponding points in the subject rather than by points outside the slice; but the necessity to unscramble the effects of other points in the same slice was not previously as clearly understood. Contrast enhancement and edge emphasis or limited deblurring could have been applied to the previous hazy images since both processes had been discussed for radiology. Mechanical scanners were not favored but there were early X-ray scanners for somewhat different purposes that solved similar equations in unusual ways (e.g., see Chapter 8). Developments in electronic technology contributed to practicality of alternatives. Digital storage of information, to be discussed later, had become more routine due to developments in solid state circuit technology, thus providing stable arbitrary resolution and storage time with fast access. Digital computers providing greater variety of processing than the alternatives were becoming relatively inexpensive, small, and simple to work with. Improvements in integrated circuits soon allowed processing of many detector channels rather easily to reduce imaging time.

The output of a computer into a television display itself was a real advantage since it readily supplied contrast enhancement that could be adjusted for best appearance while watching, and which was quite stable; though well known before, it was inherent in these systems so that everyone could see and partake of the advantages. The possible display of a small range of CT numbers was part of this, and advanced the earlier ideas of making radiology more quantitative. A major contribution was that someone took the trouble to engineer a complex system to completion thus making it available to potential users; the familiar story of the industrial versus the university model for developments does not always apply but seems to have entered here. Once success was demonstrated many groups worked hard to make units available, and the method's time had come.

These rather large and relatively expensive medical imaging systems perform a cost-effective and sometimes unique function but do raise financial questions that are sometimes debated. At an early time there were more CT X-ray scanners in California than in all England where EMI (no longer producing scanners) was located. In 1979 specific legislation was proposed in California to prohibit the installation there of "dynamic spatial reconstructors", that is, large multiple

FIGURE 4.15. Examination by CT of wood, shell, metallic and stone objects. At upper left is a section of a tree branch and at right a modern chambered nautilus shell. Center: A stony fossil after sectioning, with the CT image to the right. Some of the inner chambers appear, as does a stone inclusion in the outer chamber. Lower: A fossil ammonite with metallic appearing replacement of the septa and clear crystals elsewhere. At the left is a CT image with the CT circle marking the region over which all rays can be effective. If this display is magnified, the television line structure is troublesome. At the right, magnification is instead increased by extending the image to beyond the screen or (lower right) by recalculation of the raw data. The CT images were made at 120 or 140 kV.

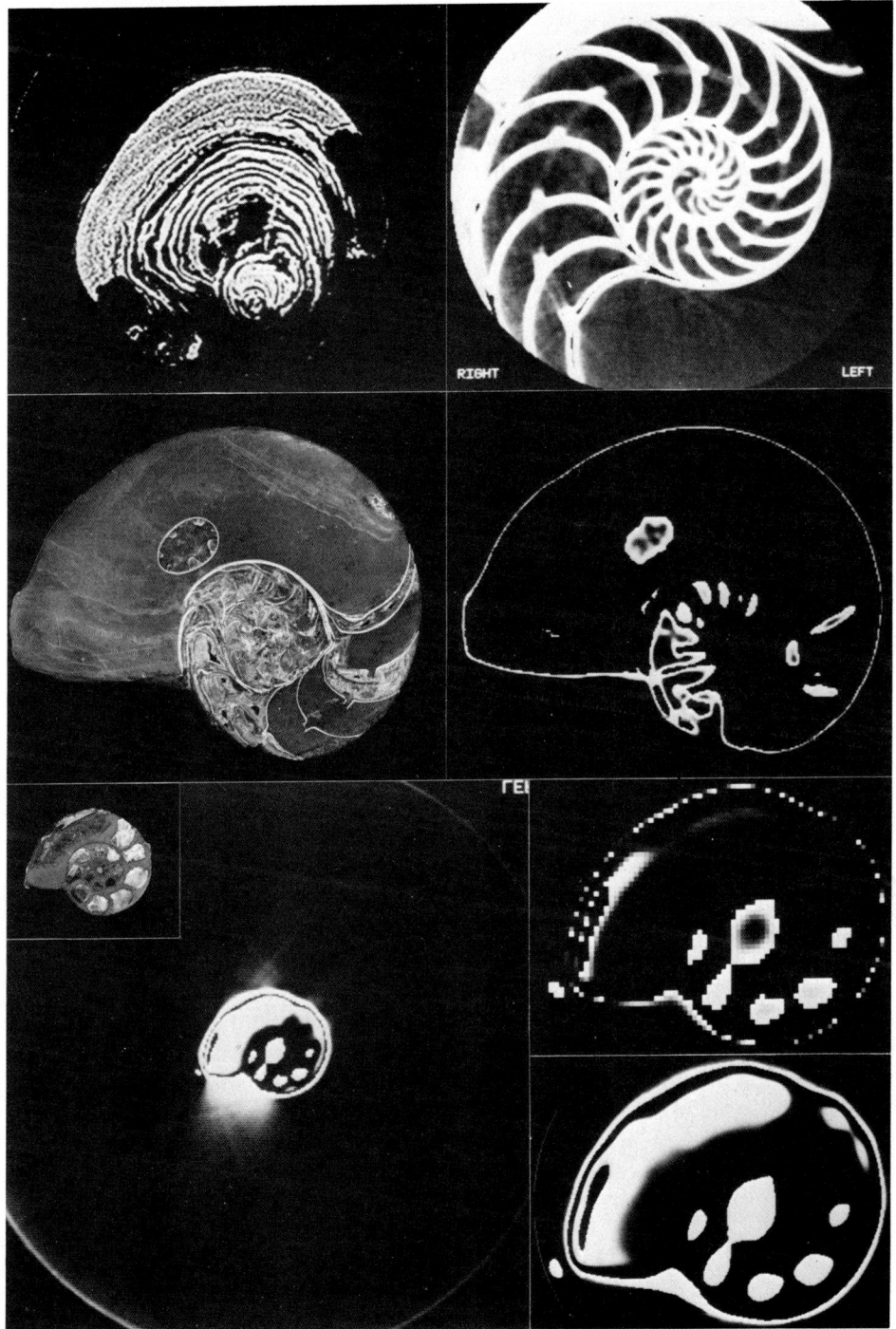

source scanners that could produce multiple images per second (Ritman et al., 1980) to give a dynamic view. This bill failed passage by a vote of 2 to 6, but is interesting in having been proposed.

The possibility of observing several people simultaneously in a CT scanner is also suggested as a cost-saving measure; unlike the above, this will appear in the April 1st issue of a journal (Reid and Dublin, 1984).

In closing this chapter we note that the same multidirectional viewing methods can be applied to objects that are not of animal tissue in order to observe small opacity differences. At the top left of Figure 4.15 is seen a cross-sectional X-ray image of a tree branch showing rings. Remaining in one position for a month after cutting often tends to make the bottom become more transparent to X-rays, apparently as part of the curing process. (A wooden board to fix the back of a person will often similarly show grain in a CT examination.) In an archeological application it was possible to assess the progress of preservation of a wooden object that we raised from a European warship that sank in 1676.

At the top right in Figure 4.15 is seen a CT section of a modern chambered nautilus shell, showing the path of the siphuncle through which all the inner chambers are emptied. The shell is of alternating layers of a crystalline form of calcium carbonate and a proteinaceous material like fingernail. Below are seen CT images of two fossil chambered cephalopods, along with the appearance when cut at approximately the same location. The small ammonite has a metallic appearance due to its septa being pyritized (deposits of iron sulfide) while the larger specimen appears rocky throughout. Similarly, we have observed some fossil fish to be less dense than their surroundings. In a display of an opaque object such as a metal hip joint, the center may appear dark (as with a transparent region) and similar artifacts can appear with these objects also.

The CT circle, within which all rays having all directions can act, is seen around the small specimen. Present clinical equipment is not ideal for such examinations, though higher Hounsfield numbers are to be displayable in commercial units announced at the time of this writing. With dense materials, penetration considerations limit specimen size, though smaller specimens suffer because of limited geometrical resolution. The final image can be enlarged though the raster scan pattern of the television display then becomes troublesome, just as looking at these images with a magnifying glass shows degraded detail. Two alternatives are seen in the final pair of images at the lower right which are of the small ammonite. Above is seen the effect of electronically spreading the image across and beyond the television monitor screen: the display scan lines are a smaller fraction of the size of the structures but the individual pixels become noticeable. Below, the same degree of enlargement is obtained by recalculation of the original raw data so that only a part of the total original field is displayed to fill the screen, giving an enlarged but smooth image.

5
X-RAYS

The discovery of X-rays was accidental, and initially their nature was not understood (Roentgen, 1895). In 1906 C. G. Barkla established the transverse wave nature of X-rays by showing they could be polarized in certain directions as a result of scattering off matter. Later it was resolved that they are electromagnetic radiation like visible light, but with a higher frequency (Fig. 2.1, Sec. 2.8). Their hazard was later realized, and their incidental production by such things as television sets is a cause for legitimate concern, but their ability to penetrate otherwise opaque materials maintains their importance both to medicine and engineering. Some of their modes of handling and image formation have been mentioned in Chapters 3 and 4 and some further information is incorporated in this chapter, hopefully facilitating comparison with other competing modalities.

5.1. SOURCES

The acceleration of a charge generates electromagnetic radiation such as X-rays. In an X-ray tube, electrons released from a hot cathode are converged into a small focal spot on an anode or target after being given a high velocity by maintaining a large difference in potential between these electrodes. The rapid slowing of the electrons after they enter the metal of the anode produces a broad continuous spectrum of X-rays that is called bremsstrahlung (German for "braking radiation"). If an alternating voltage is applied to the tube, X-rays are produced only on that half cycle which withdraws electrons from the hot cathode. During that half cycle there is a steadily changing voltage, which produces a steadily changing distribution in X-ray energies. Even if the applied voltage is rectified and filtered to give a steady direct current, a spread in wavelengths results. Though an electron can effectively give up its energy in several steps, no photon can have an energy or corresponding frequency that is higher than the energy an electron can acquire in falling through the full voltage applied to the tube. Thus there is a minimum wavelength in centimeters that is 12.4×10^{-5}/voltage of tube, which is produced when the entire electron energy goes into a single photon. (The bottom two lines of Fig. 2.1 incidentally give the minimum wavelength produced by a voltage applied to an X-ray tube.) Most photons are somewhat less energetic than this, and the wavelength corresponding to the maximum intensity is about 1.5 times the above. A thin sheet of metal over the tube can serve as a filter to reject the longest wavelengths which may not be wanted, if they are not already removed by the air and tube casing. The distribution and wavelengths from one particular filtered source are shown in Figure 5.1. This ancillary medical image is an example of what can be accomplished by an optical system, in this case the system of Figure 5.2. The more pulses of a given height that come from the electronic detector whose output is proportional to energy, the farther to the right in Figure 5.2 (which is turned to be up in Fig. 5.1) will the film be blackened at the position corresponding to that energy. Digital circuits would normally be used for recording an energy distribution, but it is a useful example of the versatility of simple optical systems. Since no pulses have a height greater than a certain amount, the film will not be darkened below a cutoff point, the energy scale in Figure 5.2 being up and down.

The electrons also can ionize atoms in the anode, creating vacancies in the inner electron shells. The rapid filling of these vacancies by transitions from outer electron shells results in the emission of a spectrum of discrete lines of "characteristic" X-rays of frequency specific to the anode material. For any particular line under consideration, the shorter will be the wavelength the higher is the atomic number (not atomic weight) of the element in the target. Above about 70 kV, radiation from the K shells of a tungsten target is released, but this is not resolved in Figure 5.1. With some X-ray tubes there can also be a small variation in wavelength across the image since the emerging rays will be partially filtered by varying thicknesses of metal in emerging from the focal spot at different angles in a tilted target.

Tubes have been built to emphasize different features. Thus in some the current is not too limited by space charge effects, and they are able to pass large currents even with a restricted voltage. When long wavelengths are of interest, a thin beryllium window allows release of very low voltage rays. In other tubes the focal spot size is sufficiently small to allow use for X-ray microscopy, with the same resolution as an optical microscope (Newberry, 1953). In the late 1940s at Berkeley we placed a negative filament near the image screen of the first commercial RCA electron microscope (for possible scanning applications) and, by thus running the beam backward, concentrated the electron beam at the specimen chamber almost to atomic dimensions (100 Å); thus the fine focus method probably can be extended.

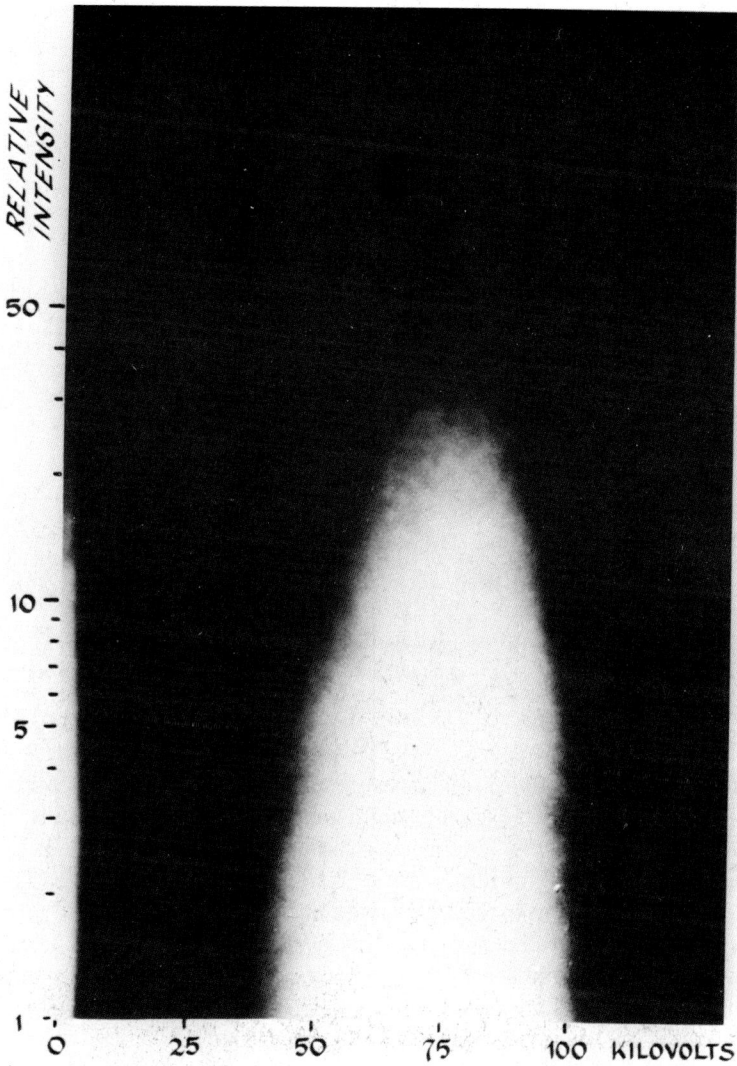

FIGURE 5.1. Application of a steady voltage to an X-ray tube yields a distribution of frequencies or energies, from the maximum value down to one limited by filtration of the beam. (From Mackay, 1961.)

A charge moving at constant speed but forced to change direction will radiate because of the centripetal acceleration. This is often called synchrotron radiation, and, indeed, such a steady dissipation of energy limits the ultimate energy of particles in circular orbit in some accelerators. Conversely, moving a charged particle in a curved path can serve as a useful potent source of radiation. One controlled source consists of a row of permanent magnets that wiggle the beam emerging from an accelerator from side to side and effectively guide the resulting

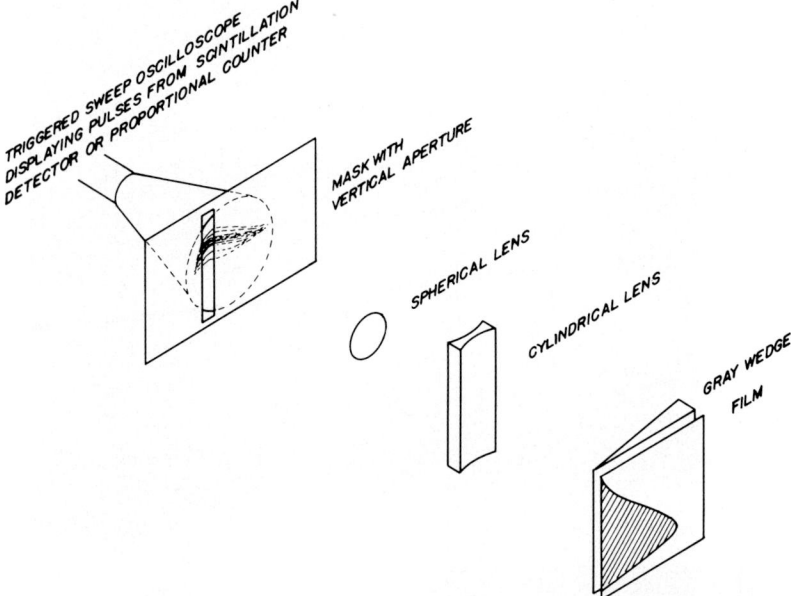

FIGURE 5.2. An example of an optical system that is directly able to produce an ancillary image in the form of a graph. The input is the pulse heights proportional to energy on an oscilloscope, the result being as seen in the previous figure. (From Mackay, 1961.)

radiation (Robinson, 1983b). Path curvature or magnet spacing and particle speed allow for the control of the resulting frequency. When the particle velocity is close to the velocity of light, this synchrotron radiation comes out tangent to the path (like the headlight beam from the front of a train). Such a specialized source may see more applications for special purposes in the future.

Laser-like coherent radiation can already be generated at frequencies corresponding to ultraviolet and bordering the X-ray region by nonlinear processes in which long wavelengths are converted to shorter wavelengths (Robinson, 1983a). Filters of different materials can somewhat restrict the range of wavelengths released from ordinary X-ray tubes, and approximations to monochromatic radiation will be mentioned in Chapter 8.

5.2. INTERACTIONS AND WHAT IS SENSED

In general the interaction of X-rays is with electrons, and thus an X-ray image is a map of the density of electrons from region to region within the subject. Electrons are not heavy, but for each electron in a neutral atom there is one proton which is relatively heavy, plus one or two neutrons in most biological materials. Thus, attenuation of X-rays is about proportional to mass of substance penetrated, and an X-ray image tends to be a map of the relative densities of matter within different parts of the subject. For a given material, absorption is directly proportional to the number of grams per square centimeter of material placed within the X-ray beam. A reduction in intensity caused by insertion of an absorber in a beam of X-radiation of given wavelength depends on the number and energy levels of electrons in the absorber and so is dependent only on the kind and number of atoms per square centimeter, and is independent of their chemical state or the combination of the atoms.

Photons can be removed from a beam of X-rays by scattering or by absorption. Total absorption is by photoionization, and scattering can be either elastic or inelastic. A scattered photon becomes troublesome because it no longer appears to emanate from the small focal spot and has "forgotten" its initial direction of travel. After scattering one or more times, a photon may have gone back into an inappropriate ray or detector; a cloud of such aimless photons can impair the contrast of the image formed by unscattered radiation. If our detector can be rendered insensitive to scattered photons either because of their reduced energy or their inappropriate direction, then a scattered photon is removed from the beam just as usefully as one that is not detected because it has given up all its energy to an electron in photoelectric absorption and vanished.

It is often convenient to speak loosely of the output from an X-ray tube as if it were effectively all at one wavelength corresponding to somewhat less than the actual applied voltage. Such an assumption is often inherent in the equations being solved in computed tomography, for example. As the wavelength is decreased by an increase in tube voltage, transmission through an absorber generally tends to increase. However, when the X-rays begin to have enough energy to knock electrons out of the K shells, the onset of this process will provide some extra attenuation and an abrupt drop in transmission with a slight increase in voltage. These "K edges," and other discontinuities corresponding to electron shells further out in the atoms, have a position on a graph of absorption versus wavelength that is different for each different element. Thus, to some extent, absorption depends upon material and wavelength. This effect can actually be used to image specific elements alone, as will be discussed in Chapter 8.

The above factors can conveniently be incorporated into a simplified equation relating absorption to the distance over which the beam traverses the material. It is common to let "absorption" refer to the actual reduction of intensity rather than to the method whereby it is accomplished. When an absorber is inserted in the path of a beam of monochromatic (single wavelength) X-rays it is found that intensity decreases exponentially with distance:

$$\frac{I}{I_0} = e^{-\mu x} \quad \text{or} \quad \log_e \left(\frac{I}{I_0}\right) = -\mu x$$

where μ is the linear absorption coefficient, x the thickness of absorber, I the outgoing intensity emerging from the absorbing material, I_0 the initial intensity, and e the base of the natural system of logarithms. This equation or law is quite plausible since it also implies a rate of change of radiation proportional to the amount of radiation, or that given thickness will obstruct the same fraction of arriving photons no matter how intense the beam. This law was given for light in 1729 by Bouguer, and again by Lambert, and by Beer, who mentioned the role of solute concentration in a mixture.

The absorption coefficient of a given element varies with wavelength, due to the changing mechanism of absorption (or more properly, attenuation), as the energy content of the individual quanta composing the radiation is changed. It is sometimes effective with a group of wavelengths to speak of one average or effective value of attenuation coefficient, which is like assuming a single wavelength. The coefficient μ enables one to calculate changes in intensity due to different thicknesses of a given absorbing material at a given wavelength, but μ is specific for the particular absorber from which it was determined and is not independent of the physical state of the absorber, though it is almost independent of temperature. (However, thermal expansion can move some matter out of the beam if there is no mechanical restraint against expansion across the beam, which can yield a small change in attenuation with temperature.) Dividing this absorption coefficient for a particular absorber by density gives a mass absorption coefficient that is the same whether the element is solid, liquid, or gaseous. This coefficient represents the sum of all processes by which radiation is reduced in intensity. In the usual range of voltages it is the sum of a photoelectric coefficient and a scattering coefficient.

The number of grams in a square centimeter across the beam indicates how much material will be encountered along the beam. From the mass attenuation coefficient (the linear coefficient divided by density of material, i.e., μ/ρ), one can conveniently calculate the mass of material required to attenuate the beam by a predetermined amount:

$$\frac{I}{I_0} = \exp\left[-\left(\frac{\mu}{\rho}\right)X_m\right]$$

where X_m is mass of attenuator per unit area normal to the beam. In units of grams per square centimeter, X_m is simply ρx, while the mass attenuation coefficient has dimensions of cm^2/g.

A consequence of these equations is indicated in Figure 5.3. When it is irradiated, an absorber having a series of equal height steps will not pass equally spaced differences in intensity. With radioactivity, a generally familiar concept is "half-life," in each of which increments of time half of the material remaining disappears. Similarly, in this drawing the steps in the object were taken as having a height of one "half-distance," traversal of which would attenuate the remaining radiation to half. Rather than equal steps in outgoing intensity, one has an exponential contour. Though this would not prevent noticing boundaries in an image, it can interfere with some comparisons within an image. Especially if subtractions of signals are involved after electronic detection, it can be desirable to pass the original signal through an amplifier with logarithmic response. For example, subtracting an image without iodine from an image after injection with iodine will give

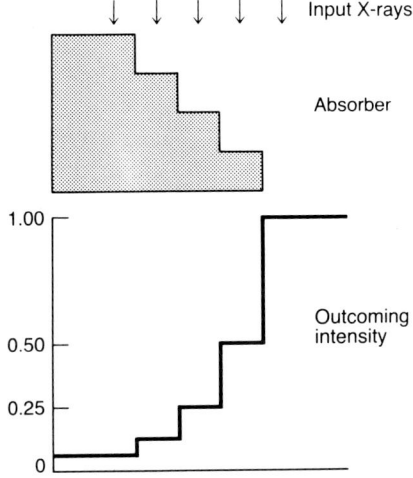

FIGURE 5.3. The exponential law of absorption for X-rays, and several other modalities, implies that a series of equal density steps in a subject will not be displayed as equal increments in the outgoing intensity. In some cases this does not matter, while in other cases it is corrected to linearity by a logarithmic electronic circuit or the logarithmic response of photographic film.

a diminished appearing amount of iodine where the subject is thicker, because the emerging intensity is lower, if this is not done to equalize the signals. The logarithmic relationship between current and voltage in many diodes can be used to produce such a response over a range of many decades in amplitude. In some cases such a compensation is taken care of by photographic film itself. Though actual photographic plates may be far from ideal, some have an optical transmittance proportional to the logarithm of the intensity of the X-ray beam which illuminated it. (The reciprocal of the fraction of intensity of light passed by a region of developed film is "opacity," and the common logarithm of opacity is "density," which can be about proportional to exposure.) In that case, the reduction in blackening of the film at each point is almost proportional to the length of the corresponding ray through a uniform density object.

Digitally computed tomography essentially assigns a numerical value of absorption to each region, and it is the distribution of such numbers that is displayed. It thus becomes convenient to express the attenuation of any tissue in terms of that of some convenient material taken as a standard. One hounsfield (H) represents 0.1% of the attenuation coefficient of water, and the scale 0 is water. Then air has H = −1000 and bone H = +1000 approximately. That is,

$$\frac{\mu \text{ (tissue)} - \mu \text{(water)}}{\mu \text{ (water)}} \times 1000 = H$$

The emphasis of differences in the display (contrast enhancement) is extremely easy by merely showing by how much each number differs from some fixed predetermined number; this is readily done with optimum adjustment for any given subject simply by turning knobs and watching the display until it appears best. But quantitative characterization of different tissue types is subject to errors that have been mentioned previously, and small machine variations can obscure or modify these numbers somewhat. Even changes in water temperature from room to body temperature can cause density changes of 5 H in this standard, which can affect calibrations and absolute CT values, as well as comparisons *in vitro* versus *in vivo*.

Attenuation takes place by redirecting photons out of the beam (Compton scattering), which takes energy away from the site of interaction, and by absorption in which energy is removed from the beam but transferred locally to the lattice as heat. True absorption or photoelectric absorption results from the absorption of a photon and consequent ejection of an electron from an atom, which results in an emission of characteristic fluorescent radiation by the absorber; if there is enough energy, the most likely interaction is with the innermost shell. In Compton scattering, the incoming photon is diverted by a free electron which recoils with some of the original energy, leaving the photon going at some new angle with less energy. The scattering coefficient depends only on the number of electrons per gram in the absorber and is otherwise independent of the material. Roughly speaking, the mass absorption coefficient varies with the cube of the wavelength and the fourth power of the atomic number, with abrupt changes at critical wavelengths peculiar to each element. Energy for "K edge" ionization is about proportional to atomic number squared. Though there are as many discontinuities as there are differences in energy levels within the atoms, light atoms such as carbon do not present them in the usual X-ray region. Energy is absorbed locally from the recoil electron of Compton interactions and from the photoelectric interactions, while scattered radiation can be produced by secondary processes such as fluorescence or bremsstrahlung as well as Compton scattering. More details are in standard texts (e.g., Evans, 1968).

On a more microscopic scale one can also consider the interaction of X-rays with matter, especially when there is a characteristic regular spacing of atoms. It is apparent from the equations for diffraction that if the wavelength of radiation is much greater than a grating space, only the zeroth order of diffraction is possible, which is equivalent to specular reflection. Since the spacing between atoms in a crystal, for example, is generally several tenths of a nanometer, visible light can only be defracted in the zeroth order. In 1912 M. von Laue conceived of using the shorter wavelengths of X-rays to probe the regularly spaced atoms within a crystal which would act as a diffraction grating. Details of structure on an atomic scale have thus been studied by observing the directions in which rays emerge from an irradiated sample. Considering an increase in wavelength up to that of light yields the usual optical laws of reflection and refraction.

For comparison with the absorption of photons or X-rays, it should be mentioned that the absorption of charged particles such as protons in a beam from an accelerator is not uniform with distance. Near the end of their range they slow and abruptly interact strongly. If the facility is available, radiograms can be formed with particle beams (Koehler, 1968), in some cases with great sensitivity and reduced subject dose over equivalent X-ray images (see also Cormack, 1980).

After death some appearances modify. For example, in the brain, white matter contains less water and more lipid than gray matter, and so its average atomic number is very slightly less. Thus it absorbs less X-radiation by the photoelectric process, and this small difference can be seen on late model CT scanners for a few hours *post mortem* until liquids shift. With some modalities there are changes *post mortem* in a subject that interfere with any successful observation but this is less true with X-rays. Images with good detail have been produced of mummies that were thousands of years old and of fossils that were millions of years old. In all cases structure is recorded as it exists, according to the above rules, and, of course, images of inanimate objects are routine for purposes ranging from engineering to art authentication.

5.3. SPEED OF EXAMINATION

If the X-ray source can be made bright enough the exposure can be very brief to minimize the effect of possible motion such as heart beat or respiration. In many cases the minimum exposure time has been limited by the ability to deliver enough photons to adequately darken a piece of photographic film. When other detectors are used to record the image then the factors that limit become those discussed in Section 3.2: supplying enough photons to limit quantum fluctuations to a degree that will adequately define the structure to be observed. In that case we saw that noise of $\sqrt{N}$ in a signal of N photons yields a ratio of signal-to-noise that is readily calculated: $N/\sqrt{N} = \sqrt{N}$. This number is made adequately large by traversing the subject with enough photons to define degree of darkening adequately. The smaller the fluctuations, the more accurately can be seen small brightness differences, while to see finer geometrical detail requires cramming the same number of photons through each correspondingly smaller region. One can trade geometrical detail for contrast detail, but some minimum number of photons must be delivered over all in order to have an adequate amount of each. This places an absolute lower limit on exposure.

When scanning is involved as in CT, overall rapidity can be limited by the speed of the scan and data recording processes. Improved technology can reduce these limits.

In an X-ray tube the more electrons that hit the cathode the more photons that will be released. The higher the tube current, the larger is the number of electrons traversing the tube and carrying this current. The longer is the time, the larger is the number of electrons that have traversed the tube. A short exposure thus requires a big tube current to provide a big blast of electrons in a brief interval. The tube voltage is first selected to be high enough so that an adequate fraction of the incident photons will penetrate and emerge from the subject, yet low enough so that the absorption will give adequate contrast with the structure in the subject; the best voltage can be predicted, or one can select a value dictated by experience (Jacobson and Mackay, 1958). A practical detail is that a change in tube voltage of 10 kV is equivalent to doubling the product of milliamperes and seconds in producing grayness of a film when operating near 80 kV, though the changes do not compensate each other in all respects.

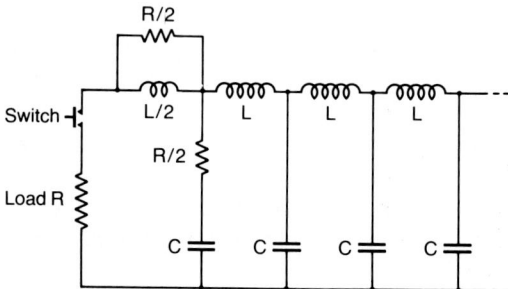

FIGURE 5.4. Discharging a delay line into a load can provide a short electrical impulse that rapidly turns on and then abruptly turns off after an interval of approximately constant voltage. There is still a Gibbs overshoot at the start of each transition which can be corrected with a suitable pair of resistors. The arrangement can be used to produce short pulses of light or X-rays, or other modalities, and in most cases the switch would be electronic in nature.

The ability of an X-ray tube to pass extremely high currents is partly limited by the ability of the cathode to liberate electrons. In the most demanding cases, rather than using a hot cathode, one can employ field emission. Here a high voltage is applied to a pointed piece of metal to generate an electric field that is strong enough to directly pull large numbers of electrons from the point. It has become almost commonplace to be able to produce X-ray images of bullets traveling down the barrels of guns or projectiles traveling along the barrels of cannons in milliseconds or microseconds.

To deliver this much energy in a short time requires considerable power, and thus it is useful to slowly store the energy over an extended period for sudden release during the exposure. This can be done with a capacitor that is charged to a high voltage and then suddenly discharged by connection to the tube. Very roughly speaking, if the tube acts like a resistance R, and the capacitance is C, then the duration of the exposure will be approximately RC. During the exposure the tube current will start high and diminish gradually, which may be undesirable.

There is a way for providing a brief discharge whose amplitude is constant during the impulse and which then abruptly drops to zero. The method is applicable as well for generating sparks that turn off abruptly as a source of visible light or underwater sparks that are a source of sound impulses and ultraviolet light. The method is to subdivide the capacitance between several smaller capacitors that successively discharge in a particular way. In Figure 5.4 the arrangement is seen. All the capacitors are initially charged gradually through a resistor (or an inductance plus diode) from a source of power that is not shown. When an impulse is desired from the device shown as a "load R," the switch is closed; the switch might be mechanical, but is probably some sort of electronic discharge. For the moment ignoring the resistors $R/2$, the capacitors C and inductances L constitute what is called a delay line in that an impulse applied at one end will travel along the combination and later come out the other end or else be reflected back from it. In the present case if $R = (L/C)^{1/2}$ then the pulse duration $T = 2\,(LC)^{1/2}$ times the number of sections. What happens when the switch closes is that the line starts to share its voltage with the load, that is, the line voltage starts dropping at its front to one-half the initial voltage while across the resistor suddenly appears a voltage that is half the initial voltage to which the line was charged. This wave of half discharge travels to the right and is reflected from the far end as a drop in voltage of one-half the initial voltage. When this wave of discharge suddenly arrives back at the left side, the voltage across the load resistor, which was half the initial voltage, abruptly drops to zero. Imperfections in the capacitors can be somewhat corrected by making the farthest right capacitor $2C$. Even with loads that change their effective resistance with applied voltage the method can be helpful since for a fixed applied voltage there is an appropriate fixed current.

However, there remains the Gibbs overshoot phenomenon mentioned in connection with the artifact of Figure 4.9, which was stated to be an inevitable accompaniment of any discontinuity. The number of sections in this delay circuit determines the highest frequency involved in passing the transient, and hence determines the width of the inevitable Gibbs overshoot of 18%; subdivision into more sections will diminish the duration of the oscillations at the start of the impulse for a given total pulse width, but they will not be totally abolished since they involve the lack of convergence of a Fourier series at a sudden step. Insertion of the two resistors shown as $R/2$ does indeed help the problem. At the first instant of the impulse, when only the left capacitor is discharging, no current will flow in the left inductance, and exactly half the initial capacitor voltage is applied to the load through the resistors. Since the impulse then starts out properly at just half voltage and eventually settles on the same half voltage, wiggles in between are minimal. Small residual variations can further be reduced by a larger resistor across the second inductance and a smaller resistor in series with the second capacitor. It may even prove possible to incorporate what we might call a "Gibbs compensator" in a feedback arrangement to correct the type of artifact seen in Figure 4.9.

In considering very short exposures, or very long ones, possible failure of the reciprocity law should be considered as it applies to film and to people. It is generally assumed that the effect on either has to do with the rate of photon arrival times time, that is, the total number of photons involved no matter how they are spread out in time. For either very long or very short times, a decreased dose rate may not exactly be compensated by a proportionately increased time. Total exposure must be changed somewhat for film under extreme conditions, but in the usual range that we are discussing, this is a minor factor. Under the usual conditions with humans, other factors cause greater variation than whether the entire dose was delivered in a very short time or not, though obviously if the dose were spread over a period such as a month some healing could occur, with repair taking place to allow a somewhat increased total exposure.

In computed tomography the overall time of an exposure is further increased by the scanning time required to collect a number of independent and statistically accurate projections. As mentioned in Chapter 4, the total dose that must be delivered to produce an adequate image increases with the size of the subject and also increases with the second or third or fourth power of the fineness of detail to

94 X-RAYS

be observed depending upon the circumstances. A large subject will not only more greatly attenuate the rays, but it will take more rays to cover the greater expanse. In some CT computations the extent to which tube voltage can be raised is quite limited, in which case an increased tube current or time can restore the number of outgoing quanta from a large subject, though at a cost of increased skin dose to the subject.

The power delivered to an X-ray tube is the product of current and voltage, and much of this energy goes into heating the target as the fast electrons slam into it. Heat collection may be more rapid than the tube is steadily able to dissipate, in which case there must be cooling periods between the storage of heat during successive exposures. A typical present CT unit takes a few seconds to collect its data, which can be stored for later slightly more lengthy computation into an image. But if repeated images are to be made, say, every 5 seconds to follow the course of a contrast injection, then these close-spaced successive images will have to be made at reduced tube current to prevent overheating. Modified tubes can store or dissipate more heat to allow more rapid successive images, so this is purely a practical problem.

5.4. BIOLOGICAL HAZARD AND DOSE

Some statistics suggest that the hazard of cancer death from low levels of radiation such as produced by living for 2 months in a stone building, or crossing the ocean by air, or living 2 months in Denver is one-in-a-million; this is the same as the risk of 700 miles of air travel (accident), 60 miles of automobile travel, 1.5 weeks work in a factory, 3 hours work in a coal mine, smoking a few cigarettes, 1.5 minutes of rock climbing, or 20 minutes of being a 60 year old man (Upton, 1982). However, it is also said that no proven body effect really establishes an increase in human cancer after doses of X-radiation below about 10 rad, though a conservative basis for risk estimation should involve an assumption of a continuous relation between dose and effect even at low doses (Webster, 1981). A major problem, still being debated, is whether low-level ionizing radiation has a cumulative effect or rather is it harmful only if it exceeds some threshold level? Certainly some repair of DNA takes place if damage is not too massive (see, e.g., Howard-Flanders, 1981). Obviously radiation to one finger is potentially less harmful than dose to the whole body, and one should preferably not X-ray the abdomen of a pregnant woman unless essential. By 1902 it was recognized that delayed development of a cancer was a possible effect of radiation. The possible effect of small doses was not considered seriously until the 1950s when E. B. Lewis inferred that cancer incidence might increase as a linear or nonthreshold function of dose.

In traversing living tissue, ionizing radiation gives up its energy through a series of random collisions and interactions with the atoms and molecules along its path, which give rise to ions and reactive chemical radicals which are able to break chemical bonds and cause other changes. Gamma rays and X-rays, which are electrically neutral, generate ions sparsely along their tracks and thus have a low rate of linear energy transfer. Charged particles such as protons and alpha particles have a higher energy transfer rate, a shallower penetration, and are generally more damaging. Improperly repaired lesions in a single molecule of DNA can be transmitted to a countless number of daughter cells as the DNA is transcribed and translated, and thus damage to chromosomes and genes appears especially important. When two breaks are close together in time and space, the broken end from one may be incorrectly joined with the broken end from the other, giving rise to structural rearrangement. The frequency of such aberrations apparently increases as a nonthreshold function of radiation dose. The probability of this being caused by low-transfer radiation is small, unless several radiation tracks traverse a given region of a cell nucleus in close succession, while a single traversal of high-transfer radiation has a high probability of generating lesions that will interact with one another.

Throughout life everyone is bathed in low-level ionizing radiation from natural sources, and medical diagnosis is estimated to contribute a further amount that approaches the dose received from the natural background. There is also a potential hazard whenever high voltages are applied to an evacuated structure. Thus an electron microscope that was delivered without proper lead glass windows created a problem, and youngsters playing with radio tubes and automobile ignition coils can be in some danger. Considerations of safety require being able to measure the amount of radiation after having decided upon an appropriate unit for describing quantity.

Several units are used in the measurement of doses and quantities of ionizing radiation. For biological purposes what is wanted is not the energy content of an X-ray beam, but rather the amount that will be utilized in the tissue, independent of wavelength. A standard recommended in 1934 essentially defined a roentgen ("r" unit) as the amount of radiation that would release one electrostatic unit of charge in one cubic centimeter of standard air, as might be measured with an ionization chamber; the definition was made more exact in 1937. It is a property of a beam and represents exposure or ionizing potential. One coulomb per kilogram of dry air is another such unit. From measured exposure at a place, dose actually delivered can be calculated from conditions and tissue types, it being higher for bone than soft tissue under some conditions, for example. The unit now often used to express dose absorbed by living matter is the rad, where 1 rad equals 100 ergs per gram of tissue. Sometimes a 100-times larger unit called the gray is used, it being 1 joule per kilogram of tissue. To make comparable the doses of different kinds of radiation, one has the roentgen-equivalent-man (rem), which is the quantity of ionizing radiation of any type that produces in humans the same biological effect as 1 r of X-rays or gamma rays, and is the dose in rads multiplied by the relative biological effectiveness of the radiation in question. The roentgen-equivalent-physical (rep) is that quantity of ionizing radiation of any kind which, upon absorption by living tissue, produces an energy gain per gram of tissue equivalent to that produced by 1 r of X-rays or gamma rays.

Either with particle irradiation or with high energy X-rays that yield recoil electrons, the dose can be higher somewhat downstream of the place of entry, and maximum dose is not at the surface but is rather where particles have slowed and interact more. Dose in a wide field can be somewhat higher than calculated for a narrow beam since photons can be scattered into a region rather than being only removed. The coefficients in the equations of Section 5.2 are to be used in evaluating removal from a narrow beam, not shielding against a broad beam where forward scattering can still allow traversal and forward exit somewhere.

Ionizing radiation is often detected by collecting the ions produced by its passage through a gas, which can be done with a pair of metal electrodes to which a small voltage is applied. What is measured in this ionization chamber is the resulting electric current, which should be about zero when no radiation is present. If the applied voltage is somewhat increased then the more rapid motion of the ions will produce other ions, resulting in a larger current in such a "proportional counter." For still higher voltages, for each ionizing event a single large impulse will take place due to the cumulative secondary ionization in the gas, and in this "Geiger–Muller counter" the size of the pulses does not depend upon the magnitude of the initial event. In the proportional counter region the original number of ions may be multiplied 10,000 times by gas amplification while in the Geiger counter region the amplification factor may be 10 billion. In these last two cases, the counting rate gives a measurement of radiation intensity.

Exposure as measured by ionization depends on both beam energy and photon density in a complicated way. Energy fluence, which is radiant energy per beam area, can be measured by recording the change in temperature produced by the beam in a few seconds in a small piece of gold or lead in plastic foam, and this can indicate the total energy that will be absorbed by a patient.

In a scintillation detector some photoelectric device monitors the glow produced in a fluorescent material. A photomultiplier tube is often used in this application because of the amplification built into it. One can either count the rate of arrival of individual flashes from individual events, or one can record the magnitude of the overall glow in order to evaluate the magnitude of a photon flux. In both cases the output can be quantitative. It was mentioned in Chapter 4, that CT scanners have employed semiconductor detectors such as CdTe and tin-loaded plastic scintillator detectors, as well as ionization chamber detectors.

One can carry an ionization chamber whose electrical discharge measures the dose received by a person. There are also thin film dosimeters such as a piece of X-ray film, or a layer of mylar or cellophane dyed with appropriate chemicals that bleach under radiation exposure.

Thermoluminescent dosimeters store energy in disturbed states. Thus a small piece of lithium fluoride if heated will momentarily release a glow that can be monitored with a photomultiplier in the dark, and which is proportional to the amount of radiation that has been absorbed (Jones, 1965). Calcium fluoride is more sensitive but shows more fading, while manganese-activated lithium fluoride shows low fading after months. Other materials also show the effect.* Before

---

*Ceramic antiquities cannot be dated by carbon-14 methods, but clay after firing starts to collect dose from natural background radiation. If a small sample is heated, the glow indicates how long it has been since it was originally formed (last fired). Competent forgers of antique pottery presumably will all own X-ray machines that are carefully calibrated.

making an observation one anneals the rod of sensitive material. After exposure, a typical observation concludes with a 6 second pre-heat at 125°C followed by a 12 second read at 285°C, during which thermoluminescence takes place. A thermoluminescent rod can be quite small, and placed on or in a person, or it can be used in or on a phantom consisting of a plastic or water model in which the distribution of radiation is to be tested for the geometry of a particular observation. An even smaller dosimeter might be made by reading the density of induced lattice defects or radiation-induced electron traps using the method of electron spin resonance (Ikeya and Miki, 1980). For monitoring during irradiation there have been built swallowable radiation measuring radio transmitters that telemeter through the body wall (Mackay, 1970).

In closing this section, it should be noted that dose reduction has many aspects. Dose will clearly be too high if almost none or almost all photons emerge from the subject. One can predict the tube voltage that will give the most contrast per unit of dose, and this often comes out to be about what experience has evolved as appropriate (Jacobson and Mackay, 1958). Once the functions of image formation and recording or display are separated, one needs detectors that are effective in using all emerging quanta (maximum detective quantum efficiency). Remaining aspects of dose reduction range from the rather theoretical to the extremely practical. As an example of the former is the replacement of an ordinary radiography process by scanning in which scan velocity is made proportional to the momentary emerging X-ray "brightness" in order that every unit of area have the same number of photons; this uniform intensity nowhere involves more quanta than needed to define structure, and a normal appearing image is produced on a film in which brightness is made proportional to momentary scan velocity (Mackay, 1958a). An example of dose reduction in practice is in one of the least glamorous reasons for introducing video magnetic tape recorders into radiology, where the instant viewing capability allows seeing whether one has what is needed and that, "experience may show that in practice the greatest saving in dose will result from taking only as many frames as needed rather than exposing extra precautionary footage before releasing the patient" (Mackay, 1959, 1961).

5.5. CONTRAST AGENTS, TRACERS, AND STAINS

Making structures in soft tissue more noticeable by placing heavy materials there is a well-known technique. Thus the swallowing of barium compounds or the injection of iodine compounds is familiar to many. Insufflated powdered tantalum has been used for outlining airways and has some advantages over barium sulfate and iodinated compounds; it has high density and atomic number and is said not to be an irritant. In some cases the difference in X-ray transmission between the structure of interest and the surrounding tissue is enhanced not by inserting a material of high atomic number, but rather by inserting a material of different physical density as in the injection of air into the spinal canal to rise and, displacing the cerebrospinal fluid, aid in the visualization of the ventricles and outer surface of the brain. New materials incorporating heavy atoms that will be carried to selected regions are increasing the applicability of the contrast enhancement and sensitivity of CT scanners, which to some extent then become a substitute for the radioactive labeling of tracers.

Such diagnostic methods may be undertaken in spite of some potential secondary side effect, as a deliberate risk in order to secure appropriate treatment for the patient. Even the introduction of air itself, as in pneumoencephalography, is not without discomfort and some hazard to the patient. Barium sulfate in water suspension is inert and none is absorbed during passage through the gastrointestinal tract but, for example, barium should not be given by mouth with symptoms suggesting large bowel obstruction. Injection of too much iodine contrast agent can yield kidney failure under some conditions, lots of fluid in advance sometimes being helpful. Some believe intravenous injections are more prone to anaphylaxis than arterial ones.

There are an estimated 15 million administrations of iodinated contrast material per year in the United States alone, with a relatively low rate of severe or fatal reactions. A typical urogram, for example, involves injection of 20,000 mg of such a "drug" directly and rather suddenly into the vascular system. The fact that not every patient has a profound reaction to such huge amounts is a testimonial to how benign these materials are. Some reactions to contrast media resemble allergic phenomena, with small amounts eliciting severe reactions by the formation of antibodies. Cross reaction with chemically related halogenated benzine derivatives in our environment presumably allow adverse reactions upon first exposure (Brasch, 1981). Since virtually all these agents are hypertonic, they can also produce dramatic and rapid shifts in body fluids, resulting in changes in cardiac output, cardiac arrhythmias, dehydration, and, in newborn infants having enemas with contrast media, hypovolemic shock. These effects and chemotoxic reactions are related to the total amount of material administered. During pregnancy the effect of contrast media (e.g., on the fetal thyroid) is uncertain. It was in part to reduce the required density of contrast agent that we explored the development of X-ray procedures that would produce images of the distribution of one element alone (Jacobson and Mackay, 1958) to be discussed in Chapter 8.

The mode of application of these materials can be important. Thus improved visualization of a solid lesion in liver, spleen, and pancreas is most likely achieved during a relatively short time span immediately after injection of a bolus during the nonequilibrium phase (Burgener and Hamlin, 1981). A myocardial infarct tends to appear as an area of decreased X-ray attenuation (cold spot image) within the first minute after injection of contrast material into a vein; later the attenuation values reverse. Observation of the heart and vascular system by observing the specified flow pattern after sudden injection is well known, and examples will be given in Chapter 11.

A number of different tracers or contrast agents have evolved recently, especially for observation by X-ray transmission computed tomography. Xenon is an inert gas of high atomic number, which readily crosses the blood–brain barrier, and which is an effective anesthetic. Successive subtraction images should allow visualization of blood flow in extremely small tissue volumes from measurements of changes with time of concentration of this nonradioactive gas. Several groups have explored the *in vivo* mapping of local cerebral blood flow using subanesthetic amounts of xenon (Drayer et al., 1980; Haughton et al., 1980; Meyer et al., 1980; Gur et al., 1981). Sulfur hexafluoride should have applications since it is a gas having a remarkably high density (about five times that of air); it is safe to breathe and lowers the voice frequency.* Sometimes iodine compounds can usefully be incorporated in different forms. Particulate contrast agents using organic compounds are accumulated in reticuloendothelial cells, with little uptake by cancer metastases, making them potentially important in scanning the liver (Violante et al., 1980). In the search for a biodegradable material for liver and spleen visualization, radiopaque liposomes were studied (Havron et al., 1981), as have been iodinated starch particles (Cohen et al., 1981) and several combinations involving metals and rare earths (Seltzer et al., 1979). Polyvinylpyrrolidone in combination with metallic salts was found to concentrate in cancerous versus normal tissue, thus increasing contrast for tumor detection (Young et al., 1981a). Fluorocarbons related to the propellants in aerosol sprays are radiodense if combined with bromine and are biologically inert. Gastrointestinal administration produced contrast enhancement in the spleen (Enzmann and Young, 1979) while injection of microemulsions caused contrast enhancement in some tumors, apparently by localization within macrophages (Young et al., 1981b). It might be mentioned that in studying some immunity reactions under an electron microscope, the presence and position of a single molecule can be observed if it is appropriately labeled by iron atoms. Other exotic combinations of materials undoubtedly will evolve for special purposes. The methods already have evolved considerably since Hascheck of Vienna in 1896 sent a picture to Roentgen in which veins had been injected with a mixture of lime, cinnabar (mercuric sulfide ore), and petroleum via the brachial artery.

---

*If altered sound is undesirable it can be somewhat corrected by an expandable periodic gap method demonstrated to improve diver or helium speech (Mackay, 1966), that mechanical method now being convenient in electronic form.

6
ULTRASOUND

One can tap out the margin of the heart just as one can locate the studs in a wall by listening to the sound of tapping. But finer detail can be resolved with a higher frequency. The range of frequencies of interest here is approximately 1–15 MHz. Frequencies much higher than this are too strongly absorbed to penetrate useful distances within the body. Acoustic microscopy of thin slices is done at frequencies up to about 2000 MHz, but this is a rather special application extending almost to the frequency of light. There seems no harm from repeated examinations, and the methods are even used to examine pregnant women. The apparatus also is taken into operating rooms to guide some procedures, and, being close to the structures of interest, the higher range of frequencies can be employed for maximum resolution of fine detail.

6.1. SOURCES

Short sharp clicks are characterized by the presence of many high frequencies of sound, though there will also be present frequencies as low as the repetition rate of the impulses. Simple sketches of sine waves compared with pulses reveal that the range of frequency extends from very roughly the reciprocal of the pulse width down to the pulse repetition rate. These frequencies travel reasonably well in tissue and so sound pulses often are used to image body structures by collecting and displaying the pattern of echoes that return after an impulse is coupled into the body. There are many ways to generate and receive these impulses, but the most common in medical applications is with the piezoelectric effect discovered by the Curies in 1880. Some materials, ranging from living bone to quartz crystals, will generate a voltage if squeezed and, conversely, if a voltage is applied will mechanically distort. (One of the most familiar examples is in the pickup arm that plays phonograph records.) If an alternating voltage is applied to such a material then steady sound will be radiated, while sudden steps in voltage or impulses applied will cause the generation of clicks; when acting as receivers, the same units will deliver electrical signals that mimic incident sound.

Especially excellent signals can be produced with crystals of lithium sulfate, though the material is adversely affected by moisture and does require a high voltage. It has become common to employ ceramic transducers (converters between sound and electricity) made of barium titanate or lead-zirconate-titanate (PZT) because their low impedance and consequently low associated voltages are well suited to transistor and other semiconductor circuits. Recently plastics have been developed that show these effects, and which have other convenient properties (Lovinger, 1983). Both ceramics and plastics can be formed in shapes that focus or have other unusual properties. There are considerable differences in the transducing capability or sensitivity of different materials, but they are not the only factors involved in choosing. As we shall see, the ability of sound to traverse the boundary between transducer material and body tissue depends upon their similarity in the product of density and stiffness, which is superior for plastics. Electric charge released by mechanical stress on these materials produces an output voltage by charging the capacitance of the structure, and thus a high dielectric constant tends to give reduced voltages but with less sensitivity to the capacitance of the connecting cables or amplifier input. Transducers often are backed with plastic loaded with tungsten powder in order to direct most of the sound forward and to rapidly damp vibrations at the end of an applied electrical impulse (Walker and Lumb, 1964; Lutsch, 1962). Double pulsing some transducers can also bring them abruptly to rest, both in reception and transmission (Mackay, 1966a).

In exploring fine structure within the body, the width of the probing beam is obviously important. A very small vibrator will radiate energy essentially in all directions, while a larger surface oscillating back and forth perpendicular to itself (in the fashion of a piston) can produce a directed beam. An example of such generation of a beam of sound is the application of an alternating voltage to the surfaces of a sheet of piezoelectric material which will cause cyclic changes in thickness, with particular efficiency if the thickness is chosen to be exactly half a wavelength at the applied frequency. The size of the piece cut from this sheet will determine the width of the beam, there being a near and a far field (Sec. 2.12).

The result is summarized in Figure 6.1 where it is seen that if the transducer is a flat disk of radius r, it will radiate a cylindrically shaped beam for a distance that is approximately equal to r^2/λ, where λ is the wavelength of the ultrasound. In this near field zone interference occurs between sound waves coming from different parts of the transducer face, and a point reflector would cause echoes of varying strength depending upon relative position. Beyond this distance the beam is no longer "collimated" but spreads spherically at an angle of approximately λ/r

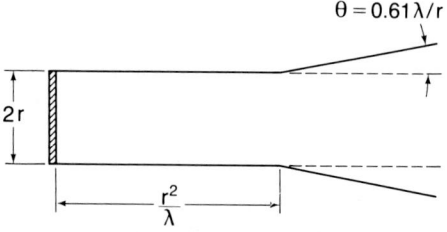

FIGURE 6.1. A vibrating surface acting like a piston releases sound energy within the approximate confines shown. The "near field" at the left is criss-crossed by an interference pattern. The transition distance and ultimate divergence of the beam depend upon piston radius and wavelength in the medium carrying the sound.

radians. There the sound field is rather homogeneous and strength variations are less of a problem. There are variations along the overall pattern, and indeed beam diameter is actually somewhat less than indicated at a position short of the transition distance, but most of the energy is contained within the lines shown. This same transducer acting as a receiver of incident sound energy will show essentially the same geometrical pattern of sensitivity.

A specific example is that a 7.5 MHz crystal (transducers traditionally being referred to as crystals because of the early use of quartz), having a diameter of 15 mm, generates a beam that remains collimated for a distance of about 28.1 cm, beyond which distance it diverges at the relatively small angle of 1.53°. If the transducer is not a circle, the transition distance will be about the same, as is derived in most advanced books on sound. The transition distance is not abrupt, and some beam pattern measurements are best made at a few times the indicated distance. The U.S. Institute (now the American National Standards Institute) arbitrarily decided and decreed that the far field for any shaped piston began at a distance of piston area/wavelength.

The far field is essentially where a diffracted wave from the piston edge and the plane wave from the surface have finally come into phase. Sound pressure beyond decreases, as in a spherical wave, in inverse proportion to distance without complicated near field structure to confuse measurements. (And energy decreases with the square of the distance.) This is the situation for distances large with respect to both piston radius and ratio of radiating area to wavelength. The same considerations apply to the other modalities. A continuous surface radiator having transverse dimensions extremely large with respect to a wavelength is formed by the partially transparent end reflector of a laser, and thus a laser can radiate an extremely narrow light beam, for example.

When a sharp impulse is being reproduced many frequencies are simultaneously involved, and the spatial distribution of energy or sensitivity can be somewhat modified from that with steady oscillation. Prediction of the transient response of a transducer is difficult; energy can be projected in directions not indicated in the figure and called "side lobes." Associated amplifiers can affect the effective overall pattern (Auphan and Dormont, 1977; Bamber and Phelps, 1977). An estimate of beam width for a transducer can be obtained at any distance by scanning across a thin nylon string suspended in saline. Near the transition from near to far field the narrowest focus will occur. Inside this distance close to the transducer a fine speckled echo pattern is commonly seen as an artifact in

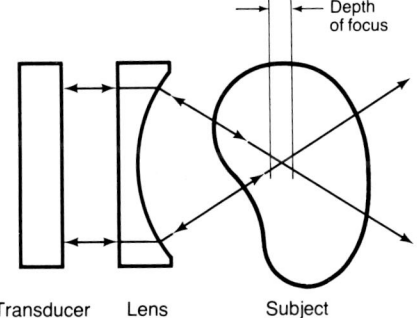

FIGURE 6.2. A broad beam can be made narrow over a limited region by a converging lens. A region of tolerable diameter in the subject is indicated. For ultrasound a converging lens is often concave. Here the transducer is shown as both the transmitter and receiver of pulses.

uniform organs, and this apparent tissue texture is at least partly due to interference phenomena.

Not only does variable transducer response produce speckle, but it also occurs in ultrasonic images because of the interference that introduces the same appearance into images made by the coherent light of a laser. Its distraction can be reduced by smoothing produced if successive clicks or "looks" are done at slightly different frequency by activating the transducer with a few cycles of a frequency that changes. At higher frequencies the images can be somewhat filled in by increased diffuse Rayleigh scattering from tissue structures.

In scanning a subject, it may be necessary to see detail finer than the width of the beam in the figure. Lenses can be made to focus sound waves, and they are often made of plastic (Sette, 1949; Szilard and Kidger, 1976). Bending occurs when there is a change in velocity across an interface. For visible light a convex lens brings rays to a focus since the velocity of light is less in glass than in air; for sound waves a converging lens is concave since the velocity of vibrations in plastic is higher than in water or tissue, though a convex converging lens can be made of rubber. A sharply converging lens can indeed narrow the distance being resolved, but at the expense of having fine resolution over only a diminished range of depths. In Figure 6.2 this is seen with a single transducer–lens combination acting for both transmission and subsequent reception of returning echoes.

Fineness of detail that can be observed along the beam depends on how short the sound impulse can be made, and if very sharp will yield echoes that precisely define the position of reflecting interfaces, while the fineness of detail transverse to the beam depends upon the width of the beam at the position in question. There are ways to compensate for some blurring that takes place in a repeatable fashion in either direction, and these will be mentioned in Chapter 10. There are also "synthetic aperture" techniques to be mentioned in Section 6.3 that allow fine detailed images to be built up after scanning with a broad beam. Unusual geometries such as a conical transducer may produce a narrow beam over a larger depth of field (McCloud, 1954) but there are problems with radiation in spurious directions that are somewhat difficult to compensate (O'Donnell, 1982). Other alternatives besides transducer modification still require exploration.

Light optical systems can be made that give images of sound wave patterns,

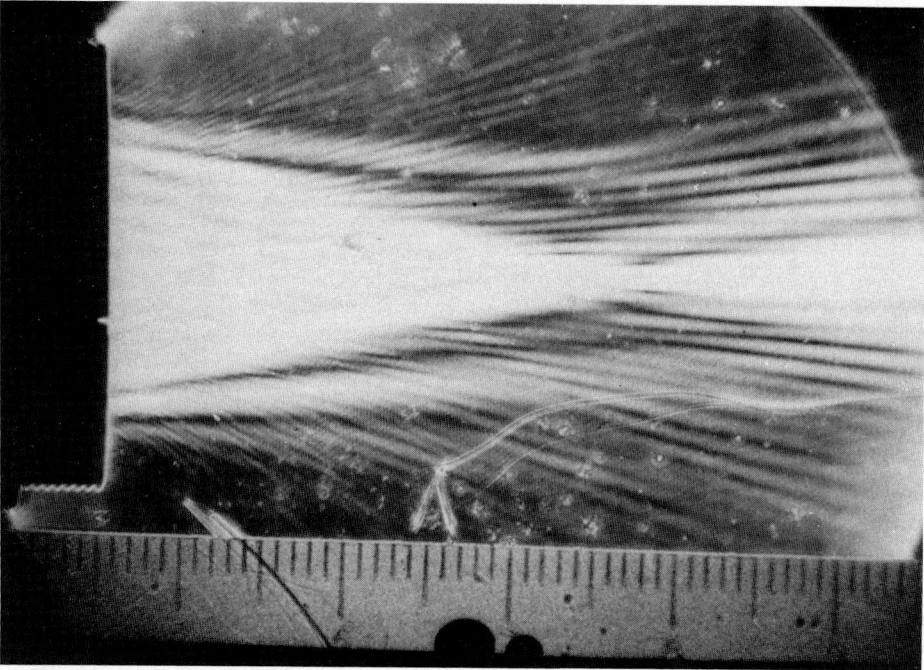

FIGURE 6.3. The pattern of sound intensity from a converging lens as made visible by a schlieren optical system. The streak at the bottom is a flaw in the camera lens used to build the light optical system. (From Mackay, 1966b.)

and the effect of fronting a transducer with a concave polyethylene lens is visualized thus in Figure 6.3. In the schlieren (German for streaks or striae) apparatus used to make this figure, parallel light is brought to focus by a lens. At the focus a small opaque spot prevents light from passing on to a camera. Any disturbance in the parallel region deviates the beam so that some reaches the camera, which is focused to photograph objects in the disturbed regions. (The optical arrangement is about as sketched in Fig. 10.9, with white light and a dot for a mask.) In the present case the disturbance was the compressions of sound in water causing changes in index of refraction, thus allowing visualization of the path of the sound transverse to the optical system. A minor imperfection in one of the camera lenses used to build the apparatus introduced the jagged artifact at lower center. Although the original vibration was largely piston-like, the well-formed sound beam is still crisscrossed by interference lines associated with the near field. The converging action of the sound lens is seen. Curved organs of the body, including the brain and the female breast, can also curve rays traversing their boundaries, which is a source of distortion in some imaging procedures. In Figure 6.4 is seen the redistribution of sound upon traversing the excised lens of a human eye such as might take place under some conditions when scanning the eye in order to build up an ultrasonic image. Changes in elastic properties of the lens might be measured ultrasonically *in situ* to serve as an index of age in animal studies.

FIGURE 6.4. The lens removed from a human eye can deform a sound pattern, as seen here using the same optical system as in the previous figure. The interference pattern at the left is quite noticeable, as is a standing wave pattern. The frequency was 15 MHz.

In the use of most of these devices an electrical impulse is applied to the transducer to generate a sharp impulse of sound, and then the same transducer is used to receive the successively arriving echoes from correspondingly deeper interfaces within the subject. Some problems in handling these impulses are indicated in extreme form by the early arrangement shown in Figure 6.5. The negative pulse is applied to the crystal whose sound output is focused by the lens upon the subject. At least when aimed at the center of the subject, the beam will encounter four interfaces from which will return four echoes in succession. These must all be allowed to return before going on to the next impulse. The echoes returning to the transducer can generate small electrical impulses which are amplified and displayed as suggested under the label "A scope." To properly use the excellent performance of the lithium sulfate crystal requires application of a pulse of perhaps 800 volt amplitude, which is seen to be applied also directly to the input of the amplifier, which must then recover in time to detect signals that are down in the 3 microvolt range. The solution to the problem shown here is to use a vacuum tube at the input which cannot be turned off more for a big input than for a medium one, and the excursion of whose anode is further limited by the diode shown. The echoes can be displayed on a standard oscilloscope and, with scanning to be discussed in Section 6.3, it is possible to display a full cross sectional image of the subject. A tendency for the crystal to "ring" and generate a positive voltage after

106   ULTRASOUND

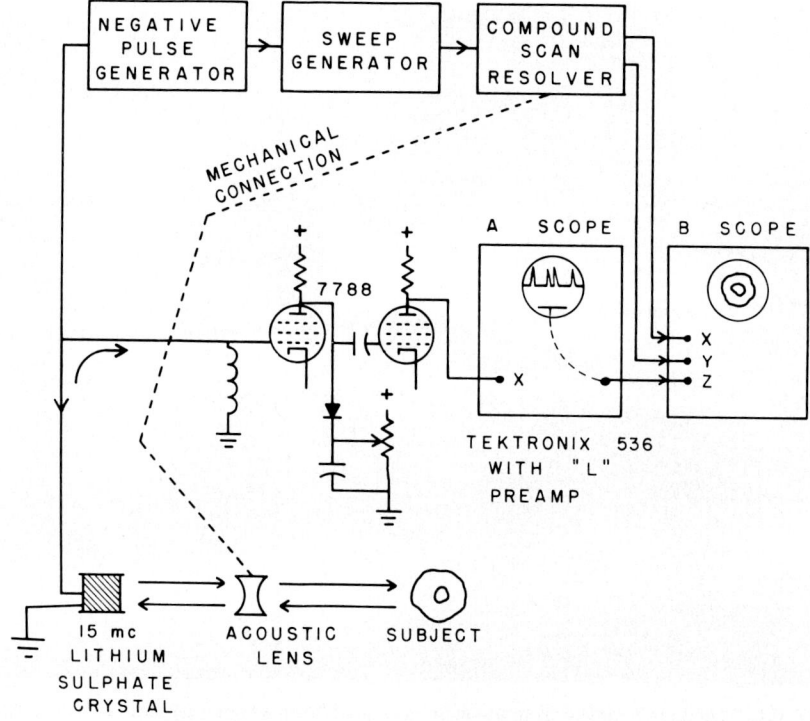

FIGURE 6.5. Pulses from the transducer are focused into the subject from which echoes are returned for display either in sequence or as a geometric image pattern. In the present case there are four echoes per impulse from two boundaries. The strong initial electrical signal is applied to the amplifier, which then must rapidly be able to respond to the tiny signals of the returning echoes. (From Mackay, 1966b).

the negative pulse can be alleviated by connecting across the crystal a large semiconductor diode that will "short-circuit" relatively large positive voltages but which will not see (be nonlinear to) the tiny voltages of the returning echoes. From the velocity of sound in tissue, it can be calculated that it takes 93 microseconds for an ultrasonic pulse to make the round trip to a target 7 cm away from the crystal, during which time this amplifier must recover from an 800 volt transient. The situation is considerably less extreme when dealing with the lower voltages involved in the use of ceramic transducers. The problem is also somewhat otherwise alleviated since, in practice, one often inserts an automatic compensation for the reduction by attenuation of the more distant echoes by simply having the gain automatically increase with time during the brief period that echoes are returning following the outgoing pulse.

The interval between pulses allows travel to the deepest parts and back before the next pulse is generated. The pulse repetition frequency is often a few thousand per second, which is much less than the ultrasound frequency, each impulse being several cycles of the sound.

Resolution in distance is higher the shorter is the radiated sound impulse but a transducer typically will vibrate for many cycles if hit with a sharp electrical impulse; this is not much reduced if there is contact with tissue of the subject since coupling is not high between these somewhat different materials. The length of the impulse can be shortened by attaching to the transducer rear a matched lossy backing of tungsten powder in plastic, as was mentioned, or by acoustically matching the transducer material to tissue by an intermediate layer of appropriate material so that energy is better transferred out. The latter is desirable also for the necessary transfer of the acoustic energy of the signals. By comparison with the familiar low reflection coatings on camera lenses, the matching layer should have a thickness of a *quarter wavelength* (at the transducer mid frequency) of material with an acoustic impedance equal to the geometric mean of that of the two materials to be matched. Gradual transition through multiple different layers may be helpful when there is a frequency spread. More about the passage of energy is in the next section.

If the outgoing pulse is long but swept in frequency, then to each instant corresponds a particular frequency, and a returned echo will indicate the distance to its reflector by the frequency it has at the time of its return. To distinguish one instant from another with a swept frequency one can either compress the "chirp" back into a sharp pulse (e.g., Lam and Szilard, 1976) or beat the returning echo with the outgoing signal to give a difference frequency that depends upon range (e.g., Kay et al., 1977).* One may also wish to produce a change in frequency in order to gain further information about a subject, as will be mentioned in Chapter 8. It is somewhat difficult to build a transducer that will work equally well over a wide range of frequencies. One can add losses either by electrical loading or by physical attachment of lossy material. This spoils the resonance so radiation is more uniform over more frequencies, though not as good at middle values. Another possibility is to clamp several slightly differently tuned disks into a single unit, with different ones taking over as the frequency shifts.

Aspects of these matters and those in the next few sections are covered in many books on sound, some being Rayleigh (1894), Wells (1969), Myer and Neumann (1972), and Hussey (1975). Some details of these matters and their history were given in Section 2.11.

6.2. INTERACTIONS AND WHAT IS SENSED

In traversing tissue a sound wave can encounter changes in velocity, absorption, scattering, and density, and a map of the distribution of any of these can provide

---

*In radio or sound ranging systems that are on rather continuously while sweeping up and down in frequency, the Doppler shift from movement averages to zero leaving distance indicated alone if the sweep up and down is symmetrical, but there is indication of motion with sawtooth frequency deviation; one can count beat cycles for increasing frequency minus those for decreasing frequency to get subject velocity, while the sum gives distance.

an image to differentiate one region from another. All the usual interactions of light with which we are familiar are operative, including reflection, refraction, diffraction, and attenuation. At high intensity sound certainly can alter tissue, but at low intensity examination seems safe. There are many ways for converting the pattern of distribution of sound intensity into an image that can be examined, one example being the above schlieren optical systems. However, many of these require more energy in the field of vibration than would be desirable, and for use with tissue or liquid-like materials some form of the piezoelectric effect is usually employed.

We are generally used to the idea of *the* velocity of sound, and tend to take it for granted that it does not depend on frequency. There is a small dependence of sound velocity on frequency in air (dispersion) but sounds of speech, for example, are usually not much distorted by this. Similarly, it has been found that the velocity of sound in tissue does not much depend upon frequency, for reasonable frequencies. The observations of many workers indicate the velocity is about the same for all soft tissues and blood, being about 1500 meters per second. There are small differences between fat, kidney, and so on, while the value for bone is several times as large because of its stiffness, and the lung and parts of the gastrointestinal tract are different where they contain gas. These changes in velocity produce refraction, as demonstrated in the previous section, and cause sound wave rays traversing the body to bend somewhat. Iterative methods in computed tomography using rays of sound are able to incorporate such bending, and this aspect was commented upon in Chapter 4.

In one cycle of oscillation a wave advances one wavelength, and thus frequency times wavelength is equal to velocity. From the measured velocities one can calculate the wavelength corresponding to any applied frequency. In an image one can resolve detail down to about a wavelength, if other factors such as ray widths do not limit resolution to a poorer value.

Absorption along a path follows the same exponential equation that was given for X-rays, and which leads to Figure 5.3. In the case of sound, the absorption coefficient by which distance is multiplied in the equation increases with frequency, being roughly proportional to frequency in some frequency ranges. For some tissues attenuation is 0.5 dB/MHz/cm. Thus lower frequencies are best for deep penetration, while higher frequencies are better for resolution of fine detail. (Tissues are not without losses even at very low frequencies, with vibrations of the eye showing a Q of about 4 at 60 Hz: see Mackay, 1962.) Different materials absorb differently, and a muscle bundle absorbs differently along and across the fibers. Bone absorbs rather strongly but one can project impulses from the side through the human skull, for example, to determine if the midline of the brain is still centered in the skull rather than having been displaced to one side by tumors or bleeding (e.g., Krogness, 1977); limited imaging has also been done through the irregularity of the skull (Erdmann et al., 1982). The skeleton of a fetus is of cartilage which is easier to look through.

Most observations presently are made by recording the reflections from the interfaces between different tissues. Characteristic of each material is an "acoustic impedance" which is the ratio of the momentary excess pressure (above atmospheric) to the resulting particle velocity. For a given material, it is simply the product of density times velocity of sound in the material. Just as one can see a reflection in a transparent piece of glass, so at an interface between different materials some sound will be reflected and some transmitted. The amount reflected is essentially proportional to the difference in acoustic impedance on the two sides of an interface divided by the sum of the two acoustic impedances, there also being some geometrical factors in the equations that relate to the angle at which the wave approaches the interface. Thus a plane wave in medium 1 moving perpendicularly toward an interface with medium 2 will have a fraction reflected back into 1 that is

$$\frac{d_2 v_2 - d_1 v_1}{d_1 v_1 + d_2 v_2}$$

where d_1 is the density of the first medium and d_2 the density of the second, while v_1 and v_2 are the respective velocities. This is the complete expression for the fraction of the pressure reflected at a large flat surface with perpendicular incidence, sound intensity being given by the square of this. One minus this fraction gives the amount transmitted into the second medium. From this expression it is seen that if two materials are absolutely identical with respect to this property of acoustic impedance then there will be no reflection of a sound at their interface, but if there is some difference then at least a small echo will be reflected back while most of the energy will go on to generate other echoes from more remote structures. A metal fragment embedded in tissue will show up well since it differs considerably in this property and thus is a good reflector. In biological cases most power incident at an interface is transmitted to the next medium; at large angles to the normal, the transmitted beam intensity can actually exceed that incident because the energy in the plane wave is concentrated or compressed by refraction. Echos can have a very large range of amplitudes that must be compressed by variable amplification before presentation to the eye, for example 100 dB reduced to 20 dB in some cases.

People have roughly the same density as water (they can swim) which is roughly a thousand times the density of air, while the velocity of sound in water is roughly five times that in air. Thus the interface between a gas and either a liquid or tissue is an extremely good reflector of sound waves, no matter whether sound is going from gas to solid or from solid to gas. Indeed, a gas–tissue interface is a much better reflector than the best of the more familiar mirrors for light. Thus one does not expect to look through a region of gas in the intestine at structures beyond. Also, a sound transducer is not merely held in the vicinity of a subject whose insides are to be explored since this would require sound to go from transducer material into air, from air into subject where it would rattle around, then go from subject to air and from air back into the transducer. (Besides the reflections, absorption by air is also greater than by water.) Transmission is quite effective if both subject and transducer are immersed in a tank of water, or

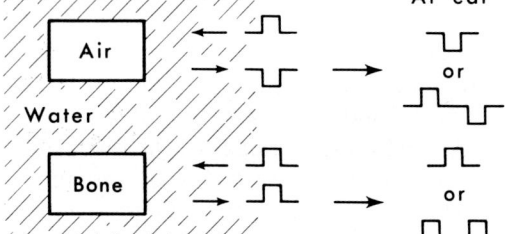

FIGURE 6.6. Good reflectors can be harder or softer than their surroundings, but reflection at a more-to-less dense interface produces a phase inversion while reflection at a more dense boundary does not produce an inversion.

preferably degassed (by boiling) saline. The other alternative is to move the transducer along the skin of the subject after smearing the transducer with some oil or gel to give a good acoustic contact, and being careful not to incorporate too many bubbles with the movement. Low reflection coatings such as are used on camera lenses can be applied to facilitate ultrasound passage, but work best with single frequencies rather than pulses.

The observation of an echo provides sensitive detection of a change in mechanical properties, namely, the product of density and stiffness. The possibility of the above expression having a negative sign has an interesting and possibly useful implication indicated in Figure 6.6. If a wave traveling in one medium comes to a more dense medium then the reflected impulse will have the same sense as that arriving. However, if the reflection is from a region that is more "acoustically dense" to one that is less dense, then the reflection will be inverted. These inversions can be displayed, perhaps as a difference in color, to further characterize a subject. We feel, for example, that a porpoise might use such information from its sonar system to rapidly distinguish the swim bladder of a fish resting upon the bottom from a less interesting smooth rock of similar size.

High frequency sound waves are not only reflected at a region of gas but can be strongly attenuated in traversing it. When gas is subdivided into many tiny bubbles some very interesting effects result that will be discussed in Section 6.7.

6.3. SCAN TYPES

In order to build up an image the sound beam must be moved, either by shifting the aim or position of the transducer or by activating a succession of separate transducers. A few explicit possibilities are described here, with their consequent performances, but other alternatives certainly exist. If the transducer is not moved, then the pattern of echoes along a particular path can be displayed as a function of time after the outgoing impulse, being shown as momentary deflections in a line on an oscilloscope as in Figure 6.5. By analogy with certain radar

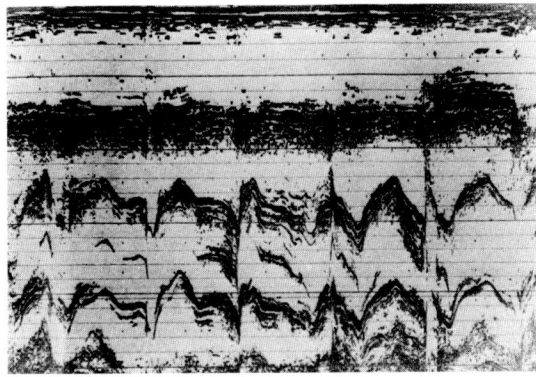

FIGURE 6.7. The M-mode recording of the position as a function of time of the interfaces in the chest, as observed from between the ribs. Increasing depth into the chest is downward, with the separation of points on the front and back of the aorta indicated. This tracing is from a sea lion, made as indicated in the next figure.

displays of World War II, this was called an "A scan." From the indicated time interval between echoes, and knowing the velocity of sound in the tissue types, it is possible to calculate the dimensions of the object. Thus an A scan can be useful in evaluating the size of a fetal head and potential difficulty with birth. The relative size of the different echoes tells something of the properties of different tissues, and how they have changed with disease, though care must be exercised in interpreting the absolute magnitude of reflections because such things as the orientation of the probe and the surface can affect the degree of reflection. An example of an A-type display will be seen at the side in the top part of Figure 9.8.

The horizontal line on the oscilloscope screen could be momentarily brightened for each returning echo rather than being deflected upward. If one of the interfaces were then to move or pulsate during successive impulses the corresponding spot on the oscilloscope would move from side to side while the transducer remained fixed in position. A sheet of film moving upward continuously would then record a wavy line whose motion duplicates that of the reflecting surface, that is, there results a graph of position as a function of time. This recording of motion is the so called "M-mode" scan. This is the basis for much of echocardiography in which echoes from successively deeper layers in the chest are displayed as successively lower lines on a tracing. An example of this is seen in Figure 6.7, where the relatively clear region is between the front and back of the aorta. The blood is uniform or structureless, and the occasional echoes within the clear areas correspond to movement of the aortic valve into the beam. This tracing is somewhat unusual, since it was made through the wet fur of a sea lion as in Figure 6.8; studies with Giavannoni and Sammons in connection with measuring dive-induced cardiac output changes will be mentioned in Chapter 9.

At the base of the aortas of seals and sea lions is a bulb (Drabek, 1977) which can be observed in these ultrasonic measurements, and which also can confuse

FIGURE 6.8. The author receiving a wet kiss from a sea lion while measuring flow in his aorta by ultrasound. The fur if wet does not interfere with adequate sound transmission.

diameter measurement. Indications of turbulence are sometimes seen in human aortic flow tracings (Chap. 9) but were not seen here. It is known that turbulent injections can create bubbles (Kremkau et al., 1969), with lesser disturbances presumably being troublesome with supersaturation. Thus the bulb may be an adaptation to smooth flow and help reduce bubble formation, thus avoiding bends in these diving animals.

A similar tracing for a human subject, with associated electrocardiogram to designate the succession of events, is shown in Figure 6.9. The top echo corresponds to the surface of the skin. Beyond the two echoes from the aorta, indicated by arrows, is the echo from the left atrium of the heart. If angled down slightly from between the ribs, the beam would instead have successively traversed chest wall, right ventricle, anterior mitral valve leaflet, and back wall of the left atrium. Further downward angling would have instead successively recorded the echoes from the chest wall, right ventricle, inter-ventricular septum, and then beyond the

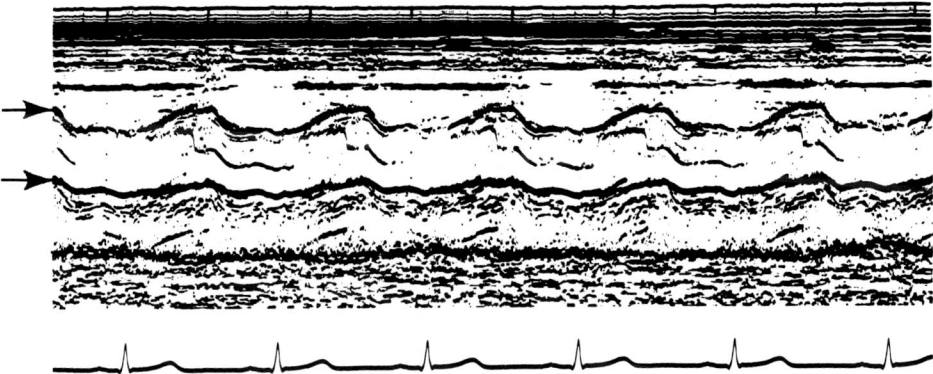

FIGURE 6.9. The M-mode recording from a human subject, again indicating the front and back of the aorta. The electrocardiogram recorded below allows correlation of various events with the cardiac cycle. The aorta is seen to shift somewhat and to pulsate slightly within the chest, while a valve leaflet periodically enters the beam.

space of the left ventricle would appear the echo from the posterior left ventricular wall. The recording of the output from a contact microphone can also be included to determine the source of certain sounds and identify irregularities associated with sounds.

If the probe is moved during the recording process then a cross sectional image can be built up depicting the distribution of material in the plane of the scan. The process can be visualized from Figure 6.10. At (a) (upper left of figure) a subject is shown consisting of a lump of soft tissue embedded in a cube of different soft tissue. This might not be well visualized using X-rays since a ray going through two kinds of soft tissue might be rather similarly attenuated to one going through the same distance of a single soft tissue. A transducer is shown having just emitted an impulse of high frequency sound. An echo returns from the front of the cube, and later in succession the echoes from the next three interfaces, while some energy emerges from the back of the subject. At (b) a dim electron beam is shown starting upward across the face of a cathode ray tube (television screen) when the click was initially emitted. As each successive echo returns to the transducer, thus generating a small electrical impulse, the beam is momentarily brightened to generate the four spots indicated. After all the echoes are collected, the transducer is moved one step to the right and the process is repeated. The figure indicates that by periodically clicking while scanning a full cross sectional image can be built up. By analogy with the radar systems of World War II that produced a map-like display using brightness modulation, this was called a "B scan."

An image of the author's eye formed in this fashion at a frequency of 15 MHz by the equipment of Figure 6.5 is shown in Figure 6.11 (Mackay et al., 1962). A tumor, for example, in the orbit behind the eye would be noticeable. But the image is quite incomplete, only one spot on the lens and three spots on the cornea being visualized. The problem is that what was said above is not strictly true.

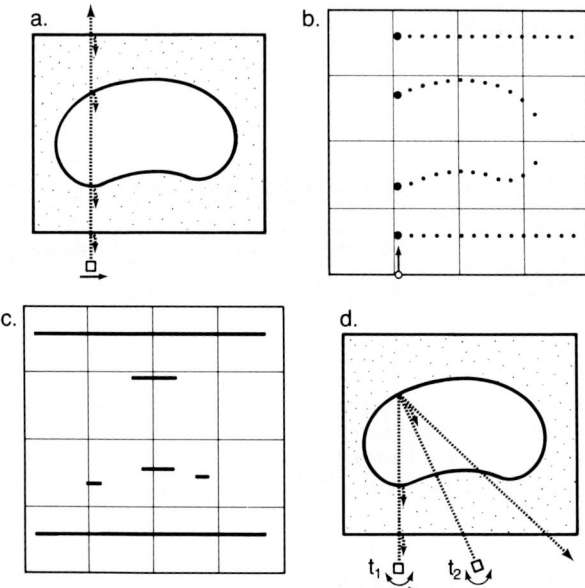

FIGURE 6.10. (a) An ultrasonic impulse from the transducer produced four echoes that return in sequence, while some sound emerges from the back, before the transducer moves on to repeat the process in a scanning movement. (b) Brightening an oscilloscope by the arrival of each echo should yield the above cross sectional image in a B-mode display. (c) What is actually recorded is more nearly this, because only echoes where the beam hits the interface perpendicular are returned to the transducer. (d) If the transducer is rapidly angled from side to side as it slowly scans to the right, then each point will be viewed from many angles, and eventually an echo will be received from each. At t_1 the echo from the third interface will not be seen since it will go off to the right, but at t_2, with the angle shown, the normal reflection will return to the transducer. There are many compound scan patterns combining a rapid and slow phase in order to give a complete B-mode image for reflections that remain in the plane.

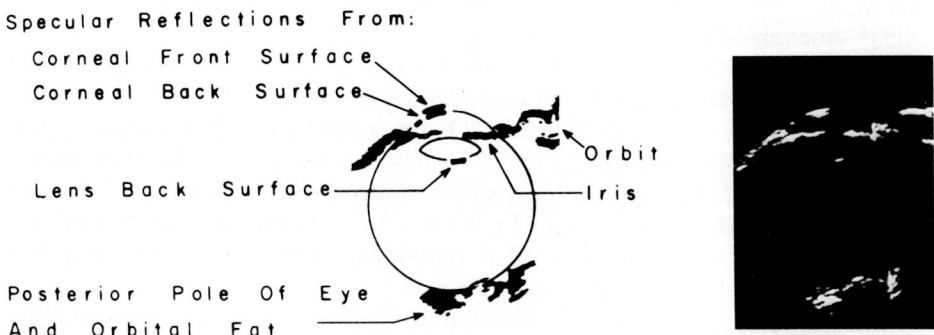

FIGURE 6.11. An ultrasonic image of the human eye and regions beyond made at 15 MHz. This was made with a sector scan only, rather than a full compound scan, to illustrate the partial display of many structures. The iris acts as if sufficiently rough to return an echo from every angle. (From Mackay et al., 1962.)

In Figure 6.10 the first echo would return as stated, to produce the first dot. The second echo would also return to the transducer slightly later, as stated, since the surface there was shown as approximately perpendicular to the beam of sound. However, the third surface is angled in such a direction as to direct the echo off to the right where it would not be collected, since with sound as in optics the angle of incidence is equal to the angle of reflection. Thus that surface would not be seen at that position. The final echo also would be seen. Thus this simple scan would produce approximately the appearance seen at (c) in Figure 6.10. In Figure 6.11 the iris is seen in its entirety since it acted as if rough by sending back sound in all directions. Many interfaces in the body act as specular reflectors in the way seen here. For example, Reid (1965) measured the reflections from excised mitral valve leaflets as a function of angle of incidence, and found a drop in echo amplitude by 50% when the beam was rotated by as little as 5° from normal incidence; a diseased leaflet was less difficult to record since it gave an appreciable echo over an angle of 30°.

One can do appreciably better by viewing all points from all directions within the plane of the scan (though a tilted surface can reflect all echoes totally out of the plane of the scan). This is indicated at (d) in Figure 6.10 where the transducer is shown as rapidly being angulated from side to side while slowly moving to the right. The true path of the third echo at (a) is indicated, making that point on the subject invisible from the first transducer position. But when the transducer happens to be angled to the left from the second transducer position (at t_2) then there is normal incidence and the echo will return to the transducer, and at that instant the point can be displayed. It is merely necessary that the electron beam moving across the screen move in the direction at which the sound transducer is momentarily aimed rather than moving straight up. This combination of two motions to view each point from many directions is called a compound scan, and thus the display of part (d) in Figure 6.10 would be called a "compound B scan." This was one of Howry's contributions, and the picture reproduced in Figure 2.10 used such a scan. In our example rapid angulation with slow translation was suggested, that is, a sector scan superimposed upon slow translation, but the same effect is achieved by rapid translation through a small distance accompanied by slow angulation around the subject. The latter was used in Figure 2.10 (as it was later used in the original X-ray computed tomography scanner of Hounsfield), while Figure 6.11 was actually produced with only a sector scan functioning while the larger scale motion was switched off. There are actually an infinite number of combinations of slow and fast motion that can be combined into a compound scan. The configuration of Figure 4.5 could provide a compound scan if pulsed and the returning echo time resolved, or it could be used for sound transmission CT; modified image processing could somewhat correct for beam spread.

In abdominal observations the transducer is often placed at the end of a mechanical arm where it can be rotated and moved by hand over the skin of the subject. The recording cathode ray tube must have its momentary path direction mimic the aim of the transducer as well as its position. Suitable sensors on the arm monitor position. The sweep of the beam is controlled by a potentiometer set by

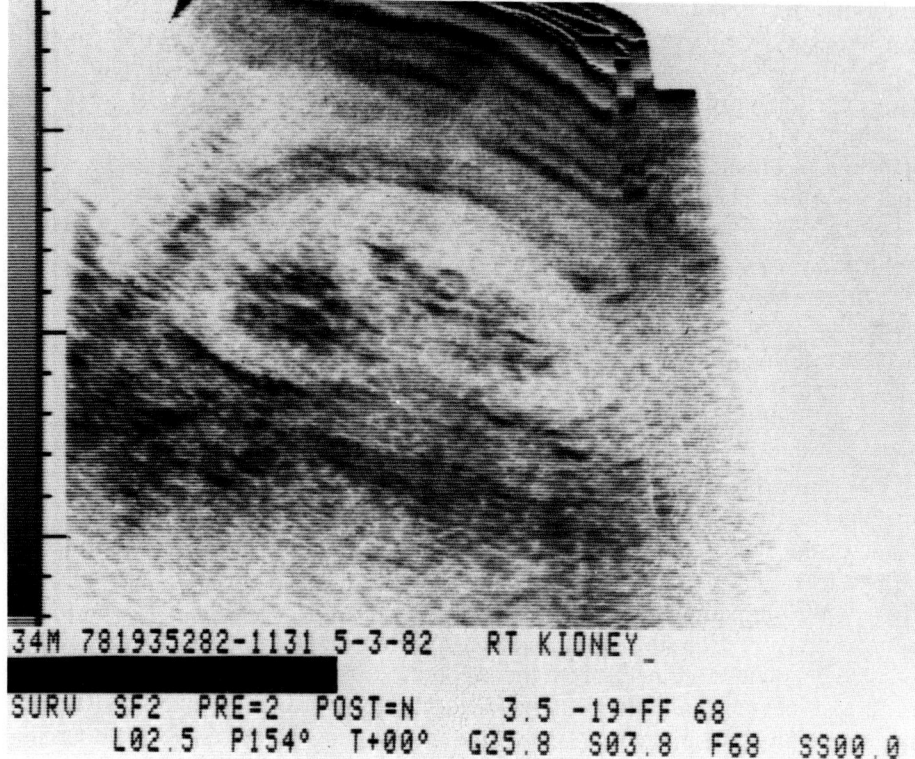

FIGURE 6.12. Compound B-scan of a human right kidney. The ultrasonic probe was guided by hand along the surface of the body and in contact with it using a liquid coupling material.

the angle of the transducer and whose outputs are proportional to the sine and cosine of the applied sweep voltage, thus giving the required momentary X and Y coordinates of the spot from instant to instant. Potentiometers often become worn with use and thus we studied other ways of converting the momentary angle of the transducer into a corresponding electrical signal; variably obstructing the light falling upon a cadmium sulfide photodetector was less effective than a variable transformer, one of whose coils moved with the sound transducer (Rubissow, 1973). Inexpensive modern integrated circuits are easily able to store information about the echoes at each point in the image plane, as the echoes are gradually accumulated. This information can then repeatedly be read out in any sequence, for example, it can be displayed on an ordinary television set scanning in its usual way, even though the data were collected in quite a different sequence. This last function is sometimes referred to as scan conversion, and in earlier times this was done in less effective special storage tubes. An image of a human kidney produced in this hand-directed fashion is seen in Figure 6.12. This is the right kidney viewed in a longitudinal scan. A transverse ultrasonic image of the same right kidney is seen in Figure 6.13. Here the liver is to the left (the patient's right) with some gall

SCAN TYPES 117

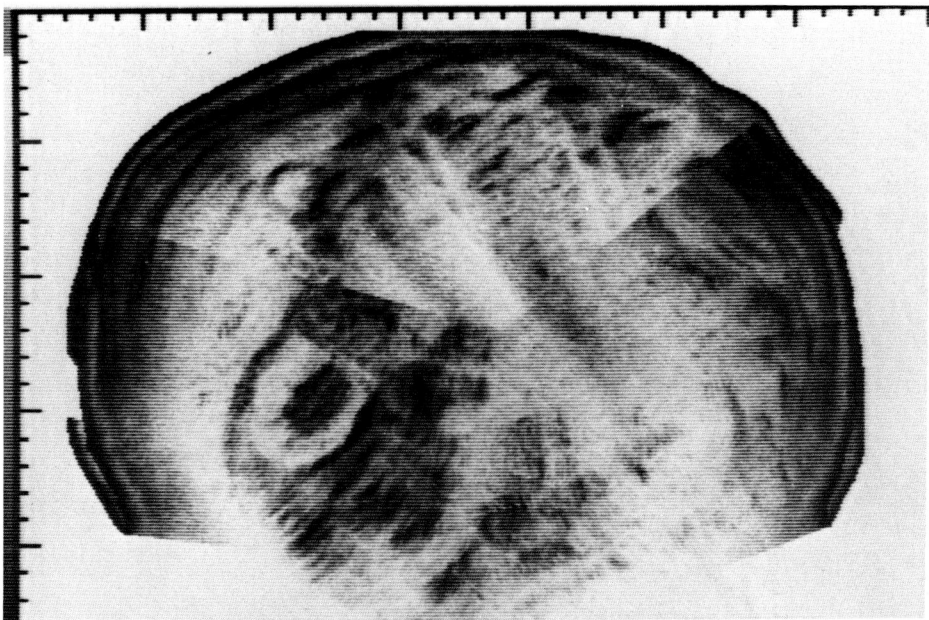

FIGURE 6.13. A different view of the kidney in the previous figure. Rather accurate distance scales can automatically be superimposed because the velocity of sound in soft tissue is not highly variable.

bladder being at the top. At the right the picture is obscured by bowel gas. The several scan lines or directions of view are especially noticeable in Figure 6.13, while the fine horizontal background lines are simply the television scan in the output display, which is continuously visible during the scanning process of building up the image. Note also the display of range or distance calibration scales that are appropriate for soft tissue where the velocity of ultrasound is approximately fixed. This is the same kidney as was previously seen by X-ray computed tomography in Figures 4.12 and 4.13 for comparison.

Scanning can be done either by mechanically moving a transducer or by successively activating a row of transducers. Modern integrated circuits allow the latter to be done much more conveniently, since each transducer requires its own individual electrical system. If there were, for example, a hundred transducers in a straight line, each could be activated in succession to give the effect of physically moving one transducer along the line a step at a time (e.g., Whittingham, 1976; Bom, 1972). But if each transducer was quite small, then its transmission and reception pattern would be unsatisfactorily broad. In that case, it is better to activate the first 10 together to give the effect of 10 times as wide a transducer, then activate 2–11, then activate 3–12, and so on, thus giving the effect of a large transducer having some holes in it moving in small steps. With such an arrangement it is not only possible to see the image form, but it is possible to produce dozens of complete images per second. This effect of a continuous moving picture

118 ULTRASOUND

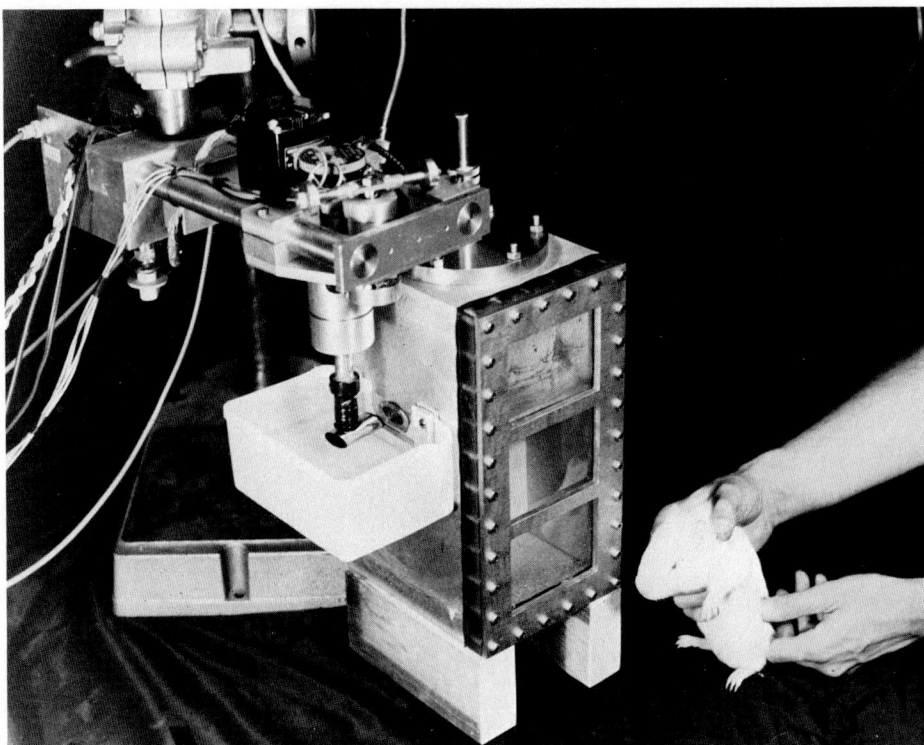

FIGURE 6.14. The pressure chamber for small animals has an acoustic window in the side consisting of a sheet of mylar. The transducer in the degassed saline opposite the window undergoes a rapid mechanical sector scan. The hydrogen thyratron for producing the high voltage pulses is seen at the left. The unit started as a sort of windshield wiper, but here has evolved to being mounted on a drill press for rigidity, which allows compound scanning in other applications. (From Rubissow and Mackay, 1971.)

being seen as it is happening is often referred to as "real-time" imaging. However, the excellent performance of a single large transducer actually can be achieved in real time if the mechanical motion is made rapid. A sector scan producing an image shaped like a piece of pie is particularly convenient, and with suitable counterbalancing the transducer can be made to vibrate from side to side about 10 times per second, giving 20 complete images per second. The unit aimed through the acoustic window of a small diving chamber in Figure 6.14 is of this sort and provided the fast phase of motion in the original compound scan for eye work (Mackay et al., 1962). If a transducer rotates steadily in one direction rather than reversing itself to go back and forth, then a mechanical scan can be somewhat faster. A further increase in image rate can be achieved by mounting several matched transducers around the periphery of a wheel and allowing each successively to provide a full image during rotation. Such an arrangement can allow

SCAN TYPES 119

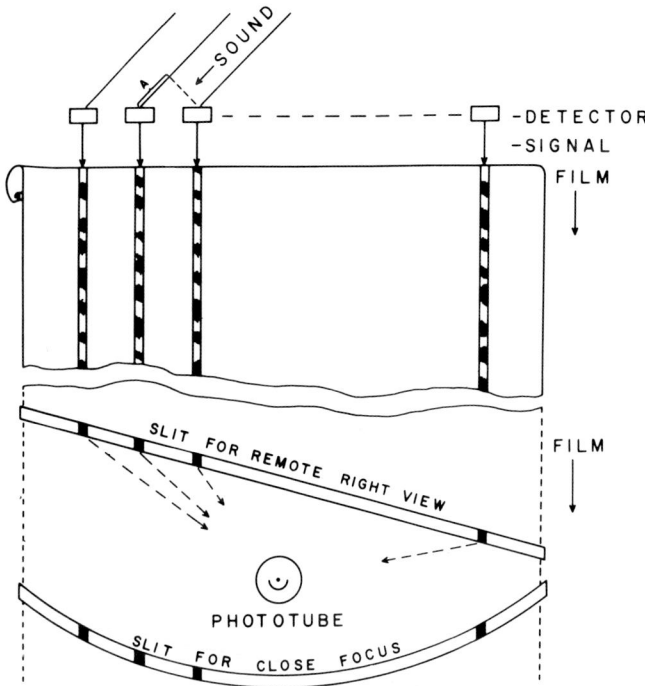

FIGURE 6.15. The scheme of a phased array and its scanning or focusing action. All echoes returning from a single click are shown being recorded simultaneously at several positions in space in the form of variable darkening on a film. Running this film after development past a slit at an angle allows one to determine the time course of echoes returning from that direction; a curved slit allows one to focus attention on a particular distance. The effect of the time delay between channels of the moving film can be produced electronically for displays in real time. (From Mackay et al., 1962.)

continuous display on an ordinary television set, or recording on an ordinary video magnetic tape recorder at the usual frame rates.

Another way of providing a rapid scan without mechanical motion, and which in addition makes possible dynamic focusing, uses the concept of a "phased array." This concept has recently attracted considerable attention, but it has a longer history. In 1960 the Mine Advisory Committee of the National Research Council convened a meeting for the National Academy of Sciences to bring together radar and sonar workers for helpful cross pollination. At that time we found the radar engineers to be somewhat familiar with phased arrays, while the sonar engineers mostly were not. (Tucker and co-workers had, however, commented on such matters in 1958.) The considerations of sonar and radar all involved scanning a beam from side to side at very distant targets, thus implying many parallel rays. The considerations were introduced to various biomedical groups with Figure 6.15 (Mackay et al., 1962; Mackay, 1966a). A number of individual sound detectors are shown simultaneously active, each recording any incident sound from all directions as a pattern of darkening on a moving strip of film. A click coming from a great distance and to the right will produce a dot first at the right hand detector, then the one next to it, then the next in line, and so on. If, after development, this film is illuminated and run past a slit tilted as shown, then for one particular angle of this slit all the dots will arrive simultaneously and produce a maximum signal in the photo tube while signals from any other direction will simply average out as noise. Thus by providing a suitable time delay between each channel one can "steer" sensitivity to any desired direction. We had already demonstrated the possibility of continuously variable electrically adjustable delay lines based upon nonlinear reactances (Kulmann, 1954), and so the time delay of the moving film was in part merely a convenient teaching aid for visualization of a steady process that could occur in real time. Phase shifting the electrical signal is a simple approximation to a delay that can provide adequate functioning for the short delays in real equipment. If the detectors instead acted as sound sources activated in sequence, then some sound would go in all directions but most of the wave from the combination would similarly be directed to one side at a particular angle depending on the time delay. Time delays for the production rather than the reception of impulses were already well known in the form of electronic circuits called monostable multivibrators, and so directing a beam in a chosen direction was a simpler process.

The figure also suggested that the same processes could focus attention on a nearby spot, which could be adjusted in position by shifting the time delays between adjacent receivers. Following an outgoing click, echoes successively return from progressively greater distances, and by shifting the time delays it would be possible to focus attention upon these steadily increasing ranges. There is thus the possibility of a dynamic focus, as if a lens were being bent progressively following any one click in order to reject noise, and so on. If one wishes only the effect of a lens free of some aberrations, without deflection from side to side, the transducers can be concentric rings to collect echoes out of the plane.

The pattern of any one transducer acting as a transmitter, or its sensitivity pattern acting as a receiver, is given by its size relative to the sound wavelength, as in Figure 6.1, while the pattern of the overall combination depends upon the overall size of the group of transducers in a similar way. Subdividing a large transducer (that could be tilted to aim to the side) into many smaller ones does, however, introduce added directions of sensitivity for the same reason that a diffraction grating produces several orders of a spectrum. This can introduce artifacts or give an impression of extra structure that is not present. In Figure 6.16 is seen a phased array image made with 64 crystals 2 cm high and of a thickness corresponding to 2.25 MHz, arranged in a row 2 cm wide. A kidney is visualized in real time while the probe is held steady. However, while maximum sensitivity is aimed straight ahead, there is some sensitivity to the side, where a "strong object" will be sensed but displayed as if straight ahead. Thus the true kidney pattern is also displayed ahead (here downward) as indicated in the drawing. This image might thus be interpreted as indicating the presence of an abscess between the kidney and liver. Similarly, in gallbladder studies, colon gas can produce

SCAN TYPES 121

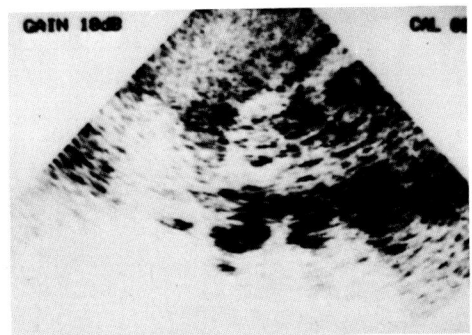

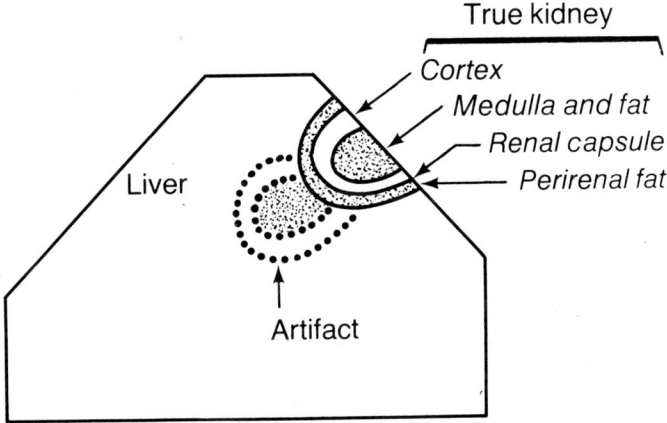

FIGURE 6.16. A real-time phased array display of the kidney in a human subject and the type of artifact that can be produced. The effective subdivision of one large transducer into a number of smaller ones can also give a noisy or grainy appearance for similar reasons.

dense echoes in an area below the liver; they may be projected into the center of the gallbladder which is normally echo free unless it contains stones. The problem is less severe than static pictures indicate because it changes dramatically with changes in transducer orientation. This same effect undoubtedly contributes some of the noise seen in phased array images. It is desirable that the transducers be spaced a half wavelength measured in subject material and have a thickness of a half wavelength in transducer material. Compact phased arrays can be built, for example, to examine the heart close up after insertion into the esophagus (Souquet et al., 1982).

One can feed back a signal to either a phased array or a sequentially activated transducer array so that the beam will follow and record a selected moving structure, the line used for tracking being interlaced with those producing a real time image.

Above a row of such phased array transducers can be stacked more rows to give a square array of points at each of which there can be transmission and

122 **ULTRASOUND**

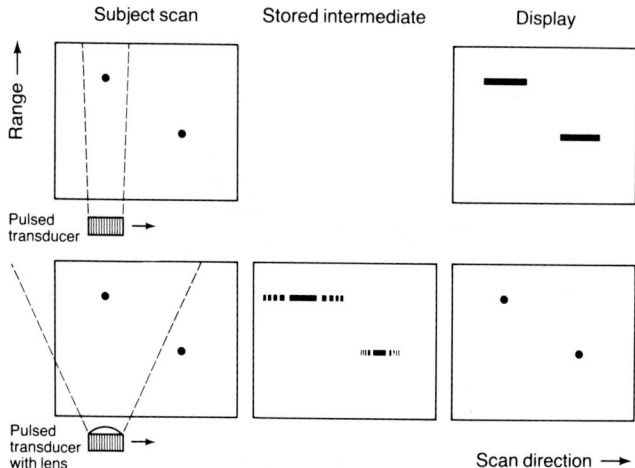

FIGURE 6.17. Combining the signals from many impulses of a scanning transducer can give improved resolution over a long distance in a synthetic aperture system. At the top each point in a B-mode display is seen as a line having approximately the width of the transducer; nothing is stored for special computation. Below the beam is shown as spread so each point returns an echo for many pulses during the motion of the transducer. A dotted pattern amounting to sections of a zone plate is stored for conversion into the final image either by computer or special optical system.

reception. Suitable phasing can then direct observation or a beam up and down as well as side-to-side, or allow focusing action in any direction. A sharp pulse (containing many frequencies) allows concentration at a given range, and frequency analysis also allows determination of how the properties of the subject material change from one frequency to another. One of the first to thus demonstrate imaging a plane across the direction of sound propagation (a sheet at a given depth in the subject) was Kay (reviewed in Kay, 1973). Complex arrays have since become practical simply because of the capacity of integrated circuits to provide complex functioning inexpensively in a small unit. One needed process is the Fourier transform, and an optical-electronic method for doing this rapidly is indicated in Chapter 10. Of course, if such an array of transducers is fronted by an acoustic lens, then activation of one transducer probes one point in the subject, and scanning can be done up and down or sideways (Beaver et al., 1977); mention was made in Chapter 3 of considering a lens as providing a Fourier transform of the energy distribution falling upon it, and the present example may help in visualizing this.

A major problem in the scanning systems has been to see detail that was fine in the direction across the beam. In the upper part of Figure 6.17 it is seen that, with short pulses, the display of a small point is blurred sideways into something that resembles a line. By comparing the response to a number of pulses from different positions one can improve this, and actually achieve resolution approaching that appropriate to a transducer as large as the path swept out, without the problem of blurring across a very wide beam. For a "synthetic aperture" system one spreads

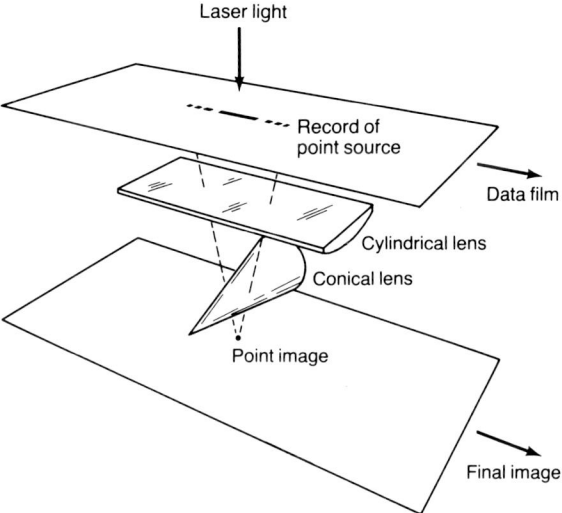

FIGURE 6.18. A simplified version of an optical system that can be used to produce final images from synthetic aperture recordings. Compression is required only along the direction of motion, and in this direction laser light interferes to produce a point. The spread in the pattern depends on the range of the subject point, necessitating use of a conical lens. Some details of the real situation are omitted, but this is the essence of the handling of the signal from one point on the subject.

the outgoing beam and pattern of returning sensitivity as shown at the lower left of the figure in order to view each point from a great variety of directions during the scan. One here has some of the advantages of a compound scan, and effort is made to record a wide range of returning intensities from strong to the weakest. The short emitted pulse is derived from a brief segment of the signal from a continuously running oscillator. The returning echo is multiplied with this continuous reference signal, making the recording alternately light and dark because sometimes the echo will be in phase and sometimes out of phase with the reference signal. The phase variation is slow when the object is in the center of the beam and becomes more rapid when the object is toward the edge of the beam. The intermediate pattern that is stored is shown in the figure, and is seen to be a slice out of the middle of a zone plate as sketched in Figure 3.1. The width of the pattern is different for points at different ranges. The system works as if it were a hologram from side to side and an ordinary pulsed ultrasound system in the distance direction. Reconstructing the stored intermediate back into a final image, in this case two points, can be done by a computer or it can be done by an unusual sort of optical system whose generalities are indicated in Figure 6.18. Along the direction of scan the recorded pattern will interfere to produce an outgoing point of light, with the conical lens taking care of different magnifications appropriate to different distances from the transducer, while the cylindrical lens maintains the point sharp in the distance direction.

Energy from a point tends to reflect in a small range of directions from many objects, giving the effect of "glint" in visible light terminology. As mentioned, in these systems one records as large a range of amplitudes as possible. Structure will not appear shifted toward bright directions since, even if only a piece of a zone plate is recorded, it will reconstruct a point at the correct spot (but with decreased resolution or greater spread) for the same reason that a piece of a hologram works. With rough surfaces there tends to be constant spatial resolution independent of distance because, as range increases, the maximum length of the synthetic aperture or the effective size of the imaging system (distance for which a point is within the illuminating beam) increases proportionately. Synthetic aperture radar systems have had considerable application in imaging large expanses of territory from moving airplanes, but they have also been mentioned in the ultrasound literature (Burchardt et al., 1974).

Camera tubes have been built that would directly convert a pattern of vibration amplitude (i.e., an ultrasonic image) directly into a signal suitable for display on an ordinary television monitor. Such a tube is fronted by a plate of quartz crystal which, in vibrating, generates voltages that can be scanned off the back by a moving electron beam in the evacuated tube, thus generating signals which can be displayed. The idea seemed to have originated with Sokolov (1936, 1939), who realized that when the outer surface of the camera face would be immersed in liquid that the plate would be damped and resonate point-by-point in sympathy with the incident ultrasound distribution, and that there would be negligible lateral spread of the thickness-mode oscillation from any point on the face of the tube. Thus an intense vibration at a "bright" spot would not spread to set the entire front of the tube vibrating vigorously. This was an unusual thought and difficult to predict. Various people have constructed such tubes, which have been limited in size without collapsing, and have used various electronic techniques for actually extracting the charge distribution (Brown and Galloway, 1976). Certain methods are able to display only changing or moving parts in a scene, while others compare the received phase of a signal with the fixed phase of a sound generator in order to give phase-contrast effects. In use one employs a converging acoustic lens to cast an image of the subject upon the front of the tube after the subject has either had a source of sound reflected from the front or transmitted through from behind. Articles on this by Turner, Smythe, and Jacobs were collected in a small volume (Mackay, 1965a) and were also reprinted into several issues of the journal *Ultrasonics*. Other camera tubes have attempted to employ thermal effects but were generally somewhat less sensitive. These image converter tubes, which change a sound image to an electrical signal, should not be confused with the scan converter tubes that were previously mentioned and which store an image gathered in one scanning process and allow it to be read off in a different sequence or at a different rate by a subsequent scanning process, thereby perhaps allowing its display on an ordinary television set after being collected in a compound scan; both the input and output of such a storage tube are electric.

A sound source driven by an oscillator is a source of steady coherent waves, and this suggests their use without lenses to produce holograms. A sound distribution can be recorded on the previously mentioned converter tubes, or it can upwardly deflect the surface of a bath of water located in an optical interferometer for observation. In practice, acoustic holography has generally been rather disappointing.

6.4. HAZARD AND DOSE

The literature indicates that sound waves of high intensity are able to promote many physical and chemical changes. Various workers have reported changes ranging from making droplets either disperse or coalesce, to making a tadpole swim out of its skin, and it probably can more rapidly age whiskey. Sound waves are able to break up kidney stones to promote passage without destruction of adjacent soft tissue. Intensities in the range 10–1000 watts/cm^2 have actually been used to perform brain surgery on humans since, depending upon duration of exposure, such ultrasound can permanently or temporarily interrupt neural paths (Fry, 1958). The effect is within seconds and, if the patient has had a piece of the bone overlying the brain removed to limit heating, such procedures can be performed on conscious cooperative subjects. He observed that white matter was more affected than gray (gray being cell bodies and white being fibers), making it possible to destroy nerve fiber tracts surrounded by nerve cell body regions. Of course, effective acoustic coupling must be provided by a suitable water bath, as sound in the air would mostly not enter the head. He also found that with a focused beam it is possible to destroy all neural components in the brain without interrupting the blood supply. Some of the earliest workers were concerned about possible genetic effects, but much of that thinking went unrecorded.

There have been many studies and reviews relating to other possible biological effects of ultrasound, the idea being that a potential for change is a potential for damage. Certainly DNA in aqueous solution can be degraded by ultrasound, though the effect could be different in sonicated mammalian tissues (Hill, 1971). It has been reported that diagnostic levels of ultrasound can produce changes in DNA strongly suggestive of unwinding of the helix or single-strand break induction, and that in one experiment tumors developed in mice at the site of injection of ultrasonically treated cells (Liebeskind et al., 1979). Others have concluded that, though chromosome aberrations and retardation of growth have been recorded, there is apparently little or no danger associated with exposure at clinical levels (Baker and Dalyrymple, 1978). In spite of the sensitivity of the eye to damage, it may be possible to control eye pressure and treat glaucoma with intense ultrasound (Mackay, 1964). Cellular attachment to a plastic substrate has been suggested as a sensitive parameter for studying the effects of ultrasound with, for example, cells cultured from human amniotic fluid being more sensitive than those derived from the kidney (Siegel et al., 1979). The effect of ultrasound on membranes might be studied by direct observation of changes in the electrical negative resistance properties of isolated nerve fibers, though the result might be different *in situ,* where a different density of naturally occurring tiny bubbles could exercise a major effect.

Causes of the various effects are still uncertain. The work of Fry (1958) seemed to rule out three possibilities: heating, local hot spots, and cavitation. Extraction of energy from the beam certainly produces heating, but the expected temperature rise is probably inadequate to explain any of the effects. If the energy were not to distribute itself uniformly but were to concentrate irregularly, then hot spots could result. This also probably is not a major factor. The cyclic negative pressures associated with sound can produce voids whose collapse can be quite destructive; such cavitation can pit the metal propellers of boats. However, irradiation of animals under increased hydrostatic pressure seemed not to raise the threshold for the effects. To the extent that these major candidates are ruled out, one must search elsewhere. If tiny bubbles are normally present in the body, then their vibration in a sound field could have drastic effects (Nyborg, 1973). The difference in susceptibility of whales and humans to the bends (decompression sickness), surprisingly, may bear on this by suggesting the normal presence of some microbubbles (Mackay, 1982). In many electron microscope pictures there appears a small unexplained artifact type that may be the residue of bubble nuclei, suggesting further studies along these lines.

The overall impression from many of these observations is that, while there are biological effects above about 1 watt/cm^2, there is little effect under about 0.1 W/cm^2. Application of ultrasound for a longer time can compensate a lesser intensity, and there seems to be a reciprocity law in which the product of time and the square root of intensity must reach a specified value to produce a given effect (Fry and Dunn, 1956). The square root of intensity is a measure of peak particle pressure or velocity amplitude or the displacement amplitude associated with the sound.

For this information to be applied usefully there must be a method for measuring the arriving intensity of sound, the assumption being made that suitable coupling is provided to allow entry into the body. Obviously a small microphone having an appropriate frequency response can compare the amplitude of an unknown sound wave with that of a wave coming from a known transducer. But ultimately the intensity of some wave must be measured in more absolute terms. Two effective ways for measuring the intensity of a sound wave will be mentioned.

If a thermocouple consisting of a pair of joined dissimilar metal wires is contained in a plastic bag of tissue-like material such as castor oil, then application of a sound beam will produce measurable heating (Fry, 1958). There is a precaution to be noted, since the recorded output from the thermocouple shows an immediate and a slower phase resulting from two heating mechanisms. The more immediate response is caused by the heating of the wires by the action of viscous forces resulting from relative motion between the wires and the imbedding medium. The slower phase is due to the heating by the absorption of the sound in the body of the medium, and it is this that is used to calculate energy in the beam. The slow phase is independent of the orientation of the wire in the sound field. Fry (1958) reports it is possible to determine the absolute acoustic intensity in a plain wave field to an accuracy of a few percent.

A sound wave exerts an alternating pressure on an interface, and in addition it produces a unidirectional pressure on any interface where there is a change in characteristic impedance or on any medium in which absorption occurs. It has long been known that the force of this net push is equal simply to the ultrasonic power divided by the velocity of sound. This gives the net force at a perfect absorber, while at a perfect reflector it is twice as large because two momentum changes occur. The effect is readily observed by placing a beam balance in a tank of water and aiming a sound transducer down at one of the pans. If the immersed sound transducer is instead aimed upward at the surface of the water, then an upward bump in the surface will be produced, or even a fountain effect. Even the small average power involved with the widely spaced pulses of medical ultrasonic equipment can be measured in this way. For example, a tiny glass bead melted onto the end of a thin glass fiber will be deflected in such a beam by the bending of the fiber, and measurement of the deflection allows absolute calibration of the sound intensity (Rubissow, 1973).

There is a precaution that is sometimes needed in making such measurements. As a vibrating "piston" moves backward, liquid comes in from all sides while during forward motion liquid is preferentially projected straight ahead. This results in some steady streaming of liquid away from a transducer. This streaming can produce an artifact by further deflecting the force measuring device if the flow is not caught by a thin plastic sheet between transducer and target, through which sound can still pass essentially unimpeded. (Small bubbles if on the target will vibrate, set up streaming toward the source, and increase the reading similarly.) Deflections can be measured either electrically or optically, and the method appears to provide better results than attempting to build an absolutely calibrated microphone to measure the alternating component of force associated with the passage of a sound wave. Ultrasound transducers can, however, be calibrated rather well with only electrical measurements (Drost and Milanowski, 1980) by an older reciprocity technique.

6.5. CONTRAST AGENTS

There are small differences in acoustic properties throughout the body that can be detected with an ultrasonic imaging system. Also, reflection systems are extremely sensitive since echoes appear on nothing (i.e., superimposed upon only a small background of noise) rather than as a small change in some general intensity as with X-rays. Since contrast is generally present, extra contrast agents are less routinely used than with some of the other modalities.

However, the introduction of a material with either different elastic properties or a different density will appear quite noticeable. Thus swallowing a glass of tap water will outline the stomach because of the many minute bubbles that such water contains. The existence of microbubbles in water can be proven in various ways. Measuring sound absorption as a function of frequency shows an increase near 200 kHz, which corresponds to a resonance of groups of microbubbles of 50 μm radius, in newly collected tap water (Iyengar and Richardson, 1958). A few hours after collection these numerous large bubbles have vanished, presumably by floating to the surface. After about 20 hours, sound absorption shows that tiny bubbles are still present, at a much smaller concentration. A resonant frequency near 1 MHz gives a calculated radius of the order of 1 μm. Applications of pressures of 10,000 lb/in.^{2} reduced the sound absorption even after the pressure was released as a consequence of the disappearance of some of the microbubbles by solution in liquid. A fraction of the microbubbles still remains in water after the pressure treatment, and they have radii around 1 μm. These authors obtained similar results from light scattering observations. Filtering water under a vacuum (Messino et al., 1965) removes solid impurities and dissolved gases, making it difficult to induce cavitation. When air is allowed to diffuse into the samples, cavitation again occurs easily, suggesting that cavitation nuclei consisting of microbubbles of about 1 μm radius, when destroyed, tend to be reestablished to an equilibrium concentration. Our methods to be discussed in Section 6.7 allow individual bubbles to be studied ultrasonically and generally confirm such observations. Minute bubbles also form the basis for contrast echocardiography, as will be discussed in Section 6.7.

In some cases a change in texture is important. Thus the region inside of a fluid-filled cyst will be relatively free of echoes just as was the blood in Figure 6.9, which can distinguish it from a tumor within which one sees a "snow storm" pattern. (If the amplification is turned too high, noise will make a snow pattern appear everywhere, however.) Similarly, ordinary saline can be injected in certain regions to outline adjacent structures. Flowing blood can display weak echoes, and control of red cell aggregation has been suggested as a way to enhance images of well perfused organs (Sigel et al., 1982). Injected gas is an exceptional reflector, but its surface may specularly direct the echoes away from the receiver, thus allowing its viewing only by its dark shadow beyond. In clinical practice, injections of gas are sometimes presently used, for example, to determine the patency of oviducts, and ultrasonic observation should be considerably more effective than the surprising methods presently used to judge the release of bubbles. Unlike groups of tiny bubbles, large masses of gas have not often proved useful, and they generally obstruct the view of structures beyond.

The passage of a broad beam of steady sound will heat tissues, as is sometimes done in therapy as a substitute for diathermy. Certainly there will be some small heating differences from region to region, even on the fine scale suggested by the differential effect on the white versus the gray matter in the brain, as mentioned above. A change in temperature is known to cause a change in the acoustic properties of material. Thus prior to each set of scanning impulses might be emitted a long pulse of ultrasound to induce temperature differences, after which one could "ping" off the thermal gradients. The effect of diathermy might be judged ultrasonically, and it might be used to generate a different set of thermal gradients.

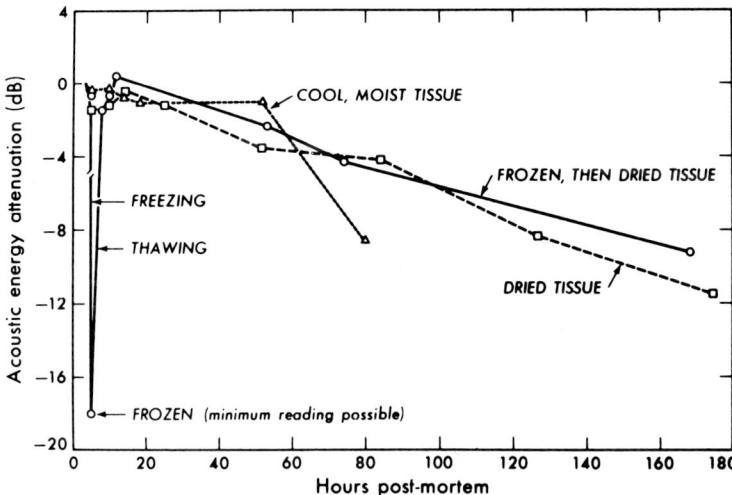

FIGURE 6.19. The modality of ultrasound is rather sensitive to the time after death and to the handling of tissue samples. These are some observations on the attenuation of acoustic energy being transmitted through samples, and changes can be considerably more rapid under some field conditions. (From Crosby and Mackay, 1978.)

6.6. POST MORTEM EFFECTS

There are changes following death that slightly alter the images produced by some modalities, but ultrasonic images can be quite strongly affected. In some early uncontrolled observations changes were noted in tens of minutes and these were attributed to the possible formation of tiny bubbles (Mackay, 1966b). Indeed, if the early workers in medical ultrasonics had employed samples from the meat market rather than using their own family members, the entire field might have been set back some years.

Before there was complete confidence in ultrasonic imaging equipment, such equipment was sometimes tested with a piece of separated tissue. Even now, the acoustic properties of different tissue types are often inferred from a piece of excised tissue. Thus it seemed desirable to have some measurements indicating the time, under controlled conditions, before which acoustically noticeable changes start to take place. A scanning transmission imager was used for the tests. The transducer was swept in frequency to allow rejection of paths longer than straight through the specimen, such as might be produced by refraction or by reflections from the side of the tank. Such longer paths would arrive later and add to the intensity of the receiver during a reasonably wide pulse if they were not rejected by having a higher frequency than the shortest path. (In transmission imaging the time measurement would allow mapping velocity changes as well as attenuation.) The general results for several tissue types are indicated in Figure 6.19 (Crosby and Mackay, 1978). The graph is for beef spleen under different modes of handling. In these same observations under somewhat controlled conditions, a sample of human uterine tissue showed no changes in transmitted acoustic energy in the first 1.5 hours, starting 15 minutes after surgical removal. Upon thawing and drying frozen tissue, its attenuation curve matched that of dried tissue that was not frozen. Moist tissue seemed to deteriorate sooner. It is suggested that fresh frozen tissue could be stored for later use and still retain at least some of the original acoustic properties. Under these moderate conditions the extremely rapid deterioration that had previously been observed was not seen. Observations of tissue *post mortem* have also been made by Hueter and by Dunn.

6.7. TINY BUBBLES

A tiny bubble in an ultrasonic beam will be forced to vibrate back and forth along the beam, and will pulsate in volume in synchronism with the pressure changes associated with the ultrasound. These vibrations cause the bubble to return more energy than would be scattered back from a heavy rigid object of the same size. In a sense, the bubble acts as if it were considerably larger than it actually is (Mackay and Rubissow, 1978). Thus bubbles much smaller than a wavelength can be noticed, localized, and measured in ultrasonic images of intact subjects (Rubissow and Mackay, 1971; Rubissow, 1973). Larger bubbles "reflect" sound but do not amplify in the same way. The first report of tissue bubbles was at a meeting on diving (Mackay, 1963b) and was quite specific:

> Much of the discussion has emphasized a fundamental need in supersaturation and bubble formation experiments for a definite indication of end point, and there may be a new approach. It is the use of ultrasound to detect the presence of tiny bubbles and the onset of the bends. In our laboratory at Berkeley, we have built pulsed 15-megacycle sound echo-exploring equipment and gas bubbles are excellent reflectors. At this frequency, the wavelength is 100 microns, and these circuits can detect and follow the progress of single bubbles that are an appreciable fraction of this dimension, or larger. In the preliminary experiment, I demonstrated the change in properties (increased opacity) to ultrasound that rat tissue undergoes upon being given vigorous decompression sickness. A focused transducer and simple scan were used, but a bulk observation with continuous sound also should have been effective. Either sound transmission or backscattering methods could be employed, and they would seem to provide an excellent tool both for *in vivo* and *in vitro* studies of the kinetics of bubble formation, for example with mixed inert gases. Some workers speak of "silent bubbles" and these should be detectable, if real. It might be mentioned that the sound intensities here are very low (about 0.001 watt/cm^2). Rapid excess inward diffusion of gas on a rarefaction half-cycle might allow intense sound to "pump up" otherwise unnoticeable small bubbles, in a fixed way, until detectable. I would imagine that a diver who was decompressing, and was hit by an intense sonar beam, might be given a case of the bends. We could investigate the effect of intense ultrasonic energy on a tissue having an overpressure with our other therapeutic equipment, but I have not yet done this; degree of supersaturation might thus be measured.

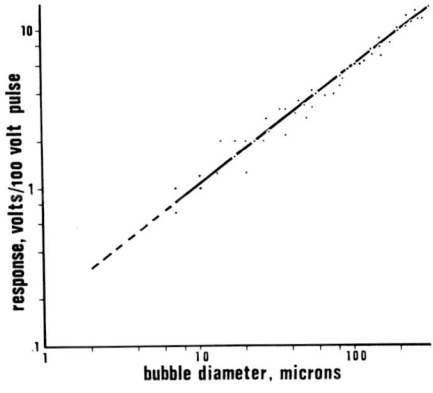

FIGURE 6.20. The measured response of an ultrasonic scanner to gas bubbles of different diameters. The frequency was 7.5 MHz, for which the wavelength is 200 microns (μm). Bubbles in the range of small sizes change or vanish rapidly. (From Rubissow and Mackay, 1982.)

This observation was made with the same equipment that had produced Figure 6.11 (Mackay et al., 1962). That equipment was modified by a number of successive students including Rubissow (1973). The frequency was reduced from 15 MHz to 7.5 MHz, and the original compound scan was inactivated leaving only the fast sector scan, in order to reduce confusing tissue echoes. The high performance of the lithium sulfate transducer still required the use of a very high voltage pulse as in Figure 6.5. The calibration data for this instrument (Rubissow, 1973) are summarized in Figure 6.20 (Rubissow and Mackay, 1982). These data were obtained by recording the echo height returning from bubbles of known size. The actual bubble size was measured by a microscope, the bubbles being either in water under a layer of gelatin or in transparent fish. Since tiny bubbles diminish or expand extremely rapidly, these data were taken by bubble collapse experiments. One watches an already formed bubble gradually diminishing in size, and at the last possible instant the push of a button simultaneously fires a strobe lamp to photograph through the microscope and also emit a sound impulse whose echo is then recorded. The bubble is centered in the thickness of the scan sheet by transducer shifts for maximum return, and hopefully does not drift partially out of the field. The smallest bubble calibrated was 7 μm, but the background noise level routinely corresponded to bubbles of about 1 μm size, this being the level to which we could measure any small rapidly changing echoes that might be recorded. From this calibration curve and the observation of a pulse height in an unknown echo (and using distance to correct for transducer pattern and tissue attenuation), the size of an unknown bubble could then be determined in the image of a subject. It should be emphasized that resonance is not involved here, and that the size of a single bubble less than 10 μm is being approximated while using a wavelength of 200 μm (corresponding to 7.5 MHz).

The general form of the scattering of a wave (sound, radio, etc.) by small objects is well known and is indicated for the present case in Figure 6.21. Obviously, near resonance a given bubble echo will provide a slightly ambiguous size indication, as is also true when bubble size becomes approximately one wave-

132 ULTRASOUND

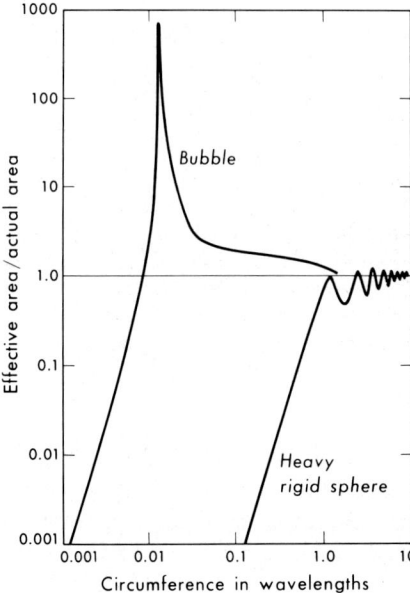

FIGURE 6.21. The well-known scattering by a bubble or by a rigid sphere as it depends upon size.

length. The resonant height in water is considerably damped if the bubble is in blood (Nishi, 1972).

Ultrasonic images of bubbles forming in the leg of a guinea pig are seen in Figure 6.22, which is made up from a series of individual frames taken from a motion picture film made with a camera whose film advance was synchronized with the mechanical scan. The scanner and chamber were seen in Figure 6.14. Because of the high solubility of inert gases in fat, many of the bubbles are seen to form in the column of fat along the leg, that here appears triangular in cross section. This image actually uses a display in which returning echoes not only cause a brightening to attract attention, but also produce a proportionate upward deflection to allow for immediate taking of quantitative data; this type of display and several others will be described in Figure 12.1 and Section 12.3. For diving studies such observations are important because they can be made without cutting into the subject and perhaps introducing bubble nuclei into the circulatory system. It was, incidentally, easy to demonstrate the presence of silent bubbles that were present but did not exert a noticeable physiological effect.

It is easy to tell which echoes in an image are coming from bubbles (pockets of gas), since an increase in atmospheric pressure diminishes the size of the bubbles and also increases their stiffness, both of which cause these echoes alone to diminish in height (Rubissow and Mackay, 1971). As long as the bubbles remain small enough so as not to be distorted from an approximately spherical shape, they will return echoes in all directions and thus be readily seen in successive images. There are several other ways for specifically sensing bubbles. The process of changing the pressure to change the echo height has a useful modification if

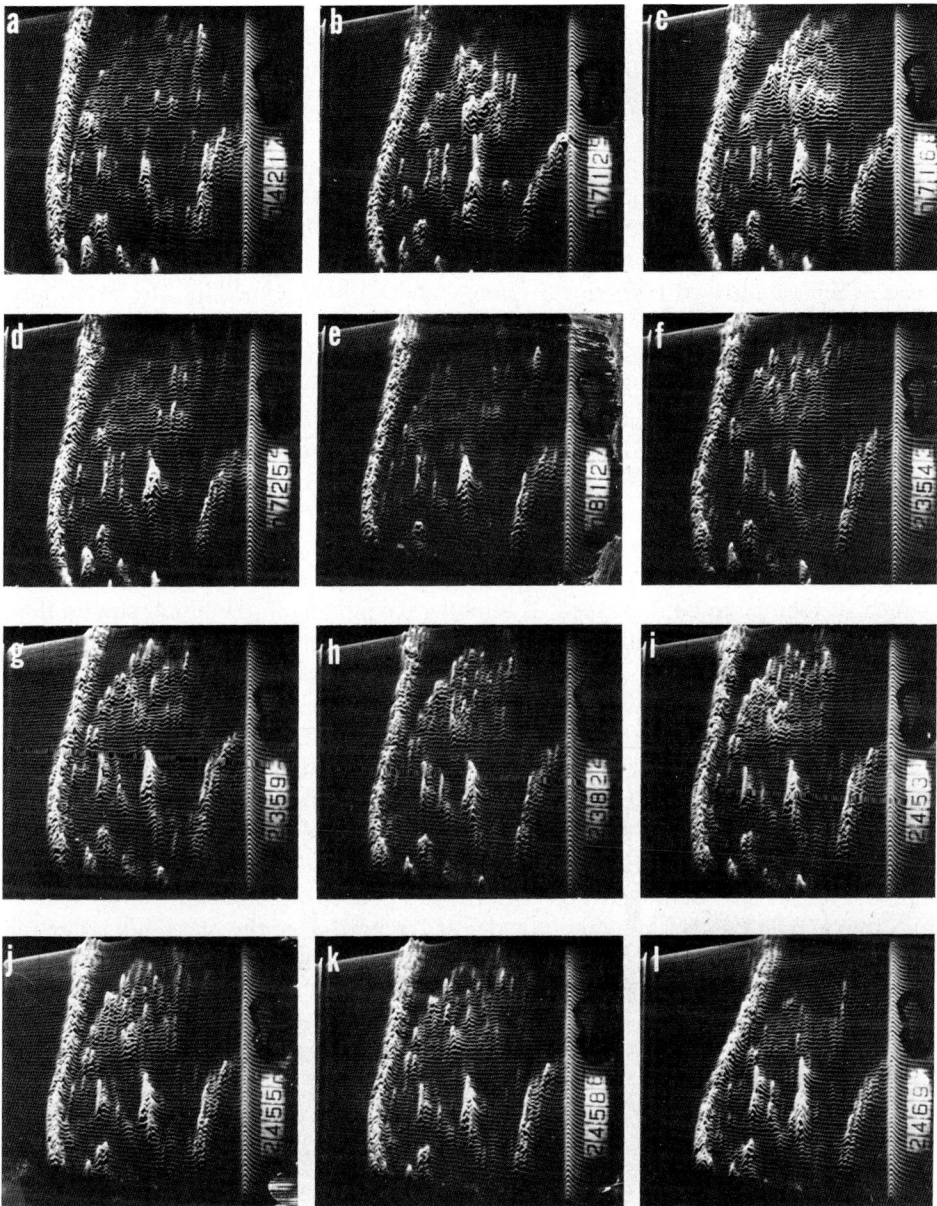

FIGURE 6.22. Conspicuous development of bubbles in the fat as observed in decompressing a guinea pig at a rate determined by ultrasonic monitoring. (a) Fat pad appears as a faint uniform triangle when sensitivity is raised before decompression. (b-c) Progressive bubble formation during linear decompression at 0.25 atmosphere/minute. (d-e) Bubble regression following recompression. (f) Echo level generally maintained, except for brief visualizing excursions, by gradual reduction of ambient pressure. (g-i) Measurable bubble growth for supersaturation of 1.1. (i-l) Survival following arrival at 1 atmosphere pressure. (From Mackay and Rubbisow, 1978.)

done cyclically and rapidly. If one cyclically changes the pressure, one can display only echoes changing cyclically (bubbles), perhaps in a different color. This vibrating of bubbles leads to a different way of considering the process. Bubbles can be set into vibration by a low frequency sound and oscillations of their surface can be detected by the Doppler shift introduced into a high frequency steady beam. Thus a 2 MHz Doppler probe to be discussed in Chapter 9 will show no coupling to a 50 kHz sound source aimed across its beam until a bubble is introduced at the intersection. (This same principle was used to detect the vibrations of the air sacs in the heads of porpoises in order to understand how they made their sounds without vocal chords and without expelling air, as will be discussed in Chapter 9.) The presence of a bubble in a reasonably intense sound beam will also cause the generation of frequencies both higher and lower than that in the beam. The detection of subharmonics can be used as an indication of the presence of bubbles (e.g., Eller and Flynn, 1969). It is also known that the presence of bubbles in a liquid or solid changes its compressibility or elasticity, and thus can drastically alter the velocity of sound (e.g., Fox et al., 1955). In this last connection it should be emphasized that dissolved gas does not affect velocity, only little pockets of it separated out as bubbles. Even a few bubbles can greatly diminish stiffness overall and reduce velocity. At frequencies above bubble resonance, velocity can be increased. If bubbles are moving in the blood stream they can be heard to give a loud sound in a Doppler flow probe (Spencer and Campbell, 1968; Nishi, 1972).

Another effect of bubbles is to absorb energy from a sound wave. When minute bubbles vibrate the gas warms and they rapidly conduct their heat to the surrounding liquid or tissue, which energy must be extracted from the source of sound. This is extremely easily demonstrated as in Figure 6.23. The liquid-filled glass at the left if tapped will give out a bell-like "clink," while the one at the right which is filled with bubbles in the same liquid will give only a dull "clunk." This dramatic effect is best demonstrated with freshly poured champagne; obviously this is a low budget figure, the white powder at the bottom of the glass indicating the presence of baking soda (sodium bicarbonate) with the liquid being only dilute vinegar. In an ultrasonic image the density of bubbles can become so great as to obscure the more remote parts. This is seen in Figure 6.24 which is a cross sectional image of a goldfish which has been given sufficiently vigorous decompression sickness to kill any mammal we have studied, but which it readily survived. The sound entered and emerged from the left in a fast sector scan, and in some of the images at the indicated times shortly after decompression the entire right side was obliterated.

It is extremely instructive to consider attempting to estimate the number of bubbles in one small region. Even if the bubbles are acting as if they were larger than their real size, the imaging system certainly can *not* image detail finer than a region approximately a wavelength on each side. All bubbles within such a region will contribute to the single echo emerging from that region. If one applies an increase in ambient pressure overall, then there will be a sudden decrease in the echo amplitude (corresponding to a decrease in size according to Boyle's Law)

FIGURE 6.23. If tapped, the left glass will give out a pleasant "clink." The similar glass on the right containing many small bubbles in the liquid gives out a dull "clunk" until all the bubbles have dissipated, after which the two sound the same. Small bubbles absorb sound, of which fact a freshly poured glass of champagne provides a good example.

followed by a rather slow further decrease in size. This slow phase is caused by the slow release of the now compressed gas of the bubble into the surrounding tissue or fluid. For a given volume of gas, ten small bubbles will have much more surface than a single larger one, and thus the slow phase will be more rapid when more bubbles are contributing to the single echo. The combination of echo height and transient response to an imposed pressure change allows one to estimate the number and sizes of the bubbles in this otherwise unexplorable region (Rubissow, 1973; Mackay and Rubissow, 1978).

This ultrasonic viewing of bubbles seems not to affect bubble formation, growth, or resolution, though we have observed sudden lower frequency sounds to initiate bubble formation in supersaturated materials, and lower frequency sounds can even pump up a bubble (e.g., Skinner, 1970). In diving studies, ultrasonic observation can control decompression rate and actually measure degree of supersaturation of tissues (Rubissow, 1973; Mackay and Rubissow, 1978). It is of interest to apply these methods to the study of bubbles in the fetus after the mother goes diving; though small, the fetus may be sensitive.

There is also relevance to clinical procedures. A heart–lung machine can introduce bubbles into the blood stream and cause brain damage. The presence of such bubbles can be monitored in these ways. Venous air embolism is unfortunately not uncommon during surgery on patients in the upright or sitting position, and early diagnosis is helpful (Michenfelder et al., 1972). In this case, or in various neurosurgical procedures, ultrasonic monitoring can play a role (Carden et al., 1970).

Indocyanine green foam used in dye dilution measurements was one of the first agents noticed to appear distinctly under ultrasound after injection (Gramiak et

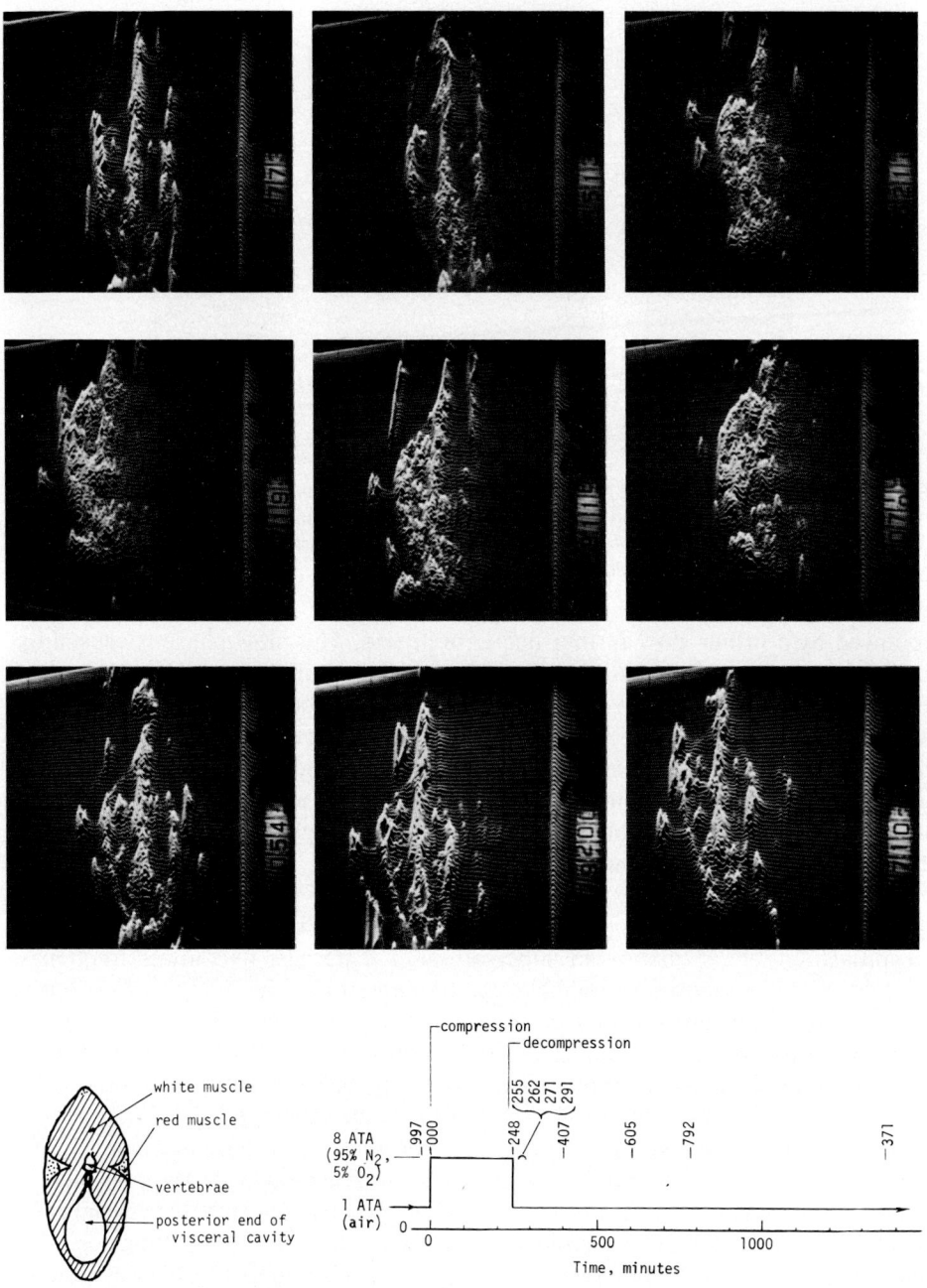

FIGURE 6.24. Ultrasonic observation of the bubbles produced in a goldfish by abrupt decompression. The fish survived despite the fact that some of the bubbling was sufficiently vigorous that it obscured the entire right side of the image of the subject.

al., 1969), but shaken saline or sonicated X-ray contrast material also show similar effects. There are many attempts being made to encapsulate tiny bubbles in more stable form. The field of contrast echocardiography (Meltzer and Roelandt, 1982) depends upon the use of aerated materials being injected to outline the cardiovascular system. The progress of single bubbles can be followed in ultrasonic images being made in rapid succession.

Washout curves might be attempted to estimate the rate of flow from the diminution in signal returning from an injection of bubble-containing material. Since for multiple incoherent scattering the reflected intensity adds approximately linearly, the detected amplitude would increase as the square root of the number of bubbles of uniform size included in the focal area of the transducer. Thus nine bubbles of 5 μm size might give three times the received signal amplitude of a single 5 μm bubble. How the signals from a group of bubbles add has been studied to some extent (Rubissow, 1973), but they can also block each other as indicated in Figure 6.24.

Other applications suggest themselves. A lump in an ultrasonic image may be either a tumor or an abscess, and it is certainly desirable to be able to distinguish these. Most abscesses may prove to contain gas, while this would not be expected of tumors, except after their death (Rankin, 1979). Thus an abscess detector might consist of the previously mentioned low frequency source for producing vibrations in conjunction with the high frequency detector of vibrations. If the sound wavelength is greater than the size of bubbles, changes in velocity caused by bubbles in an abscess might cause a lens-action to distort images of things beyond. Both natural and artificial bubbles remain interesting subjects for ultrasonic study.

7

NUCLEAR MAGNETIC RESONANCE*

An abbreviated view of nuclear magnetic resonance was given in Section 2.4. Since a mental picture of electrons moving in little orbits in atoms seems to help in remembering or predicting certain processes, a similar mechanistic picture of the effect and what is required for imaging will be given here.

7.1. THE EFFECT AND ITS OBSERVATION

The idea of electrons perpetually circulating in orbits around a nucleus involves little difficulty for most people. Thus the idea that protons act as if they are forever spinning on an axis should not be disturbing. Being charged, they are effectively a perpetual current loop, producing a small magnetic field along the axis of spin. The neutron also spins and has a magnetic field along the axis of spin, as does the earth which is considered neutral, spins, and has a magnetic field along the axis of spin; a discussion of possible structural reasons for this need not be included here.

If a pair of identical permanent magnets is dropped upon a table, they will tend to come together with opposite poles touching, that is, with a north and a south pole together at each end. This tends to cancel their magnetic effect in interacting with the rest of the world. Similarly, the magnetic effect of a pair of protons or of a pair of neutrons will cancel because of quantum mechanical effects (which like-

---

*Much of this material was preprinted into *Advances in Medical Imaging 1982*.

THE EFFECT AND ITS OBSERVATION 139

FIGURE 7.1. The toy gyroscope on the right is hanging by a thread along the direction in which it is urged, namely, the direction of gravity. The gyroscope at the left has been deflected to the horizontal and is precessing in the horizontal plane around the direction of gravity. If the outer end is weighted to effectively increase the twist of gravity, the gyroscope does not merely fall more rapidly to the final position, but it precesses more rapidly. This is an analogy of the action of some nuclei having spin when placed in a magnetic field that tends to orient them along its direction.

wise restrict pairs of electrons). However, the nucleus of an atom having either an odd number of protons or an odd number of neutrons (or both) must have some field uncancelled. There is also net spin, which gives such nuclei some of the behavior of a gyroscope. The usual atoms of many elements do not show this effect, for example, ^{12}C or ^{16}O, but most elements involved in biological systems have a useful isotope, for example, ^{23}Na, ^{14}N, ^{31}P, ^{19}F, ^{13}C, ^{17}O, and hydrogen nuclei (protons) ^{1}H themselves.

In a strong magnetic field the nuclei will tend to orient themselves along the field. Note that a compass needle tends to point north, but does not tend to move to the north pole of the earth. Similarly, in a reasonably uniform field, the nuclei orient rather than trying to migrate.

Further analogies may be helpful. In Figure 7.1 are seen two toy gyroscopes, the one at the right hanging in its ultimate equilibrium orientation along the direction of gravity. The other gyroscope is deflected to 90° and is seen precessing around the vertical thread holding up one end. In an applied magnetic field a proton, for example, is twisted so it tends to orient along the field just as the gyro at the right tends to hang along the direction of gravity. However, if a gyro is hung from a thread and oriented as at the left it will precess in a horizontal plane around the vertical support and will not fall to the vertical unless buffeted by the surroundings. The right hand gyro in the figure can be encouraged to assume the left orientation by shaking the lower tip back and forth at the frequency of precession; this action could be called applying a 90° pulse. If shaking is continued longer the gyro would rise up to face against gravity (180° pulse). A stronger orienting field does not pull the spinning object more rapidly into alignment but merely causes it to precess proportionately faster. For protons in a 1000 gauss field, this Larmor frequency is observed to be 4.2578 million cycles per second, or the gyromagnetic ratio (also called magnetogyric ratio) divided by 2π is said to be 4.26 MHz per kilogauss. (A thousand gauss is also called 0.1 tesla or a weber/square meter.) An equation stating frequency to be proportional to magnetic field and not containing Planck's constant h implies one can apply classical concepts. The constant relating frequency to field has a specific value for each different element.

Deflecting a proton, or other nucleus, from side to side can be accomplished by applying a small alternating magnetic field crosswise to the main orienting field. Likewise, when many nuclei are all sweeping around similarly oriented in synchronism across the main magnetic field, a small alternating magnetic field can be picked up externally by a coil of wire with axis pointed across the main field. This process can be observed with protons as at the top of Figure 7.2. If the magnet is turned on, the protons, for example, in a glass of water placed in the magnetic field, will gradually fall into alignment with the field; if disturbed they will similarly gradually come back into alignment after a characteristic time T_1 that depends on the environment.

This so-called longitudinal or spin-lattice relaxation time, for example, is known to be changed in fast growing tumors. If electric currents of resonant frequency appropriate to the magnetic field are applied to the transverse coil for enough time then the protons will sweep around the magnetic field in synchronism, just as the gyroscope in the photo sweeps around the gravity direction. However, they will gradually get out of synchronism in their precessing due to magnetic interactions between nuclei, and only while acting together do they induce a significant voltage in the coil, this signal being of the same frequency as that earlier applied to the coil. This time to get out of synchronism is called the transverse or spin-spin relaxation time T_2. Once the protons are precessing out of phase, one must wait until they realign with the outside field before an observation can be started again. One can observe for T_2 and then wait for T_1 to resume; where T_2 is much less than T_1 the experiment is off most of the time, giving a weak average. In continuous processes of observation the transverse magnetization disappears after T_2 but the nuclei are not readied for another cycle until T_1 and so most are unable to respond. For liquids T_1 and T_2 are similar while for solids T_1 is long and T_2 short. (The motion of atoms in a liquid causes their fields, responsible for spin-spin relaxation, to average out, while in a solid with essentially fixed nuclei this decay may be too rapid to notice; however, solids do not generate much of the right frequency to cause transitions back to thermal equilibrium and so T_1 is long.) For water T_1 and T_2 are both about 3 seconds while for ice T_1 is minutes and T_2 microseconds, for example. (C. J. Gorter in the late 1930s failed an

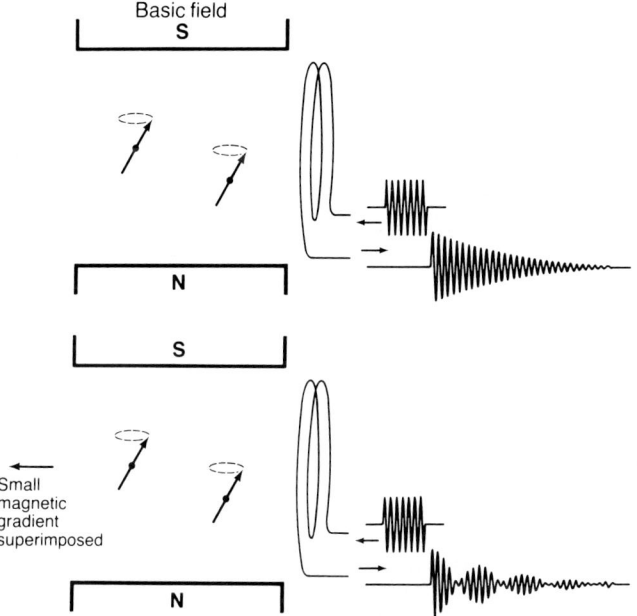

FIGURE 7.2. Top: A transverse alternating magnetic field of the proper frequency will deflect a group of aligned nuclei and start them precessing at the same frequency. They will gradually get out of step and will also fall back into alignment with the main field, thus diminishing the output signal induced into the coil from which was sent the original deflection signal. A specific example might be a group of protons if a glass of water is placed between the poles of a magnet. Bottom: If a small change in magnetic field is superimposed upon the main field, then protons at two different places will respond with two different corresponding frequencies which will beat with each other to give the type of output signal shown. A frequency analysis of this signal or Fourier transform converts the amplitude as a function of time into an amplitude that is a function of frequency to indicate the presence and relative amounts of protons at the two places.

attempt to observe NMR in a solid with long T_1.) Thus, in a complex biological structure one tends to observe mobile protons in water or fat rather than hydrogen nuclei in proteins.

Absorption of energy or stimulation of activity takes place at the same frequency as the subsequent radiation. At the top of Figure 7.2 all protons are seen acting together to return a gradually diminishing signal. If all protons (or other activated nuclei) in the entire subject or object act together, there is no information about the spatial density or distribution of these nuclei. However, if the magnet is not everywhere of the same strength, then a given type of nucleus will respond at a different frequency at different positions. At the bottom of the figure a small magnetic field is applied in such a way as to make the overall field slightly stronger on the right than on the left. A brief alternating current applied to the coil will contain many frequencies and stimulate nuclei everywhere. The two shown will continue to precess at slightly different frequencies, appropriate to their local magnetic fields, and will induce into the coil an outgoing signal consisting of two

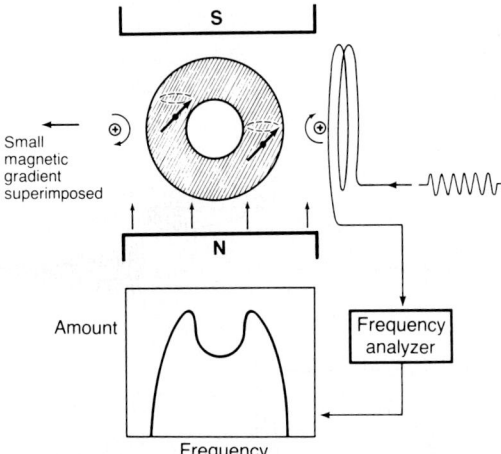

FIGURE 7.3. A somewhat more complex shaped object containing protons or other resonating nuclei will, if pulsed, return a more complex signal during free induction decay. Here the gradient is shown being supplied by inward current along the wires at right and left. Frequency analysis of the returned signal gives an amount proportional to the height of the sample at each value of magnetic field, which is just one projection of the sort used in computed tomography. With several wires the gradient can be shifted to any angle, and the projection will always be along that gradient.

different frequencies. Just as listening to two slightly differently tuned tuning forks causes one to hear a beat note, so the resulting signal here will cyclically change its amplitude. An analysis of this outcoming signal will reveal the presence of two frequencies and hence indicate the presence of nuclei at two particular positions corresponding to these frequencies. With many nuclei situated at many different values of magnetic field there will be many frequencies in the output, but a frequency analysis of this signal by a set of electric tuned circuits will show which frequencies are present, and the relative amounts of each, and thus indicate the relative number of nuclei at different positions. The process of evaluating the relative amounts present of different frequencies, that is, converting a signal of strength versus time into one of strength versus frequency, is called making a Fourier transformation.

In Figure 7.3 this process is seen applied to a hollow object uniformly composed of nuclei that will resonate. The small magnetic gradient is imposed by running a current in along the wires at right and left, which adds to the main field on the right and slightly subtracts from it at the left. The returned signal after frequency analysis will be strongest for the frequency where the column of nuclei at the corresponding field strength is longest, as shown. The resulting pattern has already been seen at the upper right in Figure 4.8. It is seen that the frequency analysis of a single signal gives the entire projection along all the vertical columns that are numbered in Figure 4.6. In the X-ray case, the amount of material along each line was judged from X-ray attenuation, while here it is evaluated by strength of radiated signal at a particular frequency. Other projections can be collected without mechanical movement by adjusting the current in other wires in order to shift the gradient of the magnetic field. Several shifts of gradient can be imposed in some sequences before the nuclei become totally disorganized, and then it will be a matter of seconds before the process can be repeated.

If changing the gradient direction gives the same projection in several directions, then the object can only be the one indicated in Figure 7.3. A neck with trachea would give more complex projections from which to compute the original form, as discussed in Chapter 4. A gradient projects the three-dimensional density of spins onto the axis line along which the gradient is applied. The object here is considered either thin or uniform perpendicular to the page; later methods can select a section.

Resonances in liquids are so sharp that frequencies differing by as little as 1 part in 10^8 are often resolved. (Responses from solids are broader.) Thus a given frequency corresponding to one place in a magnetic gradient should be localized to less than the resolvable distance of the system and this should not be a source of blurs. (The effect of spread of resonance could be somewhat compensated in the display if needed.) In actual imaging, T_2 as reduced by field nonuniformity often limits the minimum linewidth, and the gradient field must produce a frequency difference between adjacent points larger than this spread. In nuclear magnetic resonance, spatial resolution does *not* depend on wavelength or frequency in the usual ways, but does depend on gradient field strength and overall field uniformity. There is a relationship between frequency and magnetic field strength, the latter also affecting signal strength and detail for a given duration examination; thus it is possible to give an equation relating wavelength to detail, but it is misleading as to mechanism. Though somewhat irrelevant, observable detail is very much less than a wavelength (Fig. 2.1) with such methods.

One further repetition of the gyroscope analogy is given in Figure 7.4 which is a view looking down upon three gyroscopes, each frame being pivoted at one end. The downward force or torque (away from the reader) on the upper gyroscope has been increased with the small weight of an added rubber band at the end away from the pivot. The gyro to the right is provided with an even larger weight. The initial precession was started with all three facing to the right. Somewhat later when the photograph was taken they had gotten out of step or phase because of their different rates of precession. This can be considered as analogous to any of several different situations. In a uniform applied field, this could represent the desynchronization produced by interactions with the local material, including interactions with adjacent spins. It could also represent the situation when a material was placed in a magnetic field that was stronger at the right than the left, that is, in a gradient. In a fixed magnetic field the nuclei of different elements precess at different rates, and so this figure could also represent the separate monitoring of different elements by their different frequencies at any specified local value of magnetic field.

One type of imaging has been described but there are others that are similar, or faster, or employ elements other than hydrogen, or emphasize the relaxation time rather than simply amounts of material. As an example of the last, if data are

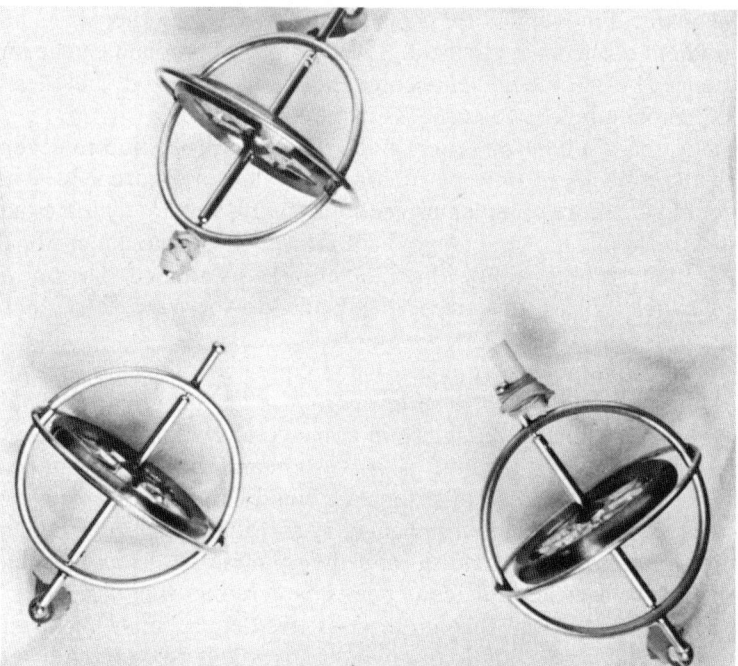

FIGURE 7.4. Looking down on three gyroscopes, each supported at one end and differently weighted to precess at different rates. If started all facing to the right, they soon get out of step as shown when this picture was snapped, just as do different nuclei in the same magnetic field, or the same nuclei in a variable magnetic field.

accumulated from any one region by sets of data taken at a high rate, tissues with a large T_1 will yield a smaller signal than those with small T_1, since the former have less chance to become fully polarized before the new accumulation is begun.

A proton image is seen in Figure 7.5 and depicts a cross section of a human upper abdomen. It was made on Philips equipment using a saturation recovery spin echo technique. Slice thickness was approximately 15 mm. Unlike Figure 2.2, which was made in a system using a superconducting magnet, this was made with a resistive magnet giving a field of 1.5 kilogauss. The dark circle above the spine is the aorta. Blood movement can carry precessing oriented protons either into or out of the active regions of sensing, and thus depending upon imaging sequences, the part of an image where flow is taking place can be either lightened or darkened.

A number of technical details warrant mention. Either absorption or emission of energy by the precessing nuclei can be observed. In a given magnetic field, the frequency of the two processes is the same, which can cause interference between transmission and reception. Absorption has been used to determine the frequency or position of the "line" associated with a given element. This has been done by slightly raising and lowering the magnetic field, and noting the changes in loading

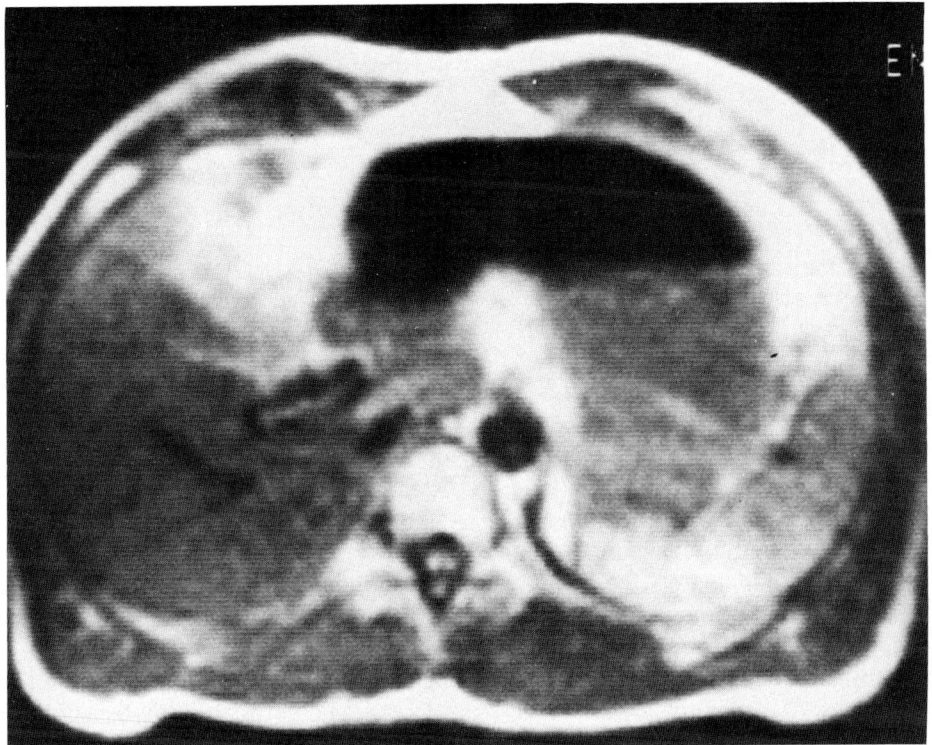

FIGURE 7.5. A transverse image of a human subject using proton imaging and a saturation recovery technique to emphasize T_1. The image was made on Philips equipment with a resistive magnet of 1.5 kilogauss. It can be compared with any of the similar X-ray slices included elsewhere in the book.

on the coil being driven by an oscillator running marginally (for a simple circuit see Singer and Johnson, 1959); this allows the observation of the change in effect upon going from an incorrect frequency to the correct one. It can also be done by placing the sample coil in a balanced bridge circuit of inductance. Instead of measuring absorption of a particular frequency, radiation can be measured in either of two ways without signal dilution and noise from the radio frequency source of stimulating power. The radio frequency can be applied to the coil briefly as in Figure 7.2, and then later when it is off, the remaining signal can be recorded as a sort of echo during "free induction decay." In this case the amplitude of the stimulating signals should be varied during the pulse in such a way that the frequencies it contains are the ones that will stimulate the atoms of interest. (Stated the other way around, to construct a wave of changing amplitude there must be present a particular collection of frequencies or Fourier components that add together at different instants to give the changing overall amplitude.)

Instead of removing background noise by waiting until the offending transmission is finished, one can instead use two different coils for transmission and reception. In Figure 7.6 are shown two arrangements of a pair of coils in which a

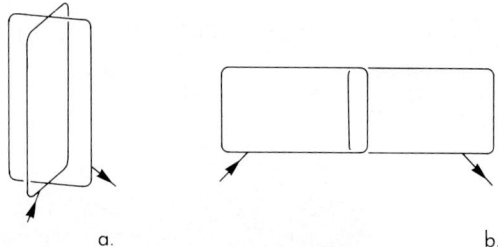

FIGURE 7.6. Coil combinations whose pairs display little direct coupling at the radio frequency applied, but which can couple input to output via nearby resonating nuclei if the magnetic field is oriented in the plane of both coils. At (a) the coils are perpendicular while at (b) they overlap and are approximately in the same plane. Intermediate angles and overlaps between these two cases can function similarly. (From Mackay, 1970.)

signal into one is not sensed by the other, that is, there is no transformer action or coupling between the coils of a pair. In the first case the coils are perpendicular and in the second case they overlap by a carefully adjusted amount in the same plane. No diluting signal, or its noise, appears at the receiver even while the transmitting coil is activated, but there is coupling between the two coils via nearby precessing nuclei which are able to take energy from the first coil and radiate it into the second.

Bridge circuits and some variable loading circuits called Q-meters separate the functions of oscillator and receiver, as do the geometrical methods such as crossed coils. Rather than sweeping the magnetic field to search for a resonance or display it, the frequencies of the transmitter and receiver can be swept synchronously, though this is difficult over more than a factor of 3, even in the geometrical case where the null depends less on frequency-sensitive components. The advantage of the separation of functions is that the noise is from the receiver and sampling circuits, not the oscillator. If the sample tuned circuit is made part of the oscillator resonant circuit then tuning is simpler but sensitivity is sacrificed. In imaging it is common to monitor a small specified frequency range after the oscillator is off, there being some problems with transients as in some ultrasonic systems.

There is a slightly different process that also has applicability and is sometimes called "free precession." An example of this involves a small bottle of water with a coil wrapped around it. A steady electric current is passed through the coil to produce a magnetic field that will orient the protons in the water along the axis of the coil. After a few seconds, this current is turned off and the protons immediately start precessing around the remaining magnetic field which is that of the earth. Assuming the bottle was not oriented exactly along this magnetic field, the movement of the protons in sweeping across the coil will induce in it an alternating current of about 2200 Hz, which provides an accurate way for measuring the magnetic field of the earth, or other fields. This observation was made some time ago (Packard and Varian, 1954) and has become an exercise for youngsters (Stong, 1968). The method can also be used to measure relaxation times (Bloom and Mansir, 1954). We built a complete NMR unit of this sort that could be swallowed but, even using averages of over fifty orienting impulses from outside the body, and providing for unclogging, it did not well serve its intended purpose (Mackay, 1965b). Physiological fluids in humans have been identified by proton resonance in the earth's magnetic field (Béné et al., 1977; Borcard et al., 1979).

We have been speaking as if all nuclei were perfectly oriented, but statistical considerations indicate this is not true. One can compare the energy of a state with the energy of thermal disruptions to obtain an estimate of the numbers of elements that will be in the state. According to quantum mechanics there are two possible states here (orientation along the magnetic field, or against it), with transitions between them being induced by the proper or resonance frequency. (Although proton spins may have only two energy levels, the Correspondence Principle allows us to assume that their average will behave classically.) The statistical result is that at an absolute temperature T in a magnetic field H, for nuclei of magnetic moment μ, the number parallel (N_1) and antiparallel (N_2) is

$$\frac{N_2}{N_1} = \exp\left(-\frac{\mu H}{kT}\right) \cong 1 - \frac{\mu H}{kT}$$

where k is Boltzmann's constant. This small majority of nuclei in the lower energy state along the field (about one in a million excess protons pointing along the field at body temperature and a field of a few thousand gauss) are the ones of interest since they are the only ones capable of being manipulated by the intrusion of energy. They produce the macroscopic or bulk magnetization that can be described as a vector that responds to the applied radio frequency.

The application of energy at the resonant frequency induces transitions in both directions, which tends to deplete the more numerous group (here N_1) and remove the signal-producing difference. If $T_2 \ll T_1$ the spin system cannot release extra energy fast enough to the surroundings (the "lattice") to maintain a signal; this "saturation" can be reduced by adding a paramagnetic material whose fluctuating local fields at the other material randomize spins to enhance relaxation or reduce T_1 so that the transverse magnetization does not disappear before the spin system repolarizes.

Very small signals are involved in these measurements. At the Larmor or precession frequencies here of a few million cycles per second, from Figure 2.1 the energy $h\nu$ of the nuclear reorientation transitions is in the range of 10^{-8} eV. A sample of 0.1 cubic centimeters of water contains roughly 10^{22} protons, and if they *all* flipped simultaneously the total energy would be about 40 ergs. This would be a few microwatts if delivered continuously at 40 ergs/sec. As indicated, not all nuclei are flipped, depending upon applied magnetic field and so on. An increased magnetic field results in a stronger signal because a larger fraction of the nuclei participates.

Many factors influence the signal and the noise in various ways. Two important factors are the magnitude of the nuclear signal itself and the coil noise, including that coupled from the subject, both of which increase with frequency. Under some conditions the signal-to-noise ratio is proportional to $f^{3/2}$ where f is the frequency, determined by the magnetic field (see, e.g., Abragam, 1961 vs. Hoult and Richards, 1976).

Strong fields somewhat modify the values of T_1 and T_2, which is involved in the suggestion that there is a possible improvement in cancer resolution at lower frequencies (Diegel and Pintar, 1975); in the cancer case, special techniques for T_1 emphasis may override these considerations.

Higher magnetic fields give stronger signals and also a higher frequency, for a given type of resonating nucleus. However, the field cannot be allowed to become too high or other problems arise. Radio signals are shielded against penetration of electrical conductors, and the body is somewhat a conductor of electricity. Thus a frequency that is very high will not uniformly penetrate into the body, or uniformly sample at different depths. There can also be differences in phase with depth. Frequencies higher than about 30 MHz are generally to be avoided, though one can pass some signal through the body at considerably higher frequencies (Sec. 2.3). Various studies have been made to characterize the electrical properties of human tissue, including some specifically dedicated to this question (Bottomley and Andrew, 1978). This factor will limit the optimum frequency, which is often assumed to be the same for all elements.

The magnetic moment of an electron is about 660 times that of a proton, which increases the magnetic effect by the same factor if concentrations are the same. The gyromagnetic ratio of a free electron is about 100 times higher than that of a proton and so higher frequencies are involved in observing electron spin resonance. Electron spin resonances have been observed at frequencies down to about 100 MHz, and thus they are most useful for small samples. There are applications for such techniques in labeling tracers.

To produce a large strong constant uniform magnetic field is one of the more demanding aspects of this equipment. A permanent magnet would be excellent but it is extremely difficult to build them adequately large. Usefully uniform fields can be produced by several types of coils: in a long helical solenoid; between two loops separated by their diameter (Helmholtz coils); in a coil wound uniformly over the surface of a sphere; in the "H" core of a cyclotron-type magnet. An approximation to the sphere case is four coils of two sizes, the two larger being toward the center. This type of coil combination is often used. Lines of force do not extend as far outside the cyclotron-type magnet to interact with metals in the building or carried by people. Since extraneous material thus has minimum influence, this form does not require an expensive special room to house the apparatus, and it allows a more convenient approach by physicians, tape recordings, wheel chairs, and so on. Power consumption by such magnets can be extremely high, and the fields difficult to stabilize adequately. Superconductors when cooled to near absolute zero display no electrical resistance, and thus a current once started can continue for an extended period, as long as adequate cooling can be provided. Filamentous superconductors solve serious problems of application, including retaining superconductivity while passing high currents in strong fields. That requires more than simply having a piece of wire of the proper element

(Schwartz and Foner, 1977; Hulm and Matthias, 1980). Niobium-titanium embedded in a matrix of copper is one combination, which is submerged in liquid helium surrounded by a vacuum chamber around which is liquid nitrogen surrounded by another vacuum chamber. Fortunately the handling of liquid gases has become rather routine in modern times.

Control of the pattern of field strength is important, an extremely uniform distribution being desirable. Similarly, the gradient fields should be uniform, and rapidly adjustable (e.g., Bangert and Mansfield, 1982). If the main magnetic field could be arbitrarily shaped in space then it could be formed to give the value for a specified frequency resonance at a point or small region, and much less elsewhere. This point could then be moved with respect to the subject to scan out the distribution of the resonating material in either two dimensions or three dimensions. Without placing magnetic material within the subject, this can be done only to a limited extent, as was described in Section 2.12. For a whole body scanner, a typical estimate of the size to which the sensitive volume could be reduced is a cube 5 cm on each side. Observations of this sort have been made (Damadian et al., 1976; Crooks et al., 1978; Gordon et al., 1980). The first two of these references relate to detection of protons while the third describes the localization of metabolites in animals using ^{31}P, for example, to diagnose liver ischemia in laboratory animals by following ATP signals.

One can shape the extent of the radio frequency field to about the same degree as either the main steady magnetic field or its gradients (Sec. 2.12), and this offers one more alternative. All fields can be shaped or just one, and there can be simultaneous gradients in the rf field and the main magnetic field. Some local spatial resolution might be achieved by passing current along a fine wire in contact with the subject or even under the skin. (Body current would tend to spread though it need not be felt if alternating.) The use of small radio frequency coils placed on the skin allows examination of tissues close to the surface of the subject (Chance et al., 1978), while several small simultaneously activated coils suitably placed allow some concentration of attention at greater depths.

If, however, the magnetic field distribution can be made to vary cyclically with time then steady resonance can be made to take place at only one point in space. Consider a set of coils oriented to produce a uniform gradient when a steady current is applied. The total applied field is a steady one plus a gradient. If the gradient coils are instead energized with an alternating current this gradient will periodically reverse itself and only at the middle will the magnetic field be unchanging. Nuclei at that region alone will steadily give out a resonance signal at a predetermined frequency, thus allowing information to be obtained from just that plane perpendicular to the gradient. If two such sets of coils are arranged perpendicular to each other, and energized 90° out of phase, then their combined field will spin in space as in familiar types of synchronous motors. Only along the axis of spin perpendicular to both fields will the magnetic field be unchanging, allowing monitoring of all nuclei along that line at one time. With three perpendicular sets of coils each activated at a different frequency, or two coils out of phase at one frequency, with a third coil at a different frequency, the resultant field (in this case field with a gradient) will spin through many directions in space, leaving the field constant only at the center point from which resonance signals can steadily add. Such movements of field have been applied to nuclear resonance imaging by Hinshaw and his collaborators (Hinshaw, 1976; Hinshaw et al., 1977). (Other aspects and applications of spinning fields are in Mackay, 1970.) The idea of spinning to emphasize a central point can be compared to the thinking of Oldendorf for X-rays in Chapter 4. If gradients move so that only a single point gives a steady signal, then one can either steadily monitor the time course of events at a single place within the subject or the subject can be moved with respect to the apparatus in order either to produce a two-dimensional cross sectional image or a full three-dimensional image.

Various combinations of applied pulses and gradients in various sequences can be used to collect information from points, lines, sections, or whole volumes at a time. The spatial rate of change of magnetic field determines how different the frequency will be from one region to the next for a given resonating type of nuclei. A larger gradient increases resolution at the expense of sensitivity. The signals are all rather weak, and noise considerations will determine how rapidly one can acquire successive pieces of data. Relaxation times also will limit the overall repetition rate. The difference between different systems is largely in selecting these various patterns or sequences, and in the various forms of information processing that can be used to reduce the collected information to final images. Some of these methods of analysis are the same as used in X-ray computed tomography (Chap. 4), and some other aspects of Fourier transformation and image processing will be mentioned in Chapter 10. The number of articles on such matters has burgeoned.

Some specific alternatives should clarify the process of scanning and data collection. One can pulse the radio frequency (rf) coil with a short pulse that will contain many frequencies (Fourier components) so that all spins within the coil will be excited, even if there are field gradients. Alternatively, the exciting pulse can be amplitude modulated in a pattern that makes present the desired stimulating frequencies to activate specified regions in a gradient, or instead a suitable modulation of a long pulse can activate a single region* (Garroway et al., 1974; Crooks, 1980). A short pulse might require the rf amplifier to deliver 10 kW for 25 μsec while a long pulse might require up to 1 kW for 5 msec, which can be provided in some cases by single-sideband radio transmitters (Holland and Heysmond, 1979). Tailored pulses can also saturate selected regions in a gradient: these regions with no net magnetization will give no response to later excitation rf pulses. One can saturate a layer, or everything but a layer, to limit the regions from which signals come. Suitable switching in different directions allows such

---

*Fourier analysis indicates that to bunch all frequencies present into a small range and thus select a specified region, the amplitude of the pulse should change with time t as $(\sin t)/t$, including a few cycles on both sides of the peak; negative rf amplitudes are obtained by changing the phase of the rf carrier by 180°. This is also a good pulse form in Doppler ultrasonic motion measurements where, for slow flow, one may need to detect frequencies shifted only slightly from the original, the original frequency being rejected by a filter (Chap. 9).

processes alone to form an image, or they can be combined with other processes. For example, after excitation, detection can be delayed while the spins precess in gradients that can be changing in order to result in a final phase angle that can be detected in the decaying signal to give spatial information along the gradient direction.

In Figure 7.3 the plane of the figure could be assumed isolated by suitable gradients, while in that plane a gradient up-and-down from one coil could combine with a gradient from side-to-side from another coil to give a resultant gradient along which there would be a steady increase in frequency to provide one projection; successively changing the relative size of these two component gradients allows swinging the angle of the projection around the subject in a plane for eventual CT reconstruction. NMR signals are received from the entire region after each pulse which helps with speed (Lauterbur, 1973).

From a group or stack of such thin section images a 3D image could be formed. Alternatively, with an adjustable third perpendicular gradient, orientation of the total gradient in three dimensions (through a hemisphere of angles) is possible with which to form projections from which a 3D image can be reconstructed. Resolution is determined by the number of projections.

There is a way of employing projections in NMR that is not available with X-rays. If nuclei along a line are precessing together and a magnetic gradient is briefly applied along the line then the nuclei will resume precessing at the same rate but they will now be spread like a spiral staircase around the line, the stronger the gradient the shorter the pitch. At the base of the line there will be no net signal if material is uniformly distributed, and a large signal if material has a periodic placement just equal to the pitch. Furthermore, the phase of the signal at the bottom indicates how far out the periodic placement starts. Recordings from a succession of increasing strength gradient pulses (duration or amplitude) can thus perform a spatial frequency analysis on material along the line. The base of this line is one point on a projection if the final recording is done with a magnetic gradient imposed perpendicular to the original line, that is, different lines parallel to the first can be simultaneously evaluated by applying a final "reading" gradient perpendicular to the lines so that whatever has resulted will give a different frequency for different original line bases that now are in a row constituting a projection. Recording is done after each of a series of preliminary gradients to give the spatial frequency distribution outward in a plane from the line onto which projection takes place. One Fourier transform gives the points along the line for each recording, while a second transform applied to the sequence produced at one point on the projection gives the distribution of material measured outward from that point; processing all points in the projection then gives the 2D image of the plane containing the two gradients. A sequence of momentary preliminary gradients in the third direction allows collecting information about all points in the three dimensions. The process, first described by Kumar et al. (1975), is to apply a 90° pulse to the sample and successively apply gradients in the x, y, and z directions, always recording the signal emitted only during the last z gradient. The phase-encoding gradient in x is successively increased from zero (rather than being rotated as in CT methods) to produce a 2D image, and if for each x gradient there is a similar sequence along y, then a three-dimensional Fourier transform yields the full 3D distribution of material. (The possible relevance of processing methods such as are in Figure 10.10 should be noted.) This might be called Fourier transformation imaging and is also sometimes called "spin warp" imaging (Edelstein et al., 1980). Again, selective excitation of the plane of interest gives a section in that plane rather than a projection of everything onto it, when a tomographic result is desired. A 3D image can be calculated from either a set of 2D projections or from a set of slice images at different levels if the calculations of the full process become too complex.

Some processes allow one to watch the image build up. Suitable rf pulses and a Z-gradient can saturate spins everywhere except in a thin section, which can then be somewhere excited in an X-gradient and then detected with frequency analysis in a Y-gradient to give an image of a strip. Successive cycles lay down successive strips in final form (Mansfield and Maudsley, 1977) to build up an image line by line as the X-gradient is progressively changed. A very fast scan could result by simultaneously exciting multiple separated parallel strips and detecting in a combination of gradients. Changing the preliminary Z-gradient values allows collection of a "stack of slices" for a full 3D image.

There are still other options. An alternating (periodically reversing) gradient along the main field, and uniform across it, gives a steady signal from all of the one plane where the gradient only reverses without field magnitude change, as was mentioned. One can apply a series of rf pulses whose result will average to zero from any region but this central plane. An alternating gradient (different frequency) perpendicular to this gives response from only one line where the two null planes intersect. A static gradient along the line gives the distribution along the line, or an alternating gradient along the line limits response to the point where three null planes intersect (Hinshaw, 1976). Image reconstruction calculations and digital data storage are unnecessary, as are great field homogeneity and gradient linearity, but scan speed is slow, limiting applicability to small regions. A sensitive line swept across a sample gives a plane image of a thin section, just as scanning a sensitive point through a subject can generate an image of a line, a section, or of a total volume.

When one speaks of the duration of a process, for example, when one specifies a relaxation time, this has a specific meaning if the process is exponential in mathematical form. In Figure 7.7 (top) is drawn an exponentially decreasing curve representing the exponential decrease of some amount or signal with time. It will be remembered that one "time constant" is the length of time it takes this function to fall to within $1/e$ of its final value, that is, it is approximately the time it will take for the value to change through about 2/3 of the amount toward which it is headed. In the next time interval of the same amount, 2/3 of what remains will be covered, again leaving about 1/3 of the way to go. The figure shows a function starting at some value and eventually approaching zero. From a few initial values one can either attempt to draw the whole curve in order to determine how long it will take to arrive at this fraction, or one can draw a straight line tangent to the initial values

THE EFFECT AND ITS OBSERVATION

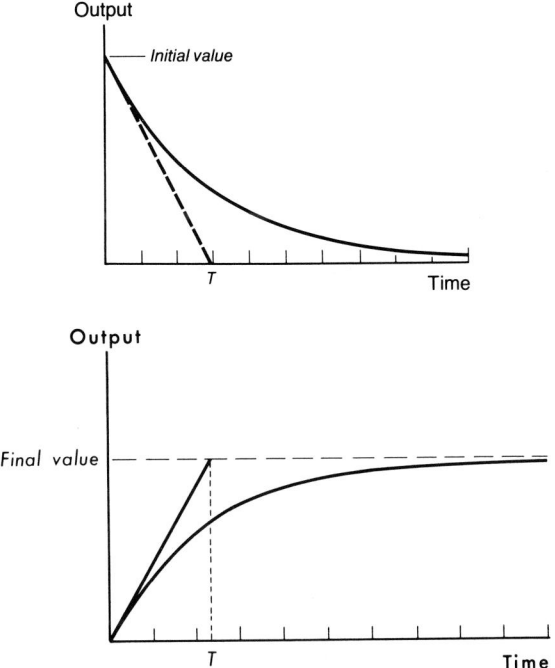

FIGURE 7.7. A decreasing or an increasing exponential are both characterized by a time constant or time T in which about ⅔ of the remaining process is completed, and which is the time taken if the process had continued at its initial rate.

as shown; this line will cut the zero axis at a time of one time constant. A "half-life" or half-time is simply 0.693 times the time constant, that is, it takes longer to complete 2/3 of the process than half the process. From a single value of diminished amplitude and time, and assuming exponential form, one can calculate a value of the characteristic T, though several values help accuracy. These same comments apply to a function that increases exponentially with time toward some final value; again a tangent drawn to the initial values will cut the line depicting the final value at a time equal to one time constant. Not all processes are truly exponential, but it is often a convenient approximation. A distribution of domains with differing T values can result in nonexponential recovery. Internal fields superimposed on the main applied field cause dephasing (as in Fig. 7.4) to produce T_2; it being a single parameter is only the result of rather special circumstances.

Two characteristic relaxation times, T_1 and T_2, have been mentioned. These can be measured and their distribution in different regions of a subject displayed or imaged, in addition to displaying the density distribution of the type nucleus doing this relaxation. Suitable pulse sequences can emphasize or selectively display nuclei with either specified ranges of T_1 (the longitudinal or spin-lattice relaxation time) or T_2 (the transverse or spin-spin relaxation time).

If T_1 is particularly relevant, it can be emphasized or measured by inverting the protons or other nuclei (a 180° pulse) and then later applying a 90° pulse. The latter will have almost full effect where the waiting time has allowed realignment, no effect where relaxation has been halfway back since the protons then are perpendicular to the main field but in all directions, and a negative effect if they have not yet gotten back halfway. The selection of pulse timing compared with the relaxation time of interest can then provide any of these effects. If two T_1 values are present in an image they can be viewed as different from each other by applying an excitation rf pulse (or "read" pulse) at a time after a 180° pulse when one is still negative but the other has relaxed back to zero or positive magnetization. In this "inversion recovery," the pulse delivered will then decay with T_2 as reduced by the nonuniform field, and it generates the image. These processes can be made quantitative by noting that, where signal amplitude changes exponentially with time t after inversion, initial amplitude of a signal is proportional to $\exp(-t/T_1)$, with T_1 being measured by applying the excitation rf pulse at one or more values of t.

Spins can be inverted by a 180° impulse but they can also be flipped by "adiabatic fast passage" in which a fast sweep in applied frequency through the resonance condition does just this in a time short with respect to T_1 or T_2. Less peak-radio-frequency power is required, and this method is less affected by inhomogeneities in the fields.

Another pulse sequence for emphasizing T_1 is the "saturation–recovery" sequence. Here periodic radio frequency pulses are applied with a spacing of about T_1 and greater than T_2 as reduced by the nonuniform field (so complete relaxation occurs during the interval); then the free induction decay signals reach an average value that depends on T_1 and on the density of the liquid-like nuclei. In areas where T_1 is long compared to the pulse interval the image intensity is weak due to partial saturation, and the image is "T_1 weighted" (e.g., Fig. 7.5). A "map" purely of T_1 values can be prepared from two images having different pulse spacings, and for soft tissues and protons, this generally shows greater differences between regions than does a density map.

If the rf pulses are so closely spaced that the signal never decays to zero then one has "steady-state free precession," which can give good images but the contributions of T_1 and T_2 are mixed.

Signal reduction associated with the desynchronization of the nuclei during T_2 has been assumed to take place in a uniform magnetic field where there is the maximum possible tendency for synchronization to be maintained. However, in many of these imaging systems gradients are set up that provide a different magnetic field in different parts of one organ. It is still possible to measure T_2 of the material, even in a nonuniform field that causes nuclei to move at different rates from their neighbors. Consider a racetrack where runners start at different speeds when a gun sounds. Soon they will be spread out, the fastest having gone farthest. When the gun sounds again at some time t, each instantly turns and runs back, the farthest moving fastest. They will remain spread out until they all suddenly come together at the starting line after $2t$, after which they again spread out on the other side of the starting line. The very same can be repeated after any reasonable time

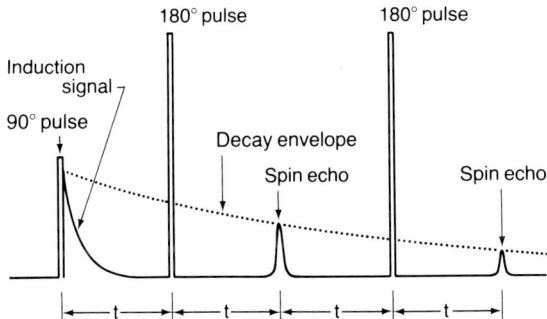

FIGURE 7.8. Even in a nonuniform magnetic field it is possible to measure the T_2 that characterizes the chemical environment by employing a suitable sequence of pulses. After the initial 90° pulse, which produces a rapidly decaying induction signal, a series of 180° pulses (either longer or greater amplitude) are applied. After successive intervals of $2t$ a series of diminishing echoes will appear whose envelope dies away with the characteristic time T_2.

to bring them together again. If some racers fall or falter then the group at the starting line will be smaller each successive time. Thus comparing successive bunchings tells something of the track conditions, that is, explores the surroundings. In Figure 7.8 the equivalent process is shown for nuclei. A 90° pulse produces a signal that dies away. At a later time t a stronger 180° pulse (either twice as long or of greater amplitude than the 90° pulse) is applied to turn over the protons where they will continue to precess in the reverse direction in whatever field they feel, producing a signal or spin echo at $2t$; it will have less amplitude than the original signal because some protons will now be out of step. Repeated 180° pulses will give a diminishing series of echoes whose heights trace out the exponential representing T_2. This spin echo sequence often involves the name Carr-Purcell, and Hahn (1950) also was involved. From the method indicated in Figure 7.7 the numerical value of this time constant can be calculated and various displays are possible. A simplified computation can be made by noting that the amplitude of the first spin echo is proportional to $\exp(-2t/T_2)$.

A distribution of densities can be displayed as brightness differences, as has been shown, as can variations in the relaxation times. It is also possible in a color display to use three different colors to simultaneously present information about nuclei density, T_1, and T_2. More will be said about displays in Chapter 12.

Neither bone nor gas obstructs either steady magnetic fields or electromagnetic fields of reasonable frequency. Thus such materials do not interfere with imaging. However, the skull region will appear in most pictures such as Figure 2.2 with most NMR proton imagers that emphasize relaxation time, not because cortical bone has a high concentration of hydrogen, but because of subcutaneous fat and yellow bone marrow. With regard to the brain itself, there is a striking differentiation between gray and white matter when "inversion recovery" is employed (e.g., Doyle et al., 1981). This is because more hydrogen in gray matter is in the form of water than is the case with white matter where a higher proportion is bound in fats and large molecules. The difference persists until about 4 hours *post mortem,* as might be expected from the earlier comments on imaging with protons or other atomic beams.

Another possibility of interest is the diagnosis of cancer by proton NMR. This has come about since the observation that relaxation times in malignant tissue can be elevated relative to the healthy host tissue (Damadian, 1971). Many groups have now imaged tumors, the discrimination with the surrounding tissue presumably being due to their elevated water content and value of T_1. Rapid growth is probably an important fact, with slow growing tumors showing lower T_1 than a rapidly growing fetus. Apparently, malignant, benign, and normal human breast tissues corrected for fat have similar values of T_1, but due to a lower fat content, tumors relax more slowly than the surrounding normal tissue (Bovee et al., 1978). This technique presumably also would show a difference in regions of myocardial infarction, and "wet lung" problems might also best be visualized this way (Hayes et al., 1982).

As mentioned, NMR scans that depend on T_1 are sensitive to the chemical environment of hydrogen or other resonating nuclei, within tissue. But, in general, there are "chemical shifts" in which the resonant frequency itself is shifted a small amount. In many of these imaging systems where the magnetic fields are not uniform this is not especially apparent, but a recording of resonances from a sample in a very uniform magnetic field indicates, for example, that in methyl alcohol there are two different resonant frequencies separated by approximately 3 parts per million arising from the methyl and hydroxyl protons. Bonding electrons screen the nucleus from the applied field by various amounts, depending on the type of chemical bond involved, and the methyl protons of methanol are more shielded from the basic magnetic field than is the hydroxyl proton. Spectroscopy by NMR of phosphorus-31 applied to a suspension of bacteria typically shows peaks from the three atoms in ATP (the molecule in which cells store energy), plus a peak from the inorganic phosphate broken off ATP when its energy is tapped, plus several other peaks. A summary of NMR of cells has been given (Schulman, 1983). An interesting case is the difference of chemical shift between the phosphorus in the two ions involved in the protonation of inorganic phosphate, and changes can be observed in response to pH changes, giving the effect of a noncontacting pH meter (Moon and Richards, 1973). Many books deal in great detail with this general subject (e.g., Lynden-Bell and Harris, 1969).

In most present cases of imaging, the field gradient across a pixel is greater than a spectral width and so nuclear resonances are considered as single lines, without chemical shifts, rather than a spectrum.

About 75% of the total body weight of an infant is water; at 2 years of age body water is reduced more nearly to the adult value of 60–65%. This is variable from organ to organ and is known to be altered by many disease states. Because of this abundance of protons, and because protons intrinsically give a relatively strong signal, the early imaging and much of the recent imaging has been done in terms of them. But certainly other elements are of extreme interest and importance. Though the signals are smaller, for example, imaging has been done of a working heart using sodium-23 (De Layre et al., 1981), and in various tissue types using phosphorus-31 (Haselgrove et al., 1983). The latter is important in investigating the metabolic state of tissues *in vivo*. The methods have been developed so that it is possible to go beyond the use of topical resonance measurements with surface coils, and it is expected that it will be possible to extend the measurements to other nuclei such as carbon-13.

One can have stroboscopic image collection to display one phase of a cyclic process, such as heartbeat, as with X-ray CT. Contrast with NMR is generally inherently higher than with X-rays when observing soft tissues.

A few general comments about magnetism may be helpful. With a strong small permanent magnet (which may show a field strength of a few thousand gauss near one of the pole faces) one can pull the bubble to one side in a carpenter's level, or one can deflect a dollar bill (Mackay, 1960). The motion of the bubble has two aspects. The oxygen gas in the bubble is paramagnetic and is attracted into the region of strong field, while the surrounding liquid is diamagnetic and is repelled out of the field, both of which effects move the bubble into the region where the field is strongest. When a molecule is placed in a magnetic field its electrons are caused to take on added circulation that produces a small opposing magnetic moment. All materials have capacity for such repulsion and, in most materials where electrons are paired and do not have any intrinsic external magnetic effect, the small repulsion of diamagnetism is observed. A paramagnetic substance outweighs this with unpaired electron spins; unlike most matter it is attracted by a magnetic field, which does this by aligning the spins and thus magnetizing the substance. Paramagnetic substances lose their magnetism when removed from the magnetic field. If the substance retains its aligned spin and its magnetic properties after the magnetic field is removed it is called ferromagnetic, iron being the most familiar such material. Groups of aligned electron spins in domains act together to give large effects.* Such materials occasionally appear in biology, as in the brains of dolphins and the radular teeth of chitons (Lowenstam, 1962).

Because of their magnetism, paramagnetic materials greatly diminish T_1 in a NMR observation. As mentioned, molecular oxygen is surprisingly strongly magnetic, it being possible to create a concentration or wind of oxygen in the air using a permanent magnet (the wind requiring withdrawal, or a heater to make two regions different). Another interesting example, with possible applications, is that oxyhemoglobin is diamagnetic while methemoglobin is paramagnetic (Paul et al., 1978), with deoxygenated blood being diamagnetic though deoxyhemoglobin has unpaired electrons. Also paramagnetic are erythrocytes in which the hemoglobin is digested by malarial parasites (Paul et al., 1981). It is actually possible to separate malaria-infected red cells from whole blood with a permanent magnet.

At the time of this writing the literature on nuclear magnetic resonance imaging is increasing rapidly. At this time the important clinical applications and methods are just starting to be revealed. In terms of the technology, the early contribution

---

*However, dc coils or permanent magnets are unable to do some things that ac electromagnets can do, including attract silver dollars (Lovell, 1946) and stably levitate metal pans, to which case Ernshaw's theorem does not apply.

of Lauterbur, who coined the term "zeugmatography" for the process, was mentioned in Section 2.4. Special mention should be made of two pioneering groups at Nottingham and Aberdeen. Some of the individuals involved were Bottomley, Edelstein, Hinshaw, Holland, Hawkes, Mansfield, and Hutchinson, whose combined output would require a long bibliography. Whether signals indicating the position and density of a nuclear species are coded as a variation in phase, or frequency, or time is a matter of the circumstances of the application involved, and it is here that the ingenuity of present designers is being expended. There are many books on magnetic resonance in most chemistry and physics libraries, and some of the early ones are still useful (e.g., Andrew, 1956; Schumacher, 1970).

7.2. SPEED OF IMAGE FORMATION

The numerous possible scanning sequences for nuclear magnetic resonance imaging impose a great variety of possible durations to the data acquisition process. Thus any comments must be necessarily of a general nature giving ultimate limits, with any particular procedure possibly being considerably slower. Mechanical movements generally do not limit these procedures, though in practice there are limits to the rate at which magnetic fields can be switched, and these can impose equivalent limits to the rapidity of the motion of a scanning process. There are also limits imposed by the fact that some observations cannot be repeated or extended in times less than about a second, corresponding to the relaxation times of the subject material. A poorly designed scanner can always require a longer time than expected. The effect of this is variable since the images formed by some methods are less disturbed by inevitable movements during the exposure than are, for example, images formed by CT-like methods where the computed indication at a point is affected by the later indications measured at other points. Present units typically require a time of the order of a few minutes to produce a large image such as a substantial portion of a human.

The active region of a sample effectively contributing to a NMR signal at any one time can be a point, a line, a plane, or a total volume. Total volume methods are most efficient in the sense that all spins contribute during each collection period, though there can be severe practical problems in implementing this. Usually a two-dimensional slice can be acquired in less time than a full 3D set of data about all the points, but the latter can yield a set of slices at leisure that would require considerable time to acquire one-by-one. The primary data from a three-dimensional array of points will often be made with equal resolution in all three directions, and thus the reconstruction problems of Figure 4.13 need not arise. A 3D collection of data about all points in a cube obviously involves storage of 1–10 million bits in memory, depending on the degree of spatial and contrast detail. The present tendency is for machines to be designed to do a plane or volume at one time.

Of the several factors that enter into the speed of an examination, one is simple geometry. In the previous section was mentioned the possibility of observing the density of nuclei, or other properties, at a point in the subject. Suppose this takes a time t because of noise or machine limitations. Then to record the information from the points along one line through the subject takes nt, if the detail is such that n points are to be resolved in the length of a line. All of a line can be monitored at once if, for example, there is a magnetic gradient along the line (and hence a distribution of frequencies), while a gradient across the line spins around it at every place along the line, so that the line provides a steady signal. Monitoring a whole line at one time provides a savings in time of n, where n is a measure of subject size measured in units of the finest detail to be observed. Even if the line is not completed in one interaction, collecting separate information from all of its points at each instant provides this savings.

This line can be stepped along through the subject to provide information about every point on a thin plane, thus forming an image of a section. If the detail is to be as fine in this new direction as along the line, and the overall region size as long, then the time required will be n^2t. Large object size or fine detail becomes expensive in terms of time required since this is n^2 times the time to collect a point of information, or n times the time to collect one line. If a whole plane of information can be acquired in the time required for any one line or point then the savings are n or n^2. Systems that collect information from all the points each time can show this savings even if, for example, the information is being collected in the form of CT projections in various directions.

Similarly, a plane of information can be moved parallel to itself through a volume of the subject to obtain full three-dimensional information. Just as a point could be moved from position to position throughout a volume in n^3 steps in time n^3t, so it is faster to acquire a line at a time, or a plane at a time, or a full volume at one time to give time savings of n^3, n^2, or n, which can be done without physically shifting the subject or apparatus.

Consider an example of producing a line of information along which could be measured 10 points. Ten parallel lines through the subject would give a 100 point image of a slice. Each point in the subject would once contribute to the corresponding point in the image. (Here a pulse sequence is assumed for which many points do not contribute to one image point, for the sake of simplicity.) If the 10 lines with 10 points each were instead 10 projections of the plane to be imaged at different angles, each object point would have contributed 10 times, having made an input to every projection. The contributions of the individual points are then separated by CT calculations. Under one set of assumptions, 10 projections with 10 points each are more valuable than 10 lines of subject information with 10 points each in producing an adequate signal-to-noise ratio since all points in the plane contribute each time. It is generally not true that each subject line would have to be repeated 10 times to do as well, but there is some advantage in time. The penalties for plane-at-a-time collection are increased storage and computation, problems with subject motion since early and later outputs are everywhere combined, and image blurring rather than geometrical or linearity distortion when magnetic field gradients are not uniform since this makes the apparent position of a point different in different projections (just as a compound ultrasonic scan can produce blurring from any irregularity in ray direction). One can not have a different frequency signal from every point in a volume at once, but methods have been indicated for obtaining some signal from each point at once. The plane and volume methods give larger signals that require sensitivity of the receiver over a big range of inputs (large dynamic range), a practical problem not unlike that of showing high contrast over a large brightness range in images.

In every case a big point (coarse detail) can give an adequate signal in less time, and the required number of these points adds to give the dimensions of the subject. A larger subject with proportionately coarser detail may go faster simply because there are more nuclei contributing signal from each little volume considered as one point, that is, the factor n above remains relevant but the original factor t may change inversely proportional to volume squared if detail is not constant but is a fixed fraction of the total extent. This situation exists when detail is specified as a fraction of a total image or field, as in the case of the specified number of lines across a television screen no matter what size object is displayed, in which case concentration on a small object requires correspondingly more signal from each little element or a correspondingly longer exposure. The specification may be instead simply a specified detail size no matter what the object size.

It should be noted that in this simplified discussion, in order to give an overall impression there was not included the fact that in going from a point to a line-at-a-time method, for example, the fraction of spins contributing at each point can change; this can considerably modify the required dwell time at any one position. A whole plane at one time may take somewhat longer than any one point alone and yet still be a significant saving. This comment applies when signals are acquired from all points in the plane at one time, even if the image is not completed on one sequence and repetitions are required.

The signal-to-noise ratio per data point can be improved by signal averaging from several repetitions. The averaged ratio will be about the original value divided by the square root of the number of repetitions. In practice there is often little necessity for improving this ratio beyond the number of gray levels or levels of spot intensity used in the final display. In going from white to black, a good practical limit is 16 uniform steps (as in Fig. 3.5).

If time is limited or fixed, one can compare the linear resolution available in a three-dimensional image of fixed contrast and signal-to-noise ratio for different chemical elements by comparing the cube roots of their sensitivity, which is just the product of natural abundance, bodily concentration, and intrinsic relative sensitivity. Thus one must average for sodium over about 1000 times the volume needed for protons, in a cube 10 times as large on a side (or 30 times as large for a slice of fixed thickness).

Extra time is involved if spin inversion pulses must be followed by waiting periods comparable, for example, to T_1 which may be a full second. In some cases repetition or the next step must wait a time T_1; in inversion recovery, a delay of $3T_1$ is needed before repetition of the 180° and 90° pair of pulses, which makes it slower than saturation recovery. This is true whether one collects information about points along lines through a subject or about points along a projection (as in Fig. 7.3). (The time course of a received signal along a column of projection need not be an exponential if materials are relaxing along it with more than one T_1, but a spot that appears weak in all lines through it after inversion recovery will be displayed differently from its neighbors as desired.) It is obviously desirable to restrict the total region observed to that which is needed, and to restrict the fineness of detail when possible by averaging over a slightly larger sample of each region of the subject. Typical proton machines presently produce an image in a few minutes, though the process must be slowed for the extremely weak resolved phosphorus spectrum signals which presently require times of an hour in a laboratory-animal-sized object (Haselgrove et al., 1983). Higher magnetic fields can reduce the time required at any one point by giving a stronger signal that is more quickly able to overcome noise, but even if more powerful magnets are available, too strong a field yields a frequency too high to uniformly penetrate or sample a full human body. Most elements are less troublesome than protons in this respect since most give lower frequencies for a given magnetic field.

7.3. HAZARDS

These procedures would not be expected to be hazardous. First of all, as indicated in Figure 2.1, the frequencies and hence energies per quantum are much lower than those values that are usually considered as able to disrupt chemical bonds, and thus this is not ionizing radiation. It is still worth considering the possible effects upon the subject of the conditions applied: the steady magnetic field, the changing magnetic fields, and the radio frequency energy.

One piece of anecdotal evidence relating to these matters can be cited from personal experience. In the early days of large circular accelerators, magnets were occasionally temporarily available free of radioactive material. It was possible to place one's head in the field, or to work in and around the magnets. First of all, there was no catastrophic result from brief exposure. But our experience was that suddenly turning the head produced a sour taste, with the angular velocities readily available. Some form of direct stimulation of receptors may have been involved, or there may have been some electrolysis of saliva. In the latter case, position of tooth fillings might even have had an effect. Certainly, as expected, changing magnetic fields can induce currents in the body, and steady magnetic fields need not be ruinous. Limits to these effects are indicated by many careful studies by many investigators. A number of these have been consolidated and reviewed by Budinger (1981), and the following suggestions for safe limits are taken from that study.

Applying radio frequency energy for nuclear exciting or inverting pulses can produce heating in a subject, just as in a diathermy machine. A crude estimate of the allowable power input can actually be obtained from the specifications for an electric blanket. These observations all take place in the near field of the transmitting coil because of the frequency at which they are made (Sec. 2.12). Here the energy is largely in the form of an oscillating magnetic field, unlike the far field where electric and magnetic energies would have equalized no matter what the form of the original transmitting "antenna." In this case some of the standard specified limits for exposure to radio can be misleading, and it should also be remembered that observations made on animals whose size approximates one wavelength can be variable and misleading (compare Fig. 6.21). It is probably true that an absorbed power of 4 watts per kilogram of body weight, which is slightly above the basal metabolic rate for man, would begin to prove troublesome if maintained for an extremely extended observation.

Steady magnetic fields of rather great strength have not yet proven dangerous, though there has been worry about possible reduction of nerve conduction velocity, and macromolecular orientation changes that could affect membrane permeability and chemical kinetics. Any movement in the field, including that of blood or the heart, will generate small voltages that are superimposed upon the natural biopotentials, but these seem not to have had a deleterious action, though there is an effect (Beischer and Knepton, 1964). Physicists in high energy laboratories are often exposed to magnetic fields as high as 10,000 gauss (1 tesla), apparently without damage, and thus this value is probably safe for a limited period.

A changing magnetic field induces a voltage which will produce electric currents in the body. These currents can stimulate excitable tissue. As an example, it has long been known that an alternating magnetic field can stimulate the sensation of light in the eye (phosphenes). Fields changing many times as rapidly as in present imagers have been reported able to stimulate peripheral nerves in the limbs (Polson et al., 1982). It is possible to develop reciprocity theorems about the currents induced by changing magnetic fields in terms of the magnetic fields detected from corresponding currents already in the body (e.g., the magnetocardiogram mentioned in Sec. 2.1). Changes in a magnetic field of a few tesla per second can induce currents of a few microamperes per square centimeter, and if maintained for a sufficient time in one direction might be expected to displace ions enough to have an effect. The stronger the fields the less the expected time requirement.

One possible worry is the production of potentially fatal ventricular fibrillation in the subject. It is known that there is a range over which the same electrical impulse will either start or stop fibrillation (Mackay et al., 1951).* It was possible to show that the property of an impulse that determined its effectiveness was approximately the total energy, rather than, for example, the total charge in the case of a capacitor discharge (Mackay and Leeds, 1953). A typical energy was of the order of magnitude of 20 watt seconds, and double-pulse experiments at that time indicated effective summation times of a large fraction of a second. There is one caution in making fibrillation observations: they should not be done on small animals (smaller than a dog), since the effect can then spontaneously disappear

---

*An important aspect in the production of fibrillation can be multiple cycles or applied impulses, those after the first being able to travel in certain directions around the heart but not all, refractory conditions from the first impulse preventing later impulses from going around the heart in both directions and cancelling at the far side; an effective stimulating frequency (Sec. 2.2) is about 60 Hz.

and may not be noticed in some cases. In a rabbit, for example, an impulse starting to travel around the small heart will arrive back at the starting point so soon that the originally excited region will still be refractory, thus terminating the uncoordinated circus activity of fibrillation. Further magnetic observation on these points is needed, keeping in mind that the current distribution undoubtedly matters, as well as the symmetry of the waveforms. If the heart is being monitored in the NMR machine, field changes will induce large voltages in the ECG leads, and there will be large motion artifacts. It is presently assumed that variations of magnetic fields as rapid as a few tesla per second are safe.

There are undoubtedly many other possible effects of pulsing electromagnetic fields, with a recent report of induced increased activity of messenger RNA and cellular transcription demonstrating that effects can be at the cellular level rather than the gross level (Goodman et al., 1983). This induction of a specific modification in normal cell function, in this case induction of transcription, need not be harmful, but it serves as a good precautionary warning that effects must be considered at all levels. In other observations (Wolff et al., 1980) it was found that no chromosomal aberrations were induced by the conditions of NMR imaging, nor was any inhibition of DNA synthesis detectable.

At this time it is probably wise to avoid imaging persons with artificial cardiac pacemakers or large metallic prostheses. Methods of Fig. 7.6 can be used to detect aneurysm clips looped closed that might be twisted in a changing or ac field. The person in Figure 4.10 would not be expected to have any problems from the surgical clips, but currents induced in the larger mass of metal could produce some image distortion and heating. (Images of the transient distortion of a changing magnetic field near a piece of metal can be seen in Mackay and Seaton, 1960.) Indeed, pushing this patient rapidly through (across) a strong field could risk tearing tissue because of the induced currents and consequent forces on the metal, even though nonmagnetic.

7.4. CONTRAST AGENTS

These methods are sensitive to one element only, and so the situation is not quite the same as the X-ray or ultrasound case where everything or anything placed in a beam contributes to the final image. Thus possibilities for useful modification of the subject include changing the amount or density of the element being recorded, or changing the response of that element already present, or changing the element being observed.

A change in state or degree of fluidity due to chemical or physical alterations can cause observable changes but the conditions may be too extreme for routine use. However, the effect of pressure on resonance signals has made a contribution to understanding the mechanism of anesthesia (Mastrangelo et al., 1979). Modifications of the gel state in the eye might be studied by modifications in the number of fluid protons. Injection of water or saline itself would be noticeable and might be used to outline structures in certain areas.

Modification of the response of protons by the addition of a small amount of paramagnetic material was a method used by the early investigators to reduce saturation of a sample. The effect of reducing T_1 is very noticeable in an appropriate image, and thus paramagnetic materials can be used as tracers. Thus amounts of manganese ions much less than lethal can produce a most noticeable effect. The paramagnetism of molecular oxygen has been mentioned, and should have applicability here. Dissolving more oxygen in the blood by elevated pressure or placing the subject in an oxygen tent should produce a noticeable change. With smaller objects the higher frequencies of electron spin resonance can be employed (e.g., Wyard, 1969). Considerable work has been done on the development and application of suitable spin labels (e.g., see the two volumes edited by Berliner in 1976 and 1979).

Imaging by a different element entirely has possible applications. For example, tritium has a larger magnetic moment than the proton (better sensitivity) and negligible natural abundance. Thus compounds labeled with it should be very noticeable. It is radioactive and has been used in various tracer studies. There is the possibility of the study of conformationally-significant couplings in multiply tritiated compounds.

Fluorocarbon liquids are an interesting possibility. As mentioned in the chapter on X-rays, they are a possible contrast agent that can appear in CT images, though here they need not be combined with other halogens such as bromine to become more noticeable. Flourine is uncommon in the body and hence any added as a tracer is noticeable by NMR. (What fluorine there is normally in the body is mostly in teeth and bone which, being solid, will give extremely broad weak signals.) Fluorine has been imaged by NMR (Holland et al., 1977). Thus the liquid can be seen with both modalities. But it also has excellent oxygen carrying properties, making it suitable as a major component of blood substitute or as the liquid for liquid-breathing experiments (Clark and Gollan, 1966). Dissolved oxygen should increasingly lower T_1, making oxymetry possible in it. Thus the various alternatives make it interesting for studying the cardiovascular system or aspects of respiratory dynamics. (However, injection of the liquid can cause gas to collect in the right heart, perhaps in part due to increasing temperature driving nitrogen from solution.) In emulsion form the liquid seems to be taken up by the reticuloendothelial system (macrophages) to appear in tumors (Young et al., 1981b); it might similarly be concentrated into an infarct. It should also enhance images of the spleen (Enzmann and Young, 1979) if administered by mouth in liquid form, or otherwise. Apparatus for protons will work with little modification with fluorine-19, since the intrinsic sensitivities and gyromagnetic ratios are similar. The apparent safety (Longo et al., 1970) and remarkable density of sulfur hexafluoride gas (which is somewhat anesthetic at elevated pressure) have already been mentioned, and under some conditions it might prove useful here.

Some relatively subtle possibilities exist in which not only gross location of a molecule is indicated, but also the properties of the immediate surroundings. An ideal spin label is a stable organic free radical having a structure and reactivity that facilitate its introduction to a specific site in a macromolecule. Nitroxides are extraordinarily stable free radicals which, because of their unpaired electron spin, display paramagnetism. Spin labeling generally refers to the use of such radicals, but other examples include nitric oxide, Mn^{2+}, other paramagnetic transition metal ions, lanthanide ions, and a few other organic radicals.

Electron paramagnetic resonance detects unpaired electrons, which often have high chemical reactivity, but with stable free radicals suitably used, observations can also be made at NMR frequencies. A spin label must resemble natural components of the system into which it is introduced, and its location on a larger biological molecule must be known. Because of the large magnetic moment associated with an unpaired electron, free radicals can cause enhanced relaxation of magnetic nuclei, depending strongly on the electron–nucleus distance. A label attached to a biological molecule will broaden all magnetic resonances in its proximity. Broadening can determine the distance from the label at a known spot to various groups on the biological molecule, whose signal can be identified by their separate resonances. By such techniques it has been possible, for example, to study enzyme functioning. Spin labeled enzymes can enhance the relaxation of water protons and allow the study of substrate interaction. It should be remembered that the distributions of paramagnetic ions can be nonuniform in a complex system. Nickel is paramagnetic and does not readily cross cell membranes, but it need not be at negligible concentration within cells since, especially when complexed, it may partition into lipid phases including membranes.

8
SPECTRAL INFORMATION

Thus far we have seen many examples of measurements of varying amounts of some property of material, usually displayed as variable darkening in a black and white image. If changes in this property can be observed as the frequency of the observing modality is changed then another sort of information is obtained, namely, the rate of change of the property with frequency. This has applicability for all modalities, or even between modalities. Differences at two frequencies can be made known by displaying the difference between two black and white images but another method of comparison is conversion into color. Indeed, the different reflecting or electron properties of different substances are observed as differences in color of reflected visible light, and thus this is the first kind of spectral information to be used (other than differences in audible sounds), and the first mentioned here.

The world is richer in information and discriminations for those with color vision. Indeed, comments about people who regularly made errors while working with colored materials are surprisingly scarce in the older writings such as the Bible; why reports waited perhaps until 1688 for Robert Boyle is unclear, though the search is an interesting story (Walls, 1956). It is important to note that in human vision, fine discriminations between different regions of the spectrum are made by comparing the response from three different kinds of receptors having very broad responses differing in different parts of the visible spectrum; these responses effectively are the Tri-stimulus curves relating to the Young–Helmholtz theory and pictured in many biology and physics texts.

Responses at different frequencies in Figure 2.1 can be compared in order to further characterize a subject, and this comparison can be done by converting the signals resulting from observation at different wavelength ranges into the small visible range. Perhaps one of the first applications suggested for such a translation of frequencies was made for the X-ray case (Mackay, 1949). The mouse image of the frontispiece to this book was produced about a year later, and these matters were discussed in various seminars at Berkeley. This eventually led to an invited lecture on this subject (Mackay, 1954), the publicity from which led to the publication in color of the mouse in the August 1954 issue of *Popular Science* magazine on page 99. The image of the hand was produced and some further details discussed (Mackay and Collins, 1957). During this period similar considerations were being developed in England (Donovan, 1951). In present times when everything from infrared images from artificial satellites to soap distributions in television hair advertisements are converted into color, such techniques are taken for granted, but at the time the ideas were somewhat unusual.

In producing the frontispiece images, three radiographs were made using three different distributions of X-ray wavelengths or energies. Each was a black and white picture depicting differences in absorption between different subject regions in one fairly broad energy distribution. (All color processes seem to have an intermediate step that can be described as devoid of color in each of several channels.) Each of these three pictures was converted to a variable thickness capable of soaking up different amounts of dye corresponding to the different degrees of darkening. The three "wash-off relief" negatives were soaked in three different colors of dye, and this dye subsequently transferred from each in register to a sheet of white paper to form the final image. Since there is no required correspondence between color and wavelength, differences in dying produced predominantly red mice and also predominantly blue mice in different prints. Changing anything photographically was extremely tedious, and quite different from the more modern equivalent of sitting before the screen of a color television set while adjusting contrast and color controls for optimum appearance. The overall result was as if the eye had receptors capable of detecting X-rays, and with three differing responses proportional to the three different wavelength distributions present. The images were not excellent but were able to demonstrate changes in color corresponding to changes in structure.

In producing the X-ray wavelength distributions it was felt desirable to have the possibility of changing the characteristic radiation as well as the continuous distribution. Changes in accelerating voltage allowed for one aspect and changes in anode material through a Wilson seal (Wilson, 1941) allowed for the other. Some modern commercial sources allow for voltage changes as rapid as 1 kilovolt per microsecond, but the process here was relatively slow. As was mentioned (Mackay and Collins, 1957), a long wavelength negative was used to mask a positive of the next shorter wavelength exposure, the similar shapes on the tails of the two X-ray distributions tending to subtract or cancel before printing as a single color, and thus making the final color somewhat less muddy. Many other alternatives were discussed including selecting the velocities of secondary electrons with velocity-

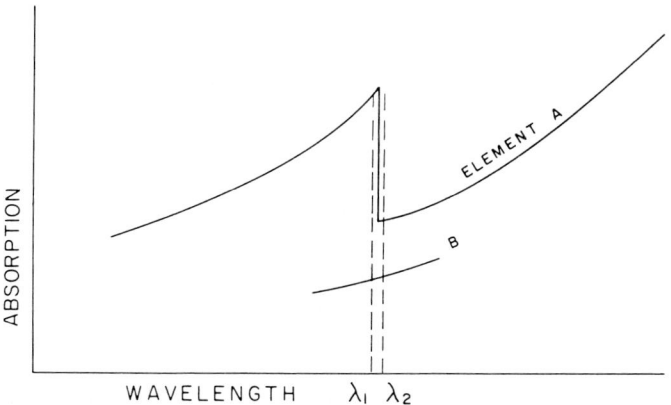

FIGURE 8.1. At the X-ray K edge a small change in energy abruptly shifts the ability of photons to interact with the inner shell of electrons of an element; thus for this element there is an abrupt change in absorption. Since the discontinuity will fall elsewhere for a different element, two images made by monochromatic wavelengths on each side of a K edge will differ essentially only where the element is present. An image depicting differences or ratios in the pair will thus largely depict the distribution of the one element. (From Mackay, 1949.)

sensitive intensifications by thin films of matter (Baker et al., 1942) to make direct color film. In connection with these first efforts it was realized that other modalities could be incorporated to extend the range, and an infrared image of the mouse was prepared with a war-surplus image converter tube which, however, introduced enough distortion so that the images could not be put in accurate register.

If the presence or distribution of a specific element is of particular interest then an alternative procedure using a property of the one element can be invoked. If images are prepared using a wavelength to the right of the K edge of an element and another by a wavelength just to the left of that edge, then the difference in the two images will depend only upon the presence of the one element, since absorptions by other elements will be about the same at these two wavelengths, as in Figure 8.1 (Mackay, 1949, 1954). It would be perfectly appropriate to display such a final image of the difference or the ratio of the two in black and white since the distribution of only one material is to be displayed even though spectral information was involved in obtaining this. This suggestion was made without realizing that such a technique had been used in quantitative microanalysis (Engstrom, 1946), though no images were formed. Jacobson had independently been considering such processes (1953), both for measurements on alternate sides of a K edge and for more widely separated measurements in a continuous region of absorption. Such methods, which are sometimes called "dichromography," depend for their success with single element imaging on having two rather monochromatic sources of X-rays at wavelengths on opposite sides of the discontinuity characteristic of the element of interest. As mentioned in Chapter 5, these occur in the useful range only for the heavier elements, of which iodine is one. The dose-dependent problems associated with an iodine contrast agent might thus be reduced, for example.

The analysis for a given element is made by measuring the intensity of transmitted radiation on either side of the absorption limit for that element. If I_1 is the intensity transmitted on the long wavelength side, I_2 transmitted on the short wavelength side, μ_1 and μ_2 the corresponding mass absorption coefficients, ρ the density of the elements in the absorber, x the thickness, and c the absorption exponent of the remainder of the material in the absorber, at the wavelength of the absorption edge, then

$$\frac{I_1}{I_2} = \frac{e^{-\mu_1 \rho x} e^{-c}}{e^{-\mu_2 \rho x} e^{-c}} = e^{-(\mu_1 - \mu_2)\rho x} = e^{-(\mu_1 - \mu_2)k}$$

where k is the amount of the element present in grams per square centimeter. Such a measurement can determine the metal content of glass or the lead content of gasoline, but here it is of interest for measuring such things as the iodine content in the body. The display of such information in the form of an image is extremely useful in judging the distribution of the element of interest, and such an image is the ratio of two original images, taken point by point. Variations in recording two separate images tend to yield erratic ratios or differences when small amounts of the special element are involved. Thus to measure the iodine distribution in a normal thyroid gland requires measuring 1 mg per square centimeter which produces about 1% difference at λ_1 and λ_2; for 10% accuracy in the difference requires 0.1% in the recordings.

The subject can be scanned with a narrow beam of X-rays which is cyclically being changed from a wavelength just above the K edge to just below, and the changing or alternating component of the received signal will be due to the difference in absorption at the two wavelengths. The alternating component of the received signal can be displayed while the scanning process proceeds, and slow changes in the detector with time or position do not affect the display. To compensate for the effect of changing body thickness, a plastic wedge can be moved in and out of the beam in such a way as to maintain the transmitted intensity of λ_2 constant. A way of assuring linearity of response in recording the amount of iodine present, for example, is to move a wedge of iodine in and out of the beam in such a way as to always return the alternating component of the signal to a fixed value, and to record wedge position in the display as the scanning process takes place.

The difference in signals, unlike the ratio, depends upon the absolute value and thus depends upon beam intensity, detector efficiency, and subject thickness. The use of wedges rather than simple recording of measurements fixes the overall outgoing intensity always to the same level.

A piece of equipment already started at the Karolinska Institute for a similar purpose was made to test this two wedge scheme in a clinical application as part of a larger project (Jacobson and Mackay, 1958). To produce a monochromatic X-ray beam, secondary radiation from an emitter bombarded by X-rays of the usual

170 SPECTRAL INFORMATION

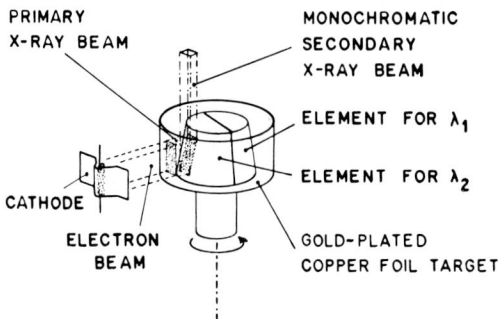

FIGURE 8.2. A special X-ray tube was constructed using rotating components to rapidly alternate the outgoing ray between two monochromatic wavelengths by the use of secondary emitters. (From Jacobson and Mackay, 1958.)

mixed wavelengths was employed; rotation of the anode and secondary emitter pair of iodine and cerium provided both for cooling and a cyclic change of wavelength (Fig. 8.2). Mechanical scan of the subject was synchronized with mechanical motion of the final photographic film past a light whose blinking rate, with constant energy per pulse, was proportional to the signal to be recorded. Stabilized servos moved the wedges, and the overall system was stabilized by taking part of the X-ray beam off as a reference into a separate detector, while the detectors themselves (a scintillator in front of a photomultiplier) were to be stabilized by automatic adjustment of their dynode voltages to make a small high frequency light signal appear as a constant amplitude (Fig. 8.3).

A recording of the iodine in a human thyroid gland is seen in Figure 8.4, where the outline of the body has been sketched in. It should be emphasized that this is the normal iodine content of that gland and is not radioactive or added for the occasion. In a normal X-ray picture the thyroid gland does not appear at all, though in the newest CT scanners it may be just visible; here *in vivo* quantitative analysis is possible. Summing the amount of iodine along the beam at each position can be made automatic (the counter at the upper right in Fig. 8.3) to give the total iodine content of the organ. In the present case the total amount of iodine was 14.2 mg, with the peak concentration in the isthmus being 5 mg iodine per cm^2 of tissue (Jacobson and Mackay, 1958).

Jacobson (1958) later noted in working with the apparatus that fat and water acted slightly differently. Two light materials thus can appear different when observed at two wavelengths separated by 6 kilovolts.

This scheme of moving wedges can be generalized to a larger number, with a correspondingly larger number of monochromatic wavelengths traversing the system in order to sample an appropriate number of absorption coefficients at the same number of wavelengths (Fig. 8.5). Thus it might be desirable to have a fat and a bone (calcium) wedge in addition to a "water" and an "iodine" wedge. If the outgoing intensities are all to be maintained constant then each wedge must move independently of the rest to indicate and compensate for the amount of its

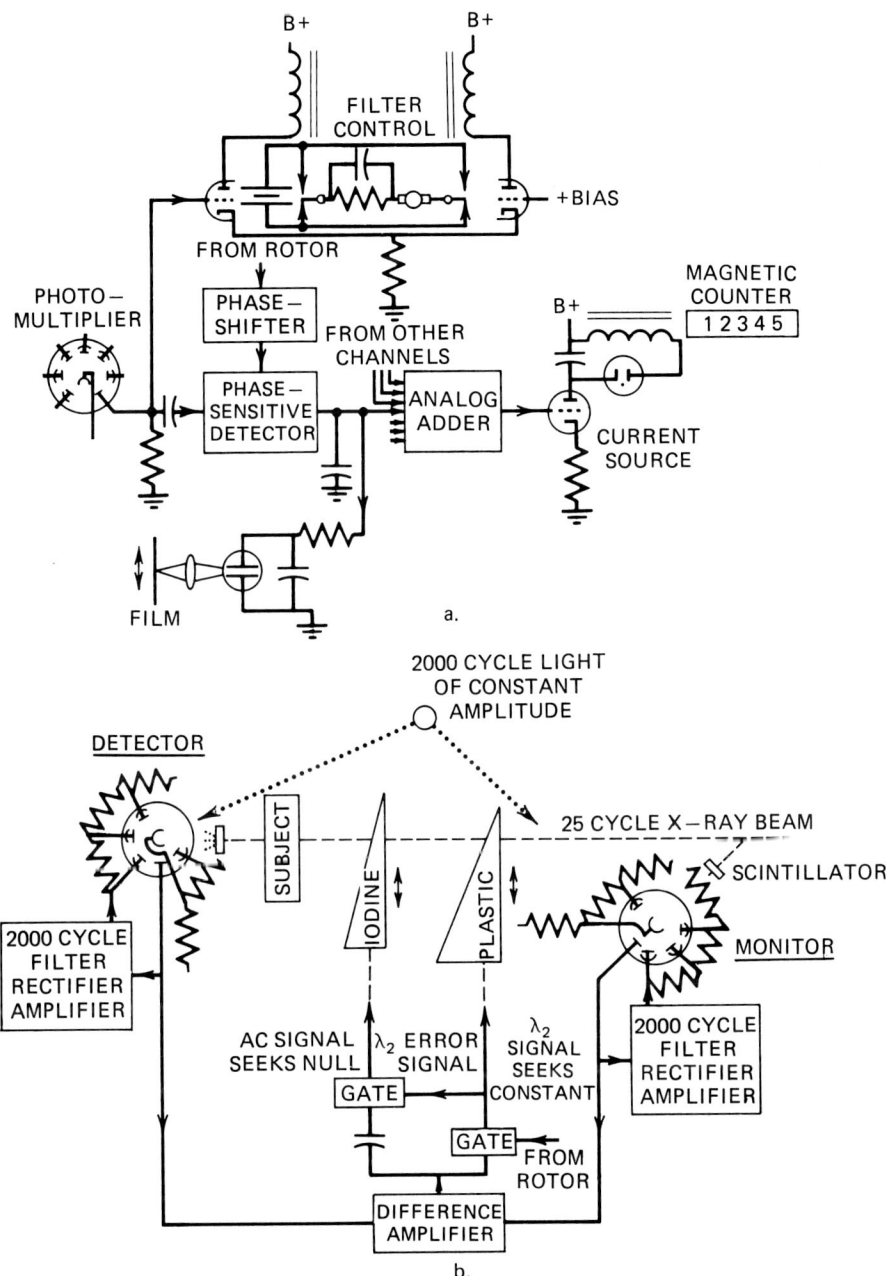

FIGURE 8.3. With everything stabilized, the distribution of iodine can be measured by the motion of a wedge of iodine to positions of different thickness in the cycling ray of two monochromatic X-ray wavelengths. Changes in overall thickness of soft tissue are compensated similarly by changes in the position of a plastic wedge. (From Jacobson and Mackay, 1958.)

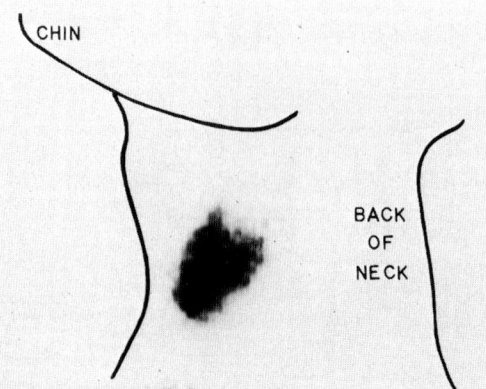

FIGURE 8.4. The distribution of normal iodine in a normal thyroid using dichromography. A sketched outline indicates the approximate position of the structure. Film density is essentially proportional to mass per unit area of iodine. The total integrated quantity of iodine is 14.2 mg determined from recordings made simultaneously with the scan. The peak concentration in the isthmus is 5 mg iodine per cm^2 tissue. (From Jacobson and Mackay, 1958.)

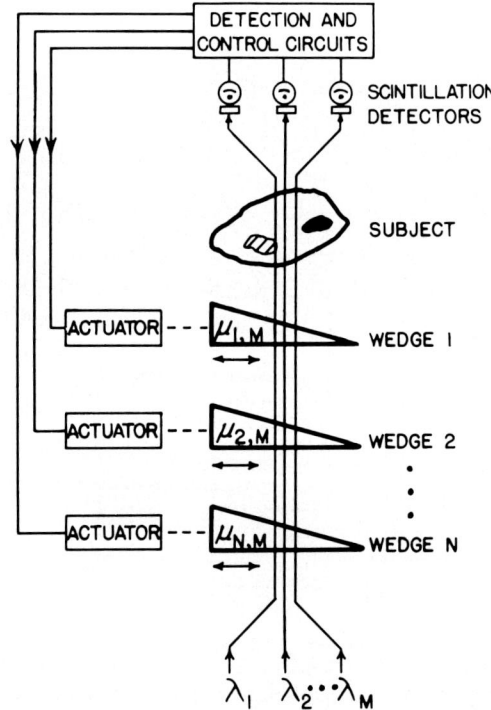

FIGURE 8.5. The moving wedge scheme can be generalized to more wedges with correspondingly more monochromatic wavelengths used to scan the subject. (From Mackay, 1958b.)

material in the subject (Mackay, 1958b, 1963a). If there are j materials and wavelengths, and if μ_{nm} is the absorption coefficient of the nth material for the mth wavelength, then there is one equation for each λ that expresses the outgoing intensity in terms of the traversed thickness of each of the materials:

$$\sum_{n=1}^{j} - \mu_{nm} t_n = \log k_m$$

Thus there are j simultaneous algebraic equations in the thicknesses of the j materials (wedge plus wedge-material in the subject). The greater the differences between materials in the way absorption varies with wavelength, the greater the expected unbalanced signals; the condition that the materials act as distinct has been mentioned (Mackay, 1958b). Certain combinations are ambiguous (just as there are two ways to obtain the color violet), but the possibility of rapidly performing some calculations during scanning could improve the guidance of the wedges. The indicated process can be considered as a sort of automatic spectral dodging, with wedge motion maintaining overall beam composition or "hardness" fixed by selectively working on different distributions within the total. The slow scan was done at a time when mechanical scans were not popular, though now the equivalent process might be speeded up by computation after measuring the alteration in composition in the emerging beam. For iodine alone, the ratio of two signals probably would now be calculated for each point in an electronically recorded pair of images. In this mechanical scanner, moving wedges effectively solved the same ray equations as in CT, there being one for each wavelength or material at every subject point rather than one for each point on each projection.

Since the discontinuity for iodine is at 33 keV, too high a skin dose might be delivered in attempting to use such a system for very thick subjects. However, the sensitivity of the method makes it applicable over a wider range than experience with ordinary beams might suggest.

There are alternative modifications of the same general scheme. Thus if lanthanum is bombarded with a broad range of X-ray energies, it will emit secondary radiation at 33.440 and 33.033 keV, which straddles the iodine absorption edge at 33.164 keV. A spinning chopper of iodine can then alternately pass both lines and one line so the differential wavelength effect can be recorded (Roy et al., 1962). A single barium secondary radiator has alpha and beta lines on opposite sides of the iodine edge and can be used with a spinning iodine chopper similarly (Jacobson and Mackay, 1958). These methods seem more practical than, for example, attempting to spin a target rapidly enough to Doppler shift lines.

One can also obtain the effect of a filter to pass a very narrow band of wavelengths by observing the difference in what passes two filters acting at different times on a broad wavelength distribution, and which cancel their effect everywhere but between their K edges (Kirkpatrick, 1944). Three of these "Ross filters" can effectively isolate two wavelengths by acting in succession in time rather than in space. If a customary polychromatic source has its distribution of wavelengths somewhat limited by standard filters then alternately placing in it a

174 SPECTRAL INFORMATION

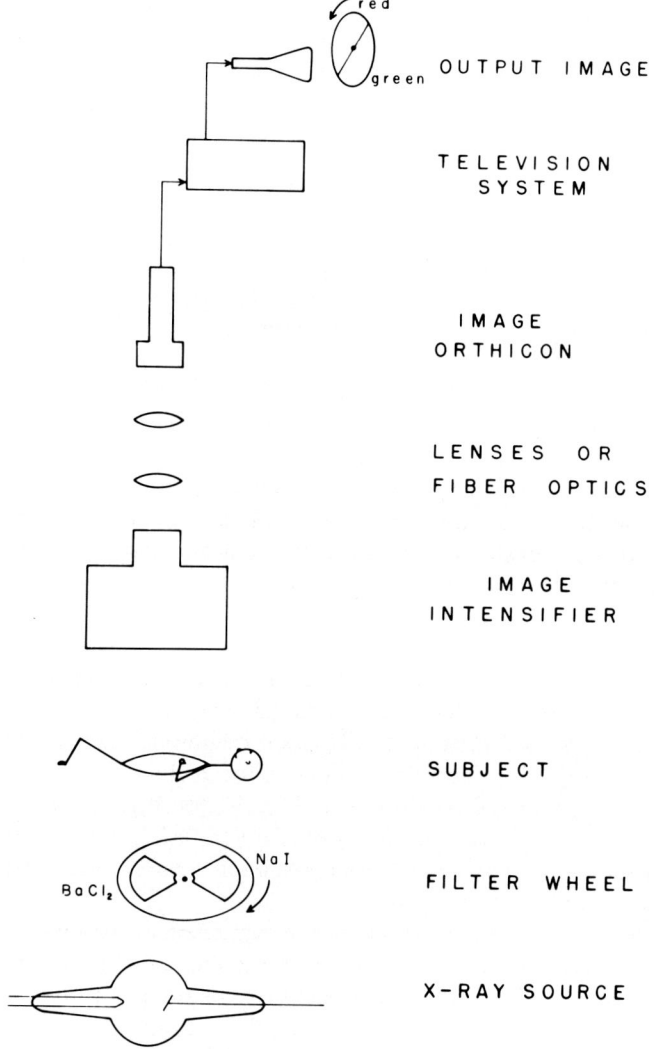

FIGURE 8.6. Changes in an image due to spinning a filter before the source of X-rays were converted to a change in color in order to detect somewhat smaller amounts of iodine than would normally be noticeable. (From Mackay, 1962.)

filter of barium and of iodine, which have K edges near each other, will result in signals whose ratio or difference largely depends upon the presence of iodine. It was hoped that the comparison of alternate signals could be done visually either by noting flicker or a shift in color for a small concentration of iodine (Fig. 8.6), and thus be rapid and require practically no extra equipment beyond a rotating filter pair. Such a filter combination was constructed as in Figure 8.7, using a

FIGURE 8.7. The rotating filter pair for X-rays used a hysteresis synchronous motor to allow easy shifting of the rotation to be in phase with the television system and the output.

hysteresis synchronous motor (having constant torque independent of speed) whose phase could be readily adjusted simply by applying a little drag to the rotor. The filter materials were in somewhat narrowed slots in the steel disk to reduce smearing due to limited transient response of the X-ray television system. Two regions can be compared for brightness equality by cyclically interchanging them and looking for flicker, which amounts to image subtraction, though color shifts here seemed less distracting. The system was tested (Mackay, 1962) and small balloons of dilute iodine solution in a container of saline were more noticeable with the combination in motion than when forming a single ordinary image.

In 1958, Jacobson, working with the apparatus of Figure 8.3 noted that water and fat act somewhat differently, as mentioned. This, of course, has nothing to do with K edges, which do not appear in the normal X-ray range for the usual biological elements. Also, if subtraction of two images made with X-ray distributions both above the iodine K edge is done, then either bone or soft tissue can be made to cancel, but not both. The shifts in color associated with such procedures as were involved in making the frontispiece mostly have to do with steady overall changes rather than with any discontinuities in absorption as a function of energy. Since any span of wavelengths can be made to correspond to any group of displayed colors, the gain of the system can be high.

As mentioned in Chapter 5, the linear attenuation coefficient of a material is the sum of two terms that represent the two major attenuating processes: Compton scattering and photoelectric absorption. The terms vary with X-ray energy E and can be expressed as $a_c f(E) + a_p/E^3$, where a_c and a_p approximately are simple functions of atomic number and atomic weight. Alvarez and Macovski (1976) calculated values of a_c and a_p from existing measurements of the attenuation coefficients of body materials at 16 energies and found that different materials grouped at different positions on a two-dimensional plot of a_p versus a_c. Soft tissue discrimination has to do mostly with the photoelectric coefficient a_p and its sensitivity to atomic number. They indicated, for example, that the average atomic number (Z) of tumors differs from that of normal tissue. Unfortunately the photoelectric cross section accounts for only a few percent of the total attenuation at useful energies. (Hydrogen density and NMR should be a better indicator of average change in Z of tissues.) The Compton coefficient has to do more with electron density or mass density. These considerations suggest making CT projections with two different source spectra so that the resulting image is not just a distribution of total attenuation μ, but of a_p and a_c themselves. These could be displayed individually, or as ratios, or as different colors. The use of a radioactive material such as gadolinium-153 as a source of the X-rays has been considered. Measurement of calcification by this method is not only of interest because of observations of bone mineral content, but because of the minute calcifications associated with breast cancer lesions. The making of observations at two different energies or wavelengths has become a popular topic, to which many workers have contributed (Rutherford et al., 1976; Di Chiro et al., 1979; Ritchings and Pullan, 1979; Talbert et al., 1980; Akutagawa et al., 1980; Dunn et al., 1980; Rutt and Fenster, 1980; Brody et al., 1981; Boyd et al., 1982; Faul et al., 1982). The comparison of this with the previous is perhaps best considered by noting that at any point on a colored picture, any color is a point on a two-dimensional chromaticity diagram relating absorption in two or three different X-ray wavelength groups; minimum detectable shifts on such diagrams have been studied extensively.

As a different aspect, ignoring the existence of spectral effects does not often distort ordinary X-ray images. But, to do the computations conveniently in computed tomography, it is assumed that there is an average or equivalent single voltage, which is not exactly the case. One result is the cupping artifact (Chap. 4). Suitable detectors can count photons of only a particular energy, but the process is not yet generally sufficiently rapid to solve this problem.

The comparison of information obtained at different frequencies can certainly be as useful with other modalities besides visible light and X-rays. In the case of radioactivity, different frequencies imply different energies, which can be helpful in isotope identification or even in determining or specifying a scattering angle (Sec. 2.9). In the ultraviolet many materials have characteristic absorptions, which have proven to act as a natural contrast effect in microscopy. Resonances at different frequencies in the infrared have provided a rich field for chemical spectral analysis, though in overall imaging it has been less important. At the bottom of the spectrum, frequency analysis of brainwave patterns from certain regions can readily indicate, for example, when a person has completed a problem in mental arithmetic; more subtle aspects have, of course, been studied in this fashion. In the case of nuclear magnetic resonance imaging, changes in frequency on a gross scale allow concentration on different elements, while on a finer scale it allows exploration of the environment of different atoms within a molecule (Chap. 7). There are many books and journals devoted to this last aspect, but from the viewpoint of application, many chemistry texts are excellent (e.g., Morrison and Boyd, 1973).

Of the major modalities, that leaves ultrasound, where a comparison of the effect at different frequencies can contribute toward tissue characterization. (Of course, at low frequency, listening to audible sounds from the body at different places does indicate something about the proper functioning of the heart or blood vessels, and even a stricture in the urethra seems able to produce a sound pattern during voiding.) Tissue texture or roughness can have a spacing of somewhat under a millimeter which means that it can well be explored by different wavelengths in this general size range. The range of frequencies often used in medical ultrasonics produces such wavelengths in soft tissue. Fat often seems to attenuate or even diffuse an image, probably more because of gross structure than because of molecular structure, for example. How the absorption and scattering of sound change as the wavelength changes can give a good indication of the nature of the structure. By analogy, the sky being light indicates the deflection of visible light down to the observer's eye, while its being blue indicates that small particles are involved rather than large fog or dust. Once again, information at two frequencies or wavelengths can be considerably more informative than an observation at one.

Melanomas, hemorrhages, and retinoblastomas exhibit scattering from internal histologic features such as cellular aggregations, small vascular channels, microscopic calcific deposits, regions of necrosis, and fine tissue septa. Spectral analysis can be used for statistical measures of volume scattering and attenuation. Tissue attenuation tends to increase linearly with frequency but the slope can change with pathology. Early work showed the difference in frequency-dependent attenuation for normal and infarcted myocardium (Lele et al., 1976). It has been shown that the slope of the attenuation coefficient with respect to frequency is highly correlated with the collagen content of tissue, and collagen is a major constituent in fibrous tissues such as scar tissue (O'Donnell et al., 1979). It also appears that many, but not all, benign and malignant tumors of the breast may be differentiated by frequency-dependent attenuation, though this may not be increased in medullary carcinomas, for example (Calderon et al., 1976). In some cases, larger structure can be determined without a sweep in frequency, for example, in observing the clear return from a fluid-filled cyst in the breast versus the snowstorm appearance of a solid malignancy *if* the amplifier controls are set so that noise is not introduced to appear everywhere. Speckle from interference appears as noise and adds distraction to images involving coherent sources such as lasers and ultrasound; observation at several frequencies can help smooth this out without disturbing real structure.

Measurement *in vivo* of attenuation coefficients for tissue segments by pulse echo ultrasound does have problems. Amplitude spectrums from time sequences of echoes can yield uncertain results since small differences can be swamped out by echoes from nearby tissues. Shortening the data segments to exclude unwanted echoes blurs the true spectrum by bringing in extra frequency components or Fourier components. The same problem occurs if one attempts to measure attenuation at two distinct frequencies with longer pulses, which results in loss of axial resolution. One solution to the dilemma is to use a signal containing many frequencies for good axial resolution and a parallel bank of receivers to monitor different components. A simpler method has used two filters and arrives at an estimate of attenuation by comparing their peak outputs (Meyer, 1981). In this case, the convenience of a frequency-modulated sinusoidal pulse called a "chirp" can be employed; commercial components for handling these are available. (These have become familiar in radar work, and also were mentioned in Chapter 6.) Some investigators have measured the frequency dependence of attenuation of soft tissue by using the transducer in an impulse-excited mode and obtained the frequency information from a spectrum analyzer (perhaps employing the Fast Fourier Transform), but the usual transducer for imaging systems is not an especially good choice because the decreasing response of the transducer adds to the increasing attenuation of the tissue to degrade the higher frequencies. Some comments on transducers for the frequency range 1–10 MHz have been made (Gammell and Le Croissette, 1978).

An object spread out transverse to the direction of observation can have any periodicity indicated by the direction of diffraction of energy from it. An object made up of an array of points that scatter incident radiation will yield a pattern that changes as the incident wavelength changes. From several such observations comes information about the distribution of points. Fourier transformation with a lens can either produce an image or retain the indication of periodicity. More will be said of such procedures in Section 10.2, and the ideas were inherent in parts of Sections 3.1 and 4.1. Wavelengths comparable to any periodicity are helpful, and the general range of 0.1 mm often seems useful, though the details of the observation will vary with the orientation of the structure, especially when the subject is not isotropic. Some texture analysis can be done for tissue characterization in an ordinary ultrasonic B-mode image without the necessity for a variety of frequencies however.

The distribution of several pieces of information can be displayed in a single colored image. An obvious possibility is T_1, T_2, and proton density in a nuclear resonance image. Attenuation at three different frequencies with one modality is in some ways a better example, since the result is not exactly equivalent to depicting the slope of the attenuation versus frequency curve in shades of gray. Any hue then corresponds to a particular trio of input values. This is perhaps best depicted on a triangular Maxwell diagram, rather than the more customary chromaticity diagram (Judd, 1935), and is probably more convenient for our purposes than points in three-dimensional space (Cohen and Gibson, 1962). In examining an image, colors in different regions can be reasonably well compared while shades of gray cannot be compared for absolute determinations. Any spread in a variable can be made to correspond to a specified color shift. However the appearance of a color can be slightly modified by the adjacent visual field, just as the brightness of one region can slightly modify the apparent brightness of an adjacent region. More will be said of color displays in Section 12.2, but it might be repeated here that rather fine discriminations are possible using detectors with rather broad responses when their outputs are compared by conversion to color.

9

MOTION AND FLOW

Changes, movement, and flow have something in common in that a comparison of two successive images or views can indicate a difference. There are a number of ways for comparing two images: for example, placing a pair of dollar bills in a stereo viewer could reveal if one were a forgery by a wavy appearance in three dimensions. Much of Chapter 8 was devoted to methods for comparing images made with different wavelength distributions in order to bring in spectral information, for example, to observe the distribution of iodine alone. We will later see that one of the main applications for the digital storage of X-ray images (digital radiography) is the display of the difference between images made before and after the injection of contrast media into blood vessels in order to remove the unchanging background structure (Chap. 11). It has been pointed out that in motion pictures one can display, or at least emphasize, only those parts that are moving by developing a positive print suitably (to a gamma of unity), and projecting the positive and negative together after shifting one forward or backward a few frames (Mackay, 1959). It is known that some doubly exposed holograms are able to show regions of motion ranging from gross motion to fine vibration. Successive positions of a place on a subject are recorded from successive ultrasonic impulses in the M-mode displays of motion in Figures 6.7 and 6.9. Flow can be determined in a moving liquid if it is not too uniform, so one region is distinct from another, whether this "labeling" of different regions is done by the presence of red corpuscles for ultrasound or the presence of oriented protons for nuclear magnetic resonance. Comparing the arrival of successive cycles in a periodic process similarly allows observation of motion as a Doppler shift. The rest of this chapter will be devoted to observation of movement and flow, rather than to changes produced by structural alterations.

The Doppler shift in frequency due to a moving source or observer is familiar to everyone who has noticed the apparent change in sound of a rapidly passing automobile. Doppler observations can be made with visible light or with radio, but we here describe it with ultrasonic examples that can be generalized. Approach of source and receiver raises apparent frequency whether the source or receiver is moving, though the equations are slightly different in the two cases (see any introductory physics text). Similarly, separation lowers frequency. If the velocity of the source or receiver v is slow with respect to the velocity of sound V, then one has the plausible result that the percentage by which the frequency f is changed (the change or Doppler frequency being Δf) is simply the percentage by which sound arrives sooner or later due to motion other than of the medium carrying the sound being superimposed upon the normal velocity: $\Delta f/f = v/V$. If the direction of movement is not along the direction of the sound beam then only the vector component of motion along the sound propagation direction will contribute to a shift in frequency, requiring multiplying by the cosine of the angle between the two directions; uncertainty in this angle contributes a major source of error in some biological and medical observations.

If the sound bounces backward off a moving reflector then the returned frequency will be shifted twice as much, the reflector first acting as a moving receiver and then as a moving source. This gives a relationship between the observed Doppler shift in a frequency, the original sound frequency, the velocity of sound in the material, and the unknown velocity to be measured:

$$\frac{\Delta f}{f} = 2\frac{v}{V}$$

This effect can be observed by sitting beside a vibrating tuning fork beyond which a sheet of cardboard is moved back and forth while reflecting sound. One will hear beat notes between the direct frequency and the frequency shifted by the motion of the cardboard, and the beat frequency is a linear measure of velocity. The stability of this observation is enhanced by the fact that a zero frequency beat note is definitely associated with no motion (no zero drift). Motion toward and away both give the same beat frequency in this simple system, and this ambiguity as to direction of motion can be removed by monitoring whether frequency is increased or decreased, as will be indicated later. There is a second way to regard this experiment which gives an equivalent result. The direct and reflected ray to the ear can be considered as interfering at every point along the path for any one reflector position, and motion of the reflector can be considered as changing the phase with which the reflected ray combines with the direct one due simply to the changed path lengths. Thus a vibration maximum may fall at the ear for one position of the cardboard while a node of vibration will be at the position of the ear for a different cardboard position; movement of the cardboard sweeps this pattern past the ear to give periodic changes in amplitude of the basic tuning fork frequency.

Rather than a single object such as a piece of cardboard, one can make similar observations on a group of many small objects, each of which will return a small signal whose phase will depend upon its position along the beam. Each little signal can be considered as a vector and the sum of all these adds to a larger vector which represents the phase and amplitude of the aggregate returned sound. One can receive a signal from a tube of flowing milk because of the scattering of sound from the fat globules, but it will not work from a tube of flowing water unless particles are added. Thus one can not measure the flow of blood this way in an octopus because it does not have formed elements in its blood.

Urine becomes observable after consuming milk and sodium bicarbonate, due to the formation of minute phosphate crystals accompanying the shift in acidity (Albright et al., 1969). Some body structures may display characteristic signals in an ordinary Doppler system. Thus an important factor in the conversion of a self-contained pocket of aberrant cells into a tumor is vascularization, and the vascular network of a malignant tumor is different from that of normal tissue. (Unlike most fish, sharks seldom have tumors, perhaps because some component of their skeleton material interferes with vascularization.) Wells et al. (1977) found that the frequency-analyzed signal over a tumor of the breast was different from a lesion such as a cyst or the signal from normal tissue on the other side. Note that in all measurements of the motion of blood, a reduced number of particles (e.g., reduced hematocrit) will lead to a more "quiet" (less intense) signal but the frequency will be unchanged.

Before giving more medical flow examples, an example of motion detection of a biological surface may be helpful. It was desired to settle the problem of how porpoises produce a variety of sounds, including simultaneous clicks and whistles, without blowing bubbles and without having vocal cords. Since the head of a porpoise, or other small toothed-whale, is largely an acoustic structure, it seemed wise to explore it with sound waves. We made M-mode recordings during their vocalization but, just as with X-ray motion pictures, nothing is sensed during the insensitive intervals between each periodic recording. Thus it was decided to determine what did and did not move in the head of a dolphin during phonation by collecting the reflected return from an uninterrupted beam of 2 MHz ultrasound. This frequency is high enough to follow individual cycles of the frequencies they produce, and also high enough to fall outside their range of normal hearing which extends to about 150 kHz.

The experiment is shown in progress in Figure 9.1, where the probe is being held against the soft part of the head in order to monitor motion of one of the air cavities therein. The probe contains two adjacent D-shaped ceramic transducers able to vibrate in thickness, one serving as the transmitter and the other as the receiver of sound at 2 MHz. Convenient sources for such continuous-wave Doppler components are the standard commercial fetal heart monitors used in obstetrics. At the receiving transducer occupying one side of the probe, two sounds combine (add) to give a resulting vibration that becomes the returned electrical

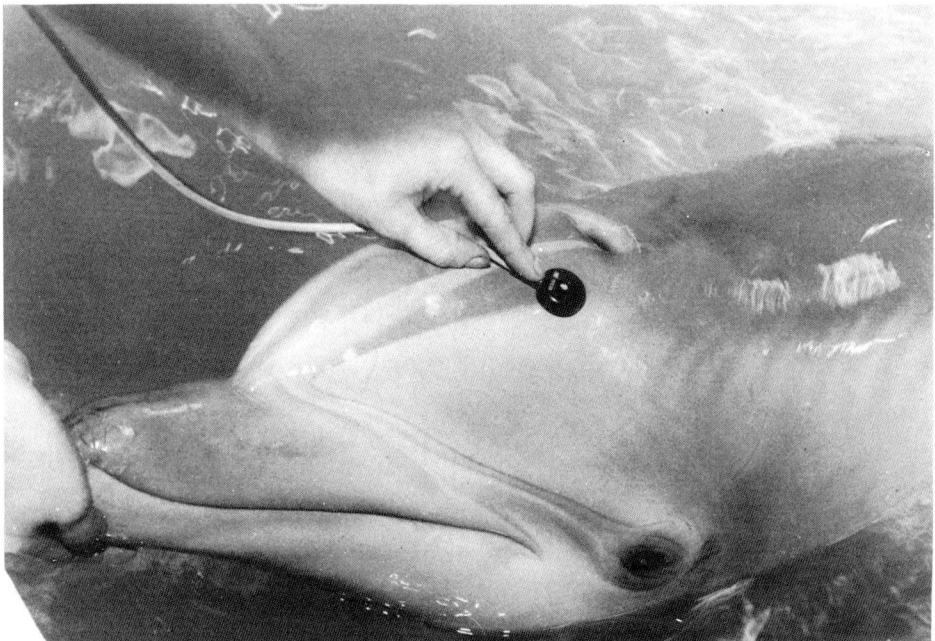

FIGURE 9.1. A continuous wave Doppler probe acting at 2 MHz being held in contact with the head of a dolphin in order to monitor movement of the air sacs during vocalization.

signal. In Figure 9.2 these two components are described as vectors having magnitude and showing the relative phase with which they combine. Sound communicating directly from the first transducer over an unchanging path is shown as e_1, while that going out into the animal's head and returning after reflection is shown as e_2. Some number of cycles plus fractions thereof will have elapsed before the return of the reflection, and this leads to the angle β being directly proportional to the distance of the reflector. As a reflector moves back and forth during a cycle of whale sound, e_2 might swing back and forth between A and C, for example. The sum of e_1 and e_2 is e_3 which is the amplitude of the 2 MHz signal. One cycle of mechanical movement from B to C and back to B will cause two

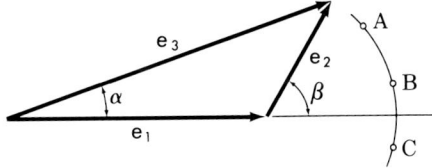

FIGURE 9.2. A simplified vector description of the Doppler process. It suggests (see text) that when monitoring small amplitude vibrations of an interface that in the detected output will appear a mixture of the vibration frequency and twice the vibration frequency.

changes in length of e_3, while a cycle of mechanical movement that causes motion from A to B and back to A will cause only one change in length of e_3. Thus a Doppler probe aimed at a vibrating reflector will return a signal that generally contains a mixture of the frequency of vibration plus twice that frequency (noted in Mackay and Liaw, 1981). (This signal of cyclically changing amplitude is processed in the way that a similarly modulated radio signal would be.) Since the angle β changes by a full 360° when the reflector moves a half wavelength distance, these considerations especially apply to small vibrations with amplitude less than a quarter of the ultrasound wavelength; if reflector velocity is constant in one direction then the usual Doppler shift frequency is predicted by the diagram. From such observations on two species of small whales we were able to show that these creatures cycle air back and forth between sacs in their head (thus not blowing external bubbles), and that the vibration of an internal structure called the nasal plug seems to generate their sonar clicks and many other of their sounds (Mackay and Liaw, 1981). However, vibrations of tissue seem not involved in their generation of whistles, as is the case with humans (and with tea kettles).

When such a device as this is aimed at a structure making large periodic excursions then a signal will be returned that varies in step with the motion. When aimed through the maternal abdomen at a fetus, for example, the motion of the little heart will return a signal that can be listened to and will have its sequence of frequencies repeated once for each cycle of heart motion. Changes in orientation of the heart can affect loudness but not interpretation of heart rate, and the cartilaginous fetal skeleton permits unobstructed observation from many directions.

One can also ultrasonically observe fetal respiratory movements in humans and animals. In work with J. Pitkin we find fetal heart slowing in seals that may be associated with these movements.

Mechanical instabilities in the transducer support must be minimized if less cyclic motions are to be recorded. Similar observations through tissue can be made with other modalities including laser light and also radio in the frequency range of a few gigahertz. In the latter case the scattered and reference signals can be combined in a hybrid junction.

Before going from the motion of solids to the flow of liquids, it should be noted that the extraction of motion information by comparing successive motion picture frames can be difficult and tedious, and there are related techniques in which a full image sequence is not recorded but only some preprocessed information. One example is electrokymography (Heyer and Boone, 1952) in which a photoelectric cell records and makes a graph of the light intensity coming from a small region of interest on a fluorescent screen as some selected structure moves back and forth in an overall X-ray image. (In an older version of the technique, a film moves vertically at uniform speed behind a narrow horizontal slit cut in a lead screen to form a line image of X-rays.) One can thus, for example, record the motion, as a function of time, of the wall of the heart as the darker margin advances and recedes across the cell. The effect of density changes of blood vessels or the heart wall can be somewhat eliminated by the use of the scanning action of a television camera, and timing and recording the interval from the start of a selected line to the border being observed; a fixed threshold trigger circuit "deciding" the level of darkening defined as the border can still see small motion where there is only an opacity change. These recordings are rather like M-mode ultrasonic recordings, which also follow pulsations or movements of selected regions, though aiming here is across rather than toward the boundary. Real-time images with any of the modalities can be handled similarly. In all these cases, one is monitoring the two-dimensional projection of the motion onto a flat surface; motion components toward and away from the surface will not be detected, while twisting of structures may be interpreted as a different sort of velocity than exists. If liquids are not uniform they can be handled similarly.

Doppler signals can be received from suitable flowing liquids, and the first such blood flow observations were reported by Satomura (1959). Attention was drawn to the method by an early, apparently independent, development of a similar method (Franklin et al., 1961). Such devices aimed through the skin allowed following such things as arteries of the leg to an obstruction (Strandness et al., 1966; 1967), or they could be implanted for more quantitative studies. (There had been other ultrasonic flow meters which transmitted an impulse diagonally along the length of a vessel to a receiver, and in which there was no change in frequency due to flow, but rather a changed arrival time; these worked but had problems and are not discussed here, nor are the electromagnetic flow meters where voltages are generated due to the flow of blood through the magnetic field of an internal unit.) Patterns of vessels can be seen in images of regions of flow produced by scanning with a Doppler system and recording a photographic intensity proportional to the Doppler shift frequency (Reid and Spencer, 1972). Perhaps the first observation through the skin of a deep vessel was that of the aorta from between the ribs (Light, 1969). Under typical conditions of flow rate and probe frequency, the shift in frequency which is directly delivered by the simpler systems falls within the human audible range and can be listened to while aiming the probe. A pulsatile sound comes from arteries, while the signal from veins often sounds more like the blowing of the wind.

Flow velocity is not the same everywhere across the diameter of a blood vessel, it tending to be higher at the center. For this reason red corpuscles do not distribute themselves uniformly, and it is their motion that is measured; in many cases this is exactly what is wanted, not the motion of the plasma. In large vessels such as the aorta, regions of reduced velocity are restricted to rather near the walls, and the velocity can be approximately the same across most of the diameter. If a continuous-wave Doppler probe is aimed across a bent vessel and the maximum shift in frequency is recorded then this signal probably originates in the region where flow is most nearly along the axis of the beam since flow elsewhere will contribute only a fraction of its speed to shifting frequency (Light, 1969). This removes uncertainty as to angle for quantitative measurements, and in large vessels a measure of maximum flow is approximately a measure of flow everywhere (the flow profile being more flat than parabolic). From a separate measurement of vessel cross section, such a measurement allows a calculation of flow volume. A

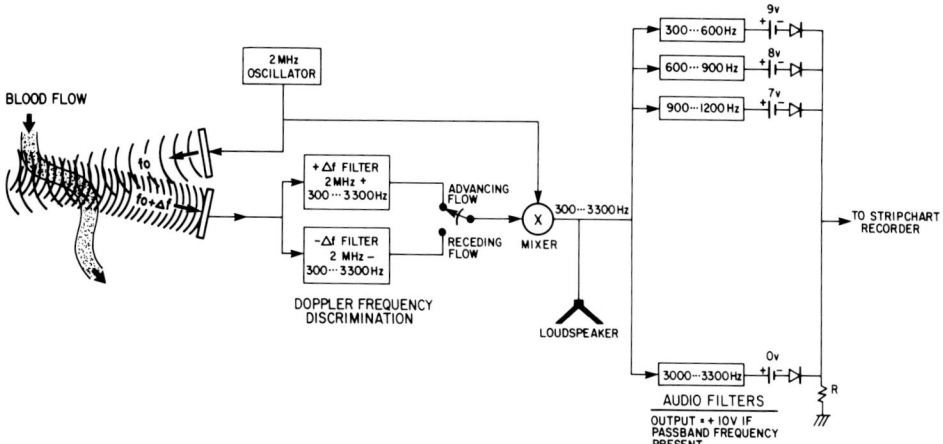

FIGURE 9.3. A continuous wave Doppler system in which flow velocity is quantitated in the region where flow is most nearly along the sound axis, which helps eliminate uncertain angular factors. Advancing and receding flow are shown being separated by a crystal filter, but the same can be done with less specialized components. Aiming of the probe is not critical and is aided by listening to the sound whose frequency is proportional to velocity.

place close to the aorta from which it is possible to aim along the aorta is the suprasternal notch, from where useful values of stroke volume could be measured (Mackay, 1972).

The general scheme is shown in Figure 9.3. The outgoing sound frequency is returned increased by the approaching blood flow. Any received signal is separated by two sharply tuned crystal filters that separate frequencies higher than the original from frequencies lower than the original, in order to discriminate advancing and receding flow. An alternative way for obtaining pure advancing flow signals or pure receding flow signals using only ordinary components has been given elsewhere (Mackay, 1972). This signal is mixed or beat with the original frequency to give an audible tone whose frequency is proportional to velocity. All signals below 300 Hz are rejected, even though loud, as being due to such things as heart wall motion or blood dragging along the walls of vessels. A group of electrical filters acting simultaneously determines the presence or absence of different flow frequencies, and the action of the diodes and batteries at the right is to deliver a voltage proportional to the highest frequency momentarily present. This will indicate blood flow velocity as determined in the position where the artery is most nearly parallel to the Doppler beam, which results in an accuracy better than 10% if the beam is anywhere within 25° of the direction of flow. A "mountain" is recorded for each beat of the heart, and the area under the curve gives the distance of effective advance of the blood on that beat of the heart in a "pipe" of the diameter of the vessel at the point of observation; curve area times vessel cross sectional area gives volume delivered. From the side of the base of the human neck one can noninvasively monitor flow in the common carotid artery.

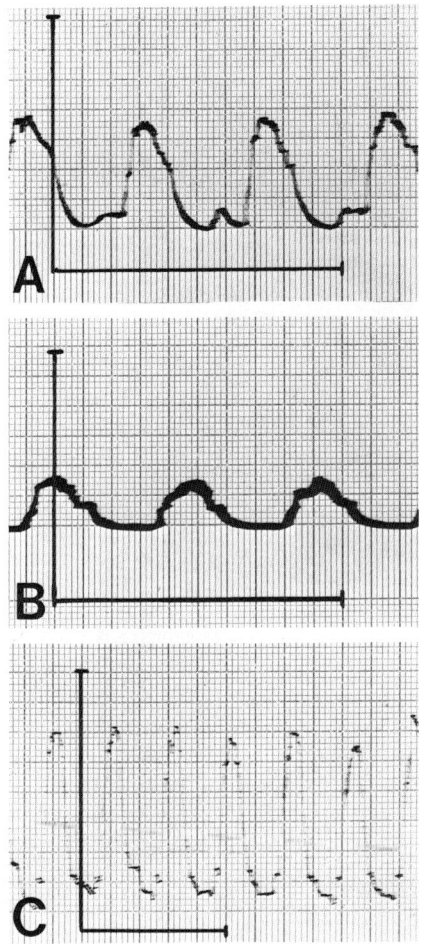

FIGURE 9.4. Velocity of blood in human fetus [transverse aortic arch at (A) and umbilical cord artery at (B)] and velocity in the descending aortic arch of a newborn at (C). The 2 MHz probe was seen in Figure 9.1. Calibration: abscissa—1 sec; ordinate—100 cm/sec flow velocity. (From Reid, Mackay and Lantz, 1980.)

The method has been used to measure cardiac output of the human fetus *in utero* (Reid et al., 1980b), the human renal artery in both normals and kidney transplants (Reid et al., 1980a), and to study dive reflexes by measuring cardiac output in sea lions as in Figure 6.8 (Mackay and Giovannoni, 1977). In the sea lion studies we found stroke volume to be about as variable as heart rate, for example, 130–590 cc per beat in one of the animals.

Three typical tracings from the fetal studies are shown in Figure 9.4. At the top is seen flow velocity as a function of time in the transverse aortic arch of a human fetus in its mother's uterus. The horizontal line represents 1 second and the vertical line represents 100 cm/sec flow velocity. The center recording is a measurement of flow as a function of time in one umbilical cord artery. The bottom recording is from the descending aortic arch of a newborn human. Such recordings are combined with an ultrasonic image that gives vessel area in order to

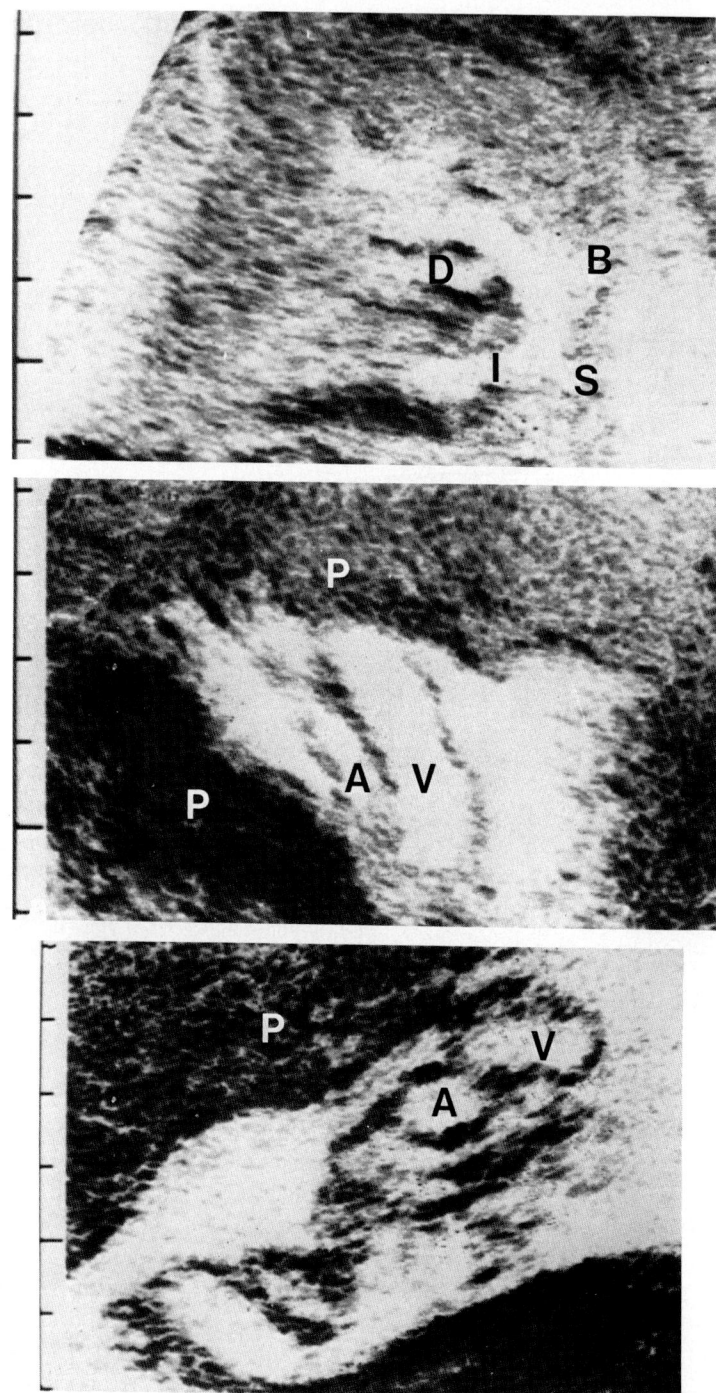

FIGURE 9.5. Top: aortic arch of human fetus with isthmus at I and ductus arteriosus at D. Center and bottom: umbilical cord of fetus with artery A, vein V, and placenta P. (From Reid, Mackay and Lantz, 1980.)

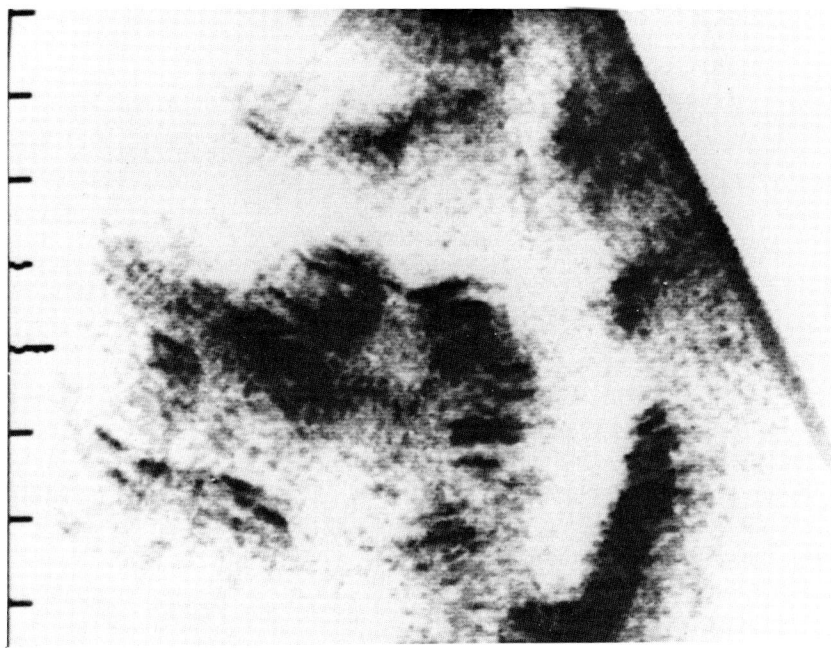

FIGURE 9.6. Ultrasonic B-scan image of adult human aortic arch, with branches, obtained from the suprasternal notch at 3.5 MHZ. (From Reid, Mackay, and Lantz, 1980.)

calculate flow volume. In the upper part of Figure 9.5 is seen an ultrasonic scan of the aortic arch in a fetus. The aortic isthmus (I) and ductus arteriosus (D) are seen along with the left subclavian artery (S) and the brachiocephalic artery (B). Below are two views of the umbilical cord of a fetus *in utero,* with an artery, a vein, and the placenta being indicated as A, V, and P. For comparison an image of the adult aortic arch parallel to the arch, with its branches, is given in Figure 9.6. Across the B-scan images can be made to appear a row of dots indicating the momentary orientation of the transducer. When these take a direction along the direction of flow, the Doppler transducer need merely be placed across the front surface of the imaging transducer in order to have approximately the correct aim, after which listening to the signal allows refinement. A major error in flow determination actually results from uncertainty in measuring vessel lumen size, the percentage error in calculated area being about twice the percentage error in the diameter measurement. From an estimate of fetal weight and assuming 15 ml/min/kg coronary flow, the fetal blood flow in the transverse aortic arch was typically found to be slightly over 100 ml/min/kg, with slightly higher values in the umbilical cord (Reid et al., 1980b).

A matter of general concern was the question raised in Figure 9.7 where a small motion of a tilted surface might be interpreted as a large motion or large velocity to a Doppler device. If the object is extremely smooth the signal will go off upward to the right and not be detected. For a real surface that is somewhat rough the

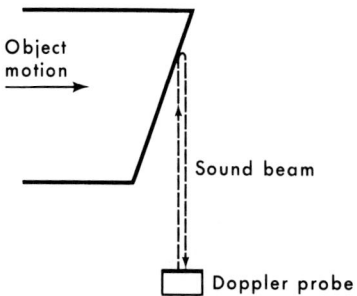

FIGURE 9.7. The surface of the reflector moves rapidly toward the transducer, and the length of the beam rapidly shortens many wavelengths, but generally there will not be a Doppler shift. Effectively, a series of small reflectors enter the beam and move straight across it rather than having a coherent pattern moving down. There will also be no Doppler shift if the surface is not rough enough to return some sound in the direction shown. (From Mackay, 1980.)

returning signal is a combination from a number of points that add to give a total response whose phase is a vector, just as in the case of an aggregate in a liquid. If this group moves toward and away from the probe, that resultant phase will shift appropriately to give a correct value of velocity or displacement (by integration). However, motion as indicated simply brings into view a new array of points of reflection which have nothing in particular to do with the previous array, thus giving a garbling or lack of signal. It would certainly be possible to design a periodic structure that would give a misleading signal in a Doppler unit, but in general it is not felt that misleadingly high readings have been produced in biological observations by this mechanism. (Similarly, aiming a radio Doppler unit up and down a street pavement or scanning across a receding landscape where everything is stationary does not indicate the presence of rapid motion.)

A few general comments on Doppler flow devices are relevant. Listening to the signal is often extremely helpful and, for example, cardiac valves have an easily recognized crisp "snap" to their audible character. This prevents their confusing the recording of nearby structures. If the full frequency analysis of the flow signal is retained, it is often possible to distinguish turbulent from laminar flow. In many cases changes in flow rather than absolute values of flow are of interest, for example, the changes produced by emotions or drugs or progressing sickness, in which case the absolute value of vessel diameter becomes unimportant if observations are always made from the same place and diameter does not change.

If the Doppler sound source is pulsed on only momentarily then the receiving circuitry can be turned on for only an instant in order to record signals returning from a predetermined distance. Such pulsed Doppler equipment has long been known in military applications, and eventually worked its way into medical applications. One prefers a short pulse in order to accurately determine or limit the distance, but one desires a pulse of many cycles in order to accurately determine the shift in frequency or the velocity. There is thus a compromise between these two factors. A single short pulse gives good range resolution but poor Doppler resolution, and a long pulse poor range resolution but good Doppler resolution. (Actually, it is not the short pulse that is important but the frequency bandwidth covered by the transmitted signal; in the chirp systems where a long frequency modulated pulse is transmitted, the range resolution is determined by the frequency sweep rather than by the pulse duration.) One should expect problems in accurately measuring velocity at a finely specified distance, and there can be particular difficulty with very low velocities. Making the transmitter pulses periodic improves the Doppler resolution since the total measurement time is increased, but at a cost of rendering uncertain at which range a point is found because it is not known which transmitted pulse caused the echo; because of attenuation more distant structures are not likely to be mistaken for nearer returns. There is a limit to the maximum velocity along the sound axis that can be recorded with periodic sampling. Two samples must be taken at the highest Doppler frequency to be passed, that is, the pulse repetition rate must be twice the highest Doppler tone desired, but yet the pulse interval must be long enough to allow the return of the deepest echoes. The maximum observation distance is inversely proportional to the recurrence frequency. For each velocity of an object in a complex field there is a corresponding frequency, which can lead to statistical complexity in some cases (e.g., Newhouse and Amir, 1983).

In spite of the various conflicting factors and compromises, such systems can be made to work rather well. In the upper part of Figure 9.8 is seen an ultrasonic image along the human neck made at 7.5 MHz. The left common carotid artery coming up from the direction of the feet is seen coming in from the right side of the image. To the right of this B-mode display is seen an A-mode display of the intensity distribution along the bright line traversing the B-mode display. The cross on this line has been adjusted into the center of the image of the artery. The B-mode scanning of the image can then be switched off, and the velocity at the position of the cross then is monitored with the pulsed Doppler mode to give the tracing in the lower part of Figure 9.8. The vertical lines mark seconds while the horizontal lines are a velocity scale. If the scanning head is not moved and if the line is oriented at an angle of 60° to the near surface of the vessel then this velocity scale indicates the frequency shift in kilohertz. The velocity recording is produced by periodic pulsing along the direction of the line indicated, and absolute values are not indicated unless the angle to the flow direction is known. However, relative flow recordings can be useful. They can indicate how long flow lasts between heartbeats and if it reverses, for example. Recordings from old people may appear "noisy," suggesting turbulent or disturbed flow in plaque disease. Moving the cross back and forth across the image of the artery allows recording of how flow changes with position across the diameter of the artery. The shape of the flow profile across an artery, and how it changes with the phase of the heart cycle, can have diagnostic significance in some cases.

For comparison, if the probe is moved up under the angle of the jaw, it is possible to view a slice across the right external carotid artery, in which is placed the cross in the upper part of Figure 9.9. Below that dark region is seen another similar dark region which is a slice across the internal carotid artery, with the common carotid artery sweeping off from them to the right. By convention, the direction of the head is to the left and the direction to the feet is to the right. The corresponding flow recording is shown in the lower part of Figure 9.9. The angle of the axis of the vessel to the Doppler axis is quite uncertain, and since the

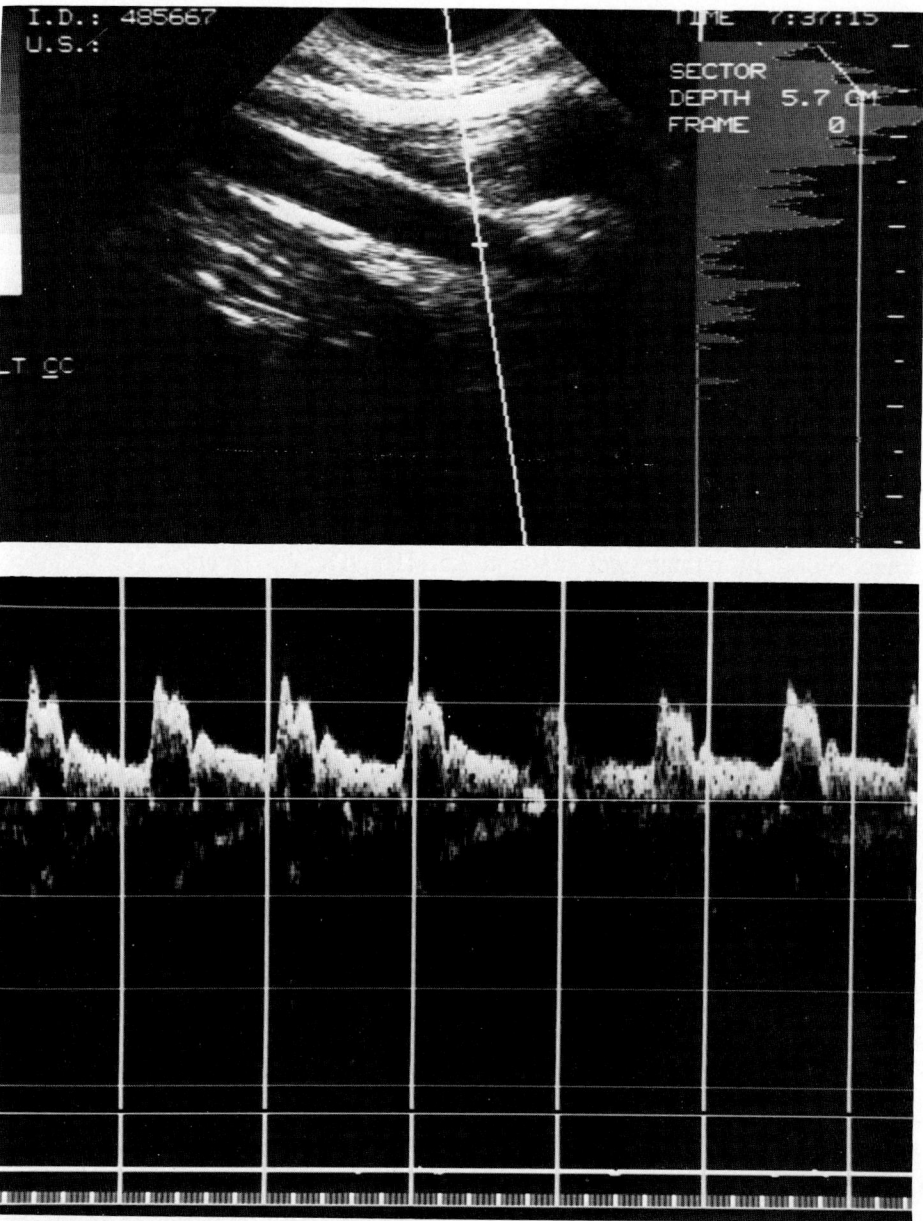

FIGURE 9.8. Top: ultrasonic image of a longitudinal section of the common carotid artery in the human neck, with a line with cross indicating the region at which the velocity component along the line will be recorded. To the right is an A-mode display of brightness as a function of position along the line. Bottom: velocity as a function of time at the point where the cross is centered, as recorded by the pulsed Doppler system.

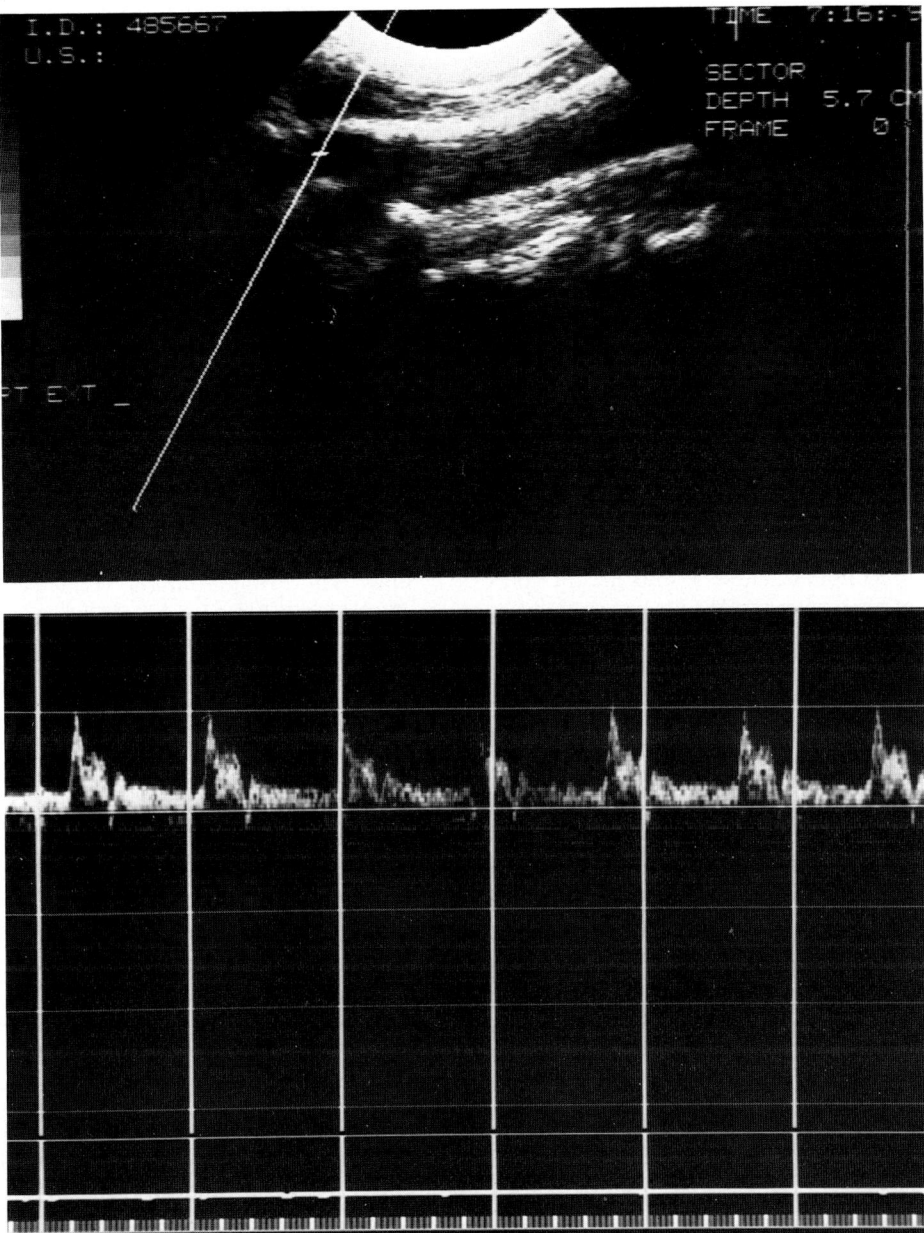

FIGURE 9.9. Recording velocity as a function of time at a point in the right external carotid artery near the bifurcation from the common carotid. Some sort of oblique slice is seen, and only relative velocities are recorded by a pulsed Doppler because of uncertainty in angle between flow direction and ultrasound axis.

component of flow across the sound axis produces no shift in frequency, this recording is not quantitative but gives relative values only.

Using a scanning imager to indicate vessel diameter along with a velocity measuring system such as shown in Figure 9.3 can have some advantages over a pulsed system. A range-gated instrument should be incorporated with a B-scanner for localization of the active region, but the former works best on arteries somewhat parallel to the beam while the latter functions better on arteries perpendicular to the beam. Thus it would have been difficult to apply a pulsed Doppler device to the previously mentioned renal artery flow measurements (Reid et al., 1980a), where observation was through the kidney of flow directly toward the probe.

Other modalities besides ultrasound provide the possibility of noninvasive flow measurement, and one of the more interesting is nuclear magnetic resonance. There are two general modes of action. Changes in the recorded signal due to flow carrying nuclei in and out of the active region can be recorded, or nuclei can be oriented at one position and the time of their arrival at a second position a known distance away then recorded. In 1951 Suryan reported experiments with flowing liquids. He found the amplitude of a nuclear absorption signal to increase with liquid velocity, this probably being because the detector had gradually saturated the sample (Chap. 7), and the motion of the fluid could bring in nuclei which had not been saturated. At higher velocities the amplitude can decrease because nuclei spend too short a time in the detector or in the region where magnetization is produced before the detector. If one is observing decay times of a signal then flow can carry precessing nuclei away from the region of detection to cause a more rapid drop; experimental arrangements can be imagined that would carry them into the detection region to apparently increase the decay time. Inversion recovery imaging can reverse results. Thus it would not be surprising to find that in an image such as Figure 7.5, regions of high flow such as the aorta might be either lighter or darker than expected, and it would be expected that in a fast image the aorta would appear differently in systole and diastole; the effect could be emphasized to measure flow. The first two to work on the noninvasive measurement of blood flow were Robert Bowman at N.I.H. (Bowman and Kudravcev, 1959) and Jay Singer (1959) at U.C. Berkeley; both had started their work earlier than these dates, and Singer's studies had been delayed by uncontrollable politics at U.C.

A book has been devoted to such matters (Zhernovi and Latyshev, 1965). Later, Morse measured blood flow by inverting protons in a time-of-flight method (Morse and Singer, 1970), and Grover determined velocity distribution patterns by recording the echo amplitude time function which is sensitive to the phase distribution of rotating nuclei that depend on the velocity of the flowing sample (Grover and Singer, 1971). Since the effect of diffusion is to attenuate an echo while the effect of flow is to shift its phase, by the use of suitable pulse sequences it is possible to separate the effects of diffusion and flow in order to measure extremely low flows (Packer, 1969). Presumably it will be possible to measure perfusion of tissues by some such method. It should also be possible to explore the spatial distribution of flow with a spin-echo experiment in a highly nonuniform magnetic field. Many other articles have appeared, and useful references include Battocletti et al. (1979) and Singer (1978).

Some direct observations can be made of human capillary flow, for example by timing the motion of red blood cells in the nailfold; this can be done through a microscope fitted with a television camera, and the results could be compared with the above.

Other tracers besides oriented protons or minute bubbles can be used to mark a small bolus of blood for travel a marked distance, or for dissipation or dilution. Radioactive isotopes of carbon, nitrogen, and oxygen can be introduced directly, uniformly, and noninvasively into the tissue of a region of interest through activation of the tissue by radiation in the multimillion volt range (Hughes et al., 1979). Tissue perfusion rate has thus been determined by them from the decay of oxygen-15 activity (positron emission) in a wash-out fashion after photon activation *in situ*. The method has been used to investigate perfusion rates in human tumors being treated with radiation. Among invasive techniques for measuring flow, dye dilution and thermal dilution are widely used.

Injections of contrast media are routine in radiology, and if this is allowable for exploratory purposes then the technique of videodensitometry can be used to compare the flow in the vessels in an image. If flow anywhere has been measured in an absolute sense, then this allows absolute calibration of all the others. In this method, a standard intensified X-ray image is viewed over a television system and probably recorded on a videotape recorder. The electrical signal corresponding to any point on the image can be taken out and separately recorded as a function of time. Injection of a standard contrast medium into a blood vessel will then have the recorded effect of a darkening that increases for a brief time. One can imagine a number of ways for combining such videodilution recordings, for example, to determine the fraction of the cardiac output that goes to the left and right kidney, and how this changes with treatment.

Consider injection of a bolus of dye into a vessel that narrows further downstream where observation is to take place. At the site of injection dye concentration will be reduced by an amount that depends on flow. In the narrowed observation region the path of dye will be elongated and moving faster, and thus spend the same duration in the monitoring spot as it would in a larger region of the vessel. The absorbing path is thinner, while the concentration and time are unchanged at a narrow spot. The recorded curve of increased darkening as a function of time thus depends on the mixing conditions at the place of injection and vessel size at the place of observation. The area under the recorded curve is measured (with a planimeter) to include the effect of the entire injection. The same curve is obtained even if, say, a third of the flow were diverted out through some branch since the length and speed of the dye column are reduced by a third, while the time, cross sectional area, and concentration are the same. However, with injection downstream of the branch, concentration would be up a third, as would be curve area. If injection of contrast material is done at two places and observation is at one, flow rates at the two sites are inversely proportional to the areas under the corresponding extinction curves, if the injected amount is the same in both cases.

Mean flow is measured if flow is pulsatile. Curve area is not affected by speed of injection nor distance to the place of observation nor the geometry of the place of injection nor the vessel shape at the place of observation. In the comparison method, vessel size at the place of observation does not matter, and thus no calibration of the detector system is required nor direct measurements on the segment of vessel on the film or television image. This procedure for measuring flow has been developed by Lantz (1975) and does not involve measuring the absolute amount of contrast medium passing the sampling area.

A weak signal resulting from the attenuation of a thick body will show less change in the image. But if the densitometer is made to have a logarithmic response to give an overall linear output rather than that suggested by Figure 5.3, the amount of tissue along the beam will not matter, and the change due to dye passage will be independent of body thickness. (The log action must take place ahead of the integration to determine area rather than after since the log of a sum is not the same as a sum of logs.) This assumes X-rays of a single wavelength since otherwise beam composition changes in traversing the subject to slightly alter the effective dye attenuation coefficient and also the ultimate interaction with the detector. One can dispense with logarithmic conversions if, rather than constant input intensity, input is adjusted so that what emerges has a preset value at the observation spot.

The fraction of flow or relative flow at two places can be converted to absolute flow by knowing the absolute flow at one of them. In some cases this itself can be done by observing the movement of radiation absorbing indicators within a vessel, that is, by measuring the mean transit time from a pair of indicator-dilution curves recorded at the upstream and downstream borders of the volume of an unbranched vessel whose size is known by being outlined by angiography (injection of a dye). Since dyes are so often injected in the normal course of radiology practice, this should prove an often useful way for quantitating flow. Such techniques have already been able to estimate flow in a single coronary artery in conscious man (Rutishauser et al., 1970). Measurement of specific flows is here done using suitably displayed X-ray images in conjunction with a routine type of injection.

10

IMAGE PROCESSING

An image, once acquired, can be systematically modified to make subtle detail more useful or noticeable. It is assumed that primary processes such as wavelength selection will have been used to make the pattern of the imaging modality as distinctive as possible, and then the secondary processes can render various aspects of the resulting image more accessible to the eye. Increasing contrast or the emphasis of edges and fine detail in an image are two such processes, and in some cases blur can be removed from a degraded image if the nature of the blurring process is known. Some of these processes incidentally emphasize the appearance of noise, while smoothing of an image can be done to reduce such distraction. In some cases quantitative analysis is helpful in extracting the maximum information possible from an image. Some displays of the final result are better than others for different purposes; these will be discussed in the final chapter, but the forms of processing mentioned here can be applied after the collection of data or signals and before final presentation in any of these ways. The processing can be done photographically, electronically, optically, or by computer, each possibility having certain advantages of speed or simplicity or flexibility in different cases. For supplementary information, Jacobson and Mackay (1958) may be helpful, though a number of new examples will be given here involving computers and computed tomography which were unavailable then.

10.1. PHOTOGRAPHIC FILM

A small piece of film can have a million to a billion pieces of information, and it can directly record a two-dimensional pattern. This means that it is not only useful for recording or displaying images, but, when processing data with an optical system, there is no medium with a larger space-bandwidth product to enter or exit an optical calculation. For some purposes the necessity to develop film, though now rapid, is a nuisance, suggesting the use of light valves for applications such as that in Figure 4.5. Although film can collect and store the effect of radiation for an extended period, unlike the eye, it is a rather inefficient detector of low levels since only a few percent of entering photons interact. A few light photons absorbed in a grain of silver halide with a volume of 1 cubic micron will produce 10^{10} silver atoms, giving an amplification by the development process of over a billion. (Thermal excitation does not immediately make all grains developable since more than one visible photon must be received within a short time.) In the X-ray case, each interacting photon can render several grains developable, though this is not equivalent in terms of ability to define structure to having less amplification and more initial interactions (Chap. 3); indeed, if one has too many developable grains per X-ray quantum one may not be able to view the very dark image that will exist if the image is statistically significant. This suggests that X-ray films could be improved in some cases by placing more silver in the beam but separating adjacent grains in a thicker emulsion so that each quantum will render *fewer* grains developable. (Films for recording an X-ray pattern often have a fluorescent screen called an intensifying screen placed against each side in order to add their glow to the direct interactions.) Electronic detectors including image intensifiers can interact with a greater fraction of incident photons, especially in the case of some of the charge-coupled devices, thus giving a higher detective quantum efficiency (Chap. 4), though the spatial resolution may be less.

The response of film to radiation is often idealized as follows. A piece of film is exposed and developed, and the percentage of a narrow beam of light of constant intensity that is transmitted by any given small area of the film is measured. The reciprocal of this is the "opacity," and the common logarithm of the opacity is the "density," which is nearly proportional to exposure (exposure being the product of time and intensity). A characteristic curve of a given emulsion is obtained by plotting density as a function of the common logarithm of the exposure. These characteristic curves are often called "H and D" curves for Hurter and Driffield (1890). Such a curve will typically show a straight line region beyond which it curves toward the horizontal at both ends, giving an S-type curve. The slope of the straight line region is usually called gamma and is different for different emulsions. If development is of longer duration or at a higher temperature, gamma will tend to be higher.

Since gamma relates the difference in light expected to come from adjacent regions on the developed film to the difference in light at the corresponding points in the original scene, gamma is a measure of how contrast is reproduced. Developing a picture to high gamma can yield an image with more contrast than the original scene, in which case the eye may be able to notice smaller initial differences. Some films intrinsically have such a high gamma that the smallest change in illumination will take them from total "transparency" to total "blackness," and they cannot be used for reproducing shades in between. (Such films are used for reproducing line drawings.)

To most senses a small change in a small amount matters about the same as a large change in a correspondingly larger amount (just as pennies may be as important to a poor person as dollars to a wealthier one), and therefore these logarithmic measures somewhat match the response of the visual system. Thus a so called Weber–Feckner curve relating sensation to stimulus on a decade scale has a straight line region at its center. Because of this photographs look natural. The slope of the sensitivity curve of the eye is such that one can just distinguish a difference in brightness of about 1% between contiguous uniformly illuminated fields. Photography often can enhance contrast by roughly ninefold—threefold by the photographic process of taking the primary photograph and threefold by the printing process—making it possible to see detail in a photograph which is invisible to the unaided eye.

The exposure range represented by the straight segment is called the latitude of the emulsion, and in some cases there will be no straight segment at all. Since both axes of this curve are logarithmic, it implies a curved relationship between brightness in and out if plotted on linear scales. A steeper slope above unity on the log scales implies a more sharply upward curved graph on linear scales, both implying more enhancement of contrast. (This will be seen in Figure 10.1 in the next section, where it is applied as well to television and other systems.)

A gamma of unity implies unchanged contrast between a scene and the resulting negative. If the negative is used to expose a second film by contact printing to make a positive, then it can be shown that the intensity of light transmitted through the positive is proportional to the original signal intensity raised to a power that is the product of the gammas of the two steps in the sequence. When $\gamma_1\gamma_2 = 1$, the intensity of the light transmitted through the positive when illuminated by a uniform light intensity is proportional to the original light intensity. (Exposing a film to a varying light intensity can be accomplished by moving a constant-intensity source across the film at changing speeds so that the product at any point of intensity and time gives the desired value of exposure.) In some applications of using optical systems to process signals, one wishes to insert into the system a light *amplitude* proportional to some signal. If intensity is the square of the desired signal, then the amplitude will be proportional to the desired signal. Thus, if a desired input signal is represented by the intensity used to expose a film, and if the gamma product of the two steps is set equal to 2, the resulting amplitude transmission function will approximately correspond to the desired signal. This is what is required for some forms of optical image processing; it must be remembered that in a sense negative values of light are not possible and only light intensity can be sensed. Electronic recording of electrical signals from detectors specifically eliminates some of these questions, and the display to be mentioned in Section 12.3 can be used to conveniently record the amplitude of a signal on film for viewing.

The structure of film itself introduces some noise into an image due to grain. Thus a microphotometer will yield a signal producing a deflection in a root-mean-square (rms) voltmeter while the small spot of light or aperture scans a sheet of uniformly exposed film. Many analyses of granularity have been made (e.g., Doerner, 1962). The use of film can produce other changes in an overall imaging process as well. For example, if the developer is not stirred, then its depletion at any boundary will make dark regions extra dark and nearby light regions extra light. This emphasis of edges to be mentioned in Section 10.3 can be desirable in some cases.

Photographic prints to be observed by reflection are somewhat different. Contrast is more inherently a property of the emulsion, and development modification is not very useful. Also, there is a difference between glossy prints and matt prints. Indeed, there is a difference between photographs reproduced in a book on rough versus smooth paper. A photographic print or a page in a book may reflect as little as 5% of the incident light or as much as 95%, which means the brightest part is perhaps 19 times as bright as the darkest. This range in brightness is considerably less than the range of 100 to 1000-to-1 that can appear on a screen while projecting a slide. A few percent improvement in the dark regions considerably increases the number of discernible steps that will fit into the total brightness range. A matt print scatters some light from dark regions in every direction, thus giving a smaller ratio of light to dark, though there is no glare. A smooth print or page reflects glare from a light source off in some direction, which must be avoided in viewing, but from other directions the blacks look blacker. Thus a glossy print can return a wider brightness range. For this reason the three-dimensional pictures built up from CT sections to be shown in Section 12.4 are best printed upon glossy paper to increase the number of sections that can be included before arriving at full light or full dark. (A glossy book page also allows finer dots for greater detail in the printing process.)

The use of photographic film is simple, compact, and cheap in experiments, but it can become expensive and time consuming in some clinical situations. It is important to remember that silver grains and photons are not the same, though one leads to the other. Thus scattered photons coming into a detector can uniformly "dilute" an image, while uniform blackening over a film from scatter can, to some extent, be overcome by viewing the film with a brighter light since contrast is not reduced by a uniform added layer equivalent to a filter; beyond the film's linear range uniform scatter gives darkening correlated with the image negative.

10.2. CONTRAST ENHANCEMENT

The eye will notice differences in brightness of about 1%, and significant differences less than that can be recorded on a photographic film or in the memory of some electronic imaging systems. By contrast we mean the percentage change in brightness at a boundary. In some cases increasing the contrast in an image will allow things to be seen that would otherwise be unnoticeable. It has been mentioned that a simple change in brightness, for example, by looking through a filter, does not change contrast, nor would increasing the intensity of illumination. The systems that can change contrast can be understood from Figure 10.1. One can study a television system, for example, by shining a light into the camera and measuring what intensity comes out of the television screen. The results for

CONTRAST ENHANCEMENT 201

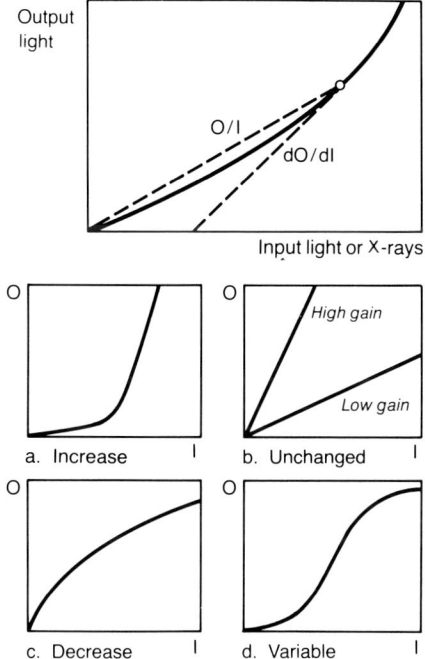

FIGURE 10.1. On a linear graph of output light intensity as a function of input, if the tangent to the curve cuts the horizontal axis to the right of the origin, then contrast will be increased in the brightness range corresponding to the point. A linear system does not change contrast even if the amplification is large. Several examples of actual characteristics are indicated.

various values of brightness could be plotted on linear scales as at the top part of the figure. If one observes a small change in brightness dO (such as might exist at a boundary of two adjacent regions in a real scene) at an average value of brightness O, due to a change in input dI, at an average value I, then to increase contrast

$$\frac{dO}{O} > \frac{dI}{I} \quad \text{or} \quad \frac{dO}{dI} > \frac{O}{I}$$

Contrast is increased if the ratio of changes or differences is increased relative to the ratio of averages. We see in Figure 10.1 that this means that the slope of the curve at the brightness of interest must be greater than the slope of the line leading from that brightness value back to the origin. This implies a curve that is concave upward relating input and output intensity, while a curve that is concave downward would be expected to decrease contrast. On a linear scale, a straight line function does not alter contrast since it modifies differences to the same extent as their averages. When the tangent to the curve cuts the horizontal axis to the right of the origin, contrast will be increased for that general range of brightness.

In the four lower parts of the figure several alternatives and their effects on contrast are indicated. The first of these can be considered as equivalent to giving no response below a certain light level, that is, subtracting away a certain amount of light overall, which is not the same as dividing all intensities by a certain factor; this can be accomplished with a suitable level of bias electrically introduced into a television system, the response being rather linear above this cutoff point. Such a system can compensate for some uniform scatter (not image-correlated), to the extent that the linearity of the detector itself is not compromised by the presence of this extra radiation. In other applications, more can sometimes be seen by aiming a television camera with contrast control down a microscope than can be seen by direct viewing.

Increasing the amplification of the television system as in the graph of part (b) does not increase the contrast since all differences would be magnified just as much as their averages would be. Many image intensifiers are linear and will not enhance contrast, and this is why an X-ray film image will often display more contrast than does the image in an image intensifier.

The brightness range in a small region can be amplified to fill the entire span from bright to black, with all changes outside that brightness range being displayed either as full black or full white. Sensitivity can also be increased until noise just becomes noticeable. This process can produce especially satisfying images if these adjustments can easily be made while watching the accompanying changes in appearance (though it has long been known that some such process could profitably be routinely applied to certain types of radiograph without preliminary observation). Such adjustments while watching are conveniently possible with computed tomography, and this is one of the main reasons for the satisfying character of its images. A small range of Hounsfield numbers can be set to span the full range from black to white, with the large range of numbers beyond simply being unable to turn an already off oscilloscope beam further off or make a fully bright one much brighter.

One can not magnify every brightness range because there would then be no way to display the total span, which must be compressed into the number of steps that can be displayed between fully bright and fully dark. Slow meaningless variations across a subject can drive the output from fully dark to fully light while there is not enough amplification to reveal small brightness changes at some intermediate level. Thus changes in overall thickness of a head can interfere with the observation of small changes in brightness associated with a blood vessel. Placing the head in a suitable bag of water can make the emerging intensity everywhere more nearly the same. In a scanning system, slow variations can electronically be rejected as having to do with large structures, allowing greater amplification of high frequencies that relate to small structures. There are several other ways of accomplishing an equivalent result.

An X-ray or other image on film can be printed onto another sheet of film while a sheet of glass is placed between the two. The blurred image on the second film will not contain any fine detail but will represent large overall variations in intensity in an inverse fashion from the original film. These two films can then be placed in direct contact and printed onto a third film which can be developed to extremely high contrast. Slow changes will thus cancel out leaving magnified changes in intensity associated with small structure. This is the so called "unsharp mask" method (Eden, 1955).

A sort of electronic equivalent is the method called LogEtronics (Craig, 1955;

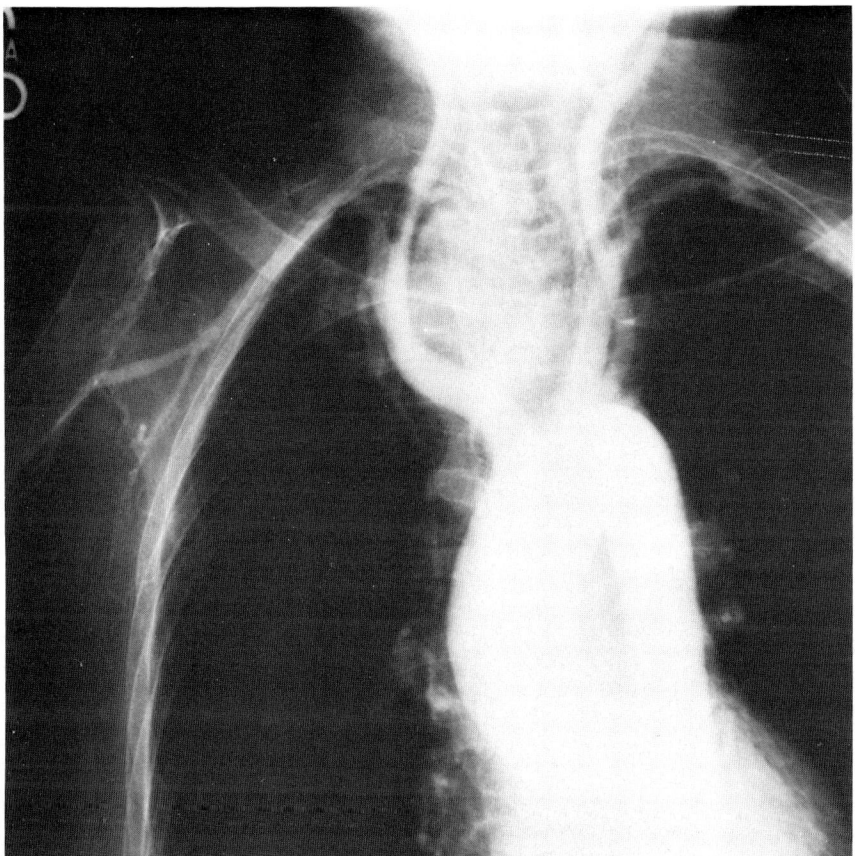

FIGURE 10.2. A radiograph made before the injection for an angiogram.

St. John and Craig, 1957). In this a radiograph is printed onto a contacting sheet of film by the scanning spot of light of a television tube whose brightness is decreased whenever a phototube at the far side of the two sheets of film sees an increase in transmitted brightness. The tendency is to cancel all variations and give a uniformly gray print, but the scanning spot is made somewhat larger than the detail of interest, but smaller than the major density masses. The feedback will cancel out slow changes but fine detail will print in the usual way, with the momentary intensity of the light source depending on the average darkness of that region. (An unsharp mask appears on the cathode ray tube.) As in the out-of-focus mask method, one can then print to a very high gamma on contrasty paper or film to emphasize detail and yet not run out of the range of the film because of slow cumulative variations leading to a region of saturation. An example of this is seen in the processing of the image of Figure 10.2. This figure shows an ordinary radiogram before the injection of dye in producing an angiogram. Figure 10.3 shows this same image after processing by LogEtronics. Of special interest are the

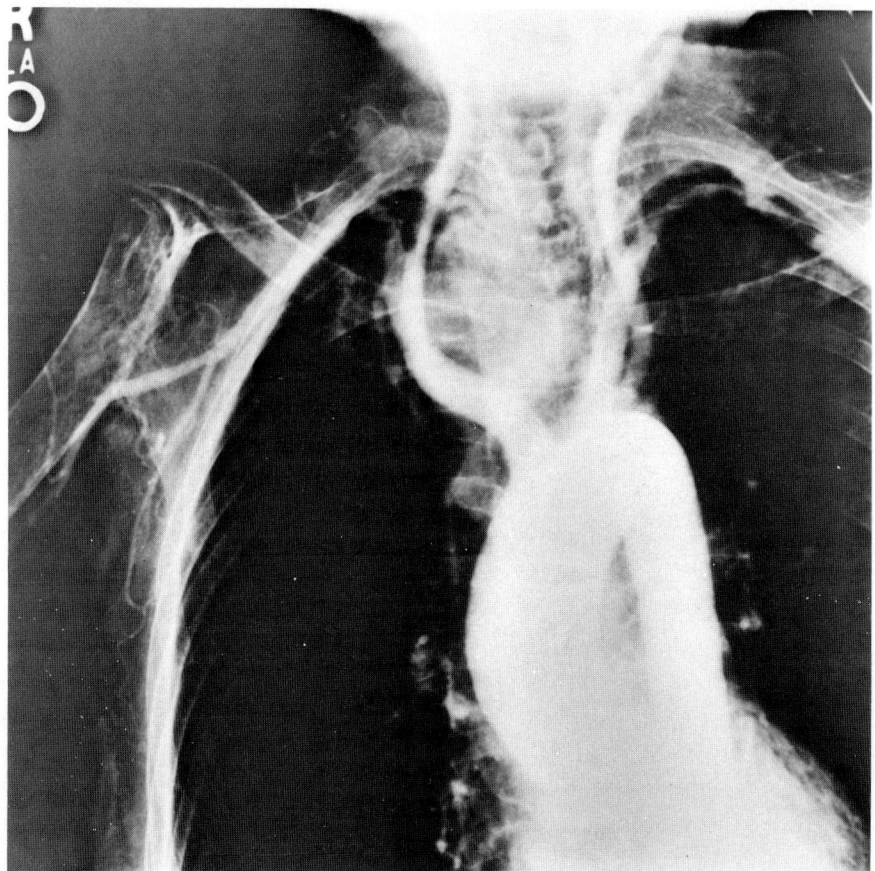

FIGURE 10.3. A more contrasty print of Figure 10.2 reveals the small vessels in the shoulder. The range from black to white is not exceeded due to slow changes across the film having been removed by LogEtronics processing, which is somewhat equivalent to the unsharp mask method.

small vessels around the shoulder seen to the left. The image of this same subject after injection of iodinated contrast agent is included in Section 10.6.

Certain types of image might routinely have their contrast enhanced, while others should not. An image whose content is difficult to discern because of noise will not be improved by contrast enhancement. There is no need for any process that enhances contrast above the level of statistical fluctuation. On the other hand, it is preferable to make certain that this level is always reached or else information may be lost. In typical films at normal density, the area having a noise value perceptible to the eye is of the order of a square millimeter. It is not profitable to attempt contrast enhancement on scenes with such fine detail. Conversely, if larger image elements appear uniform they can contain unnoticed information, and perhaps should be subjected to some form of contrast enhancement to ensure

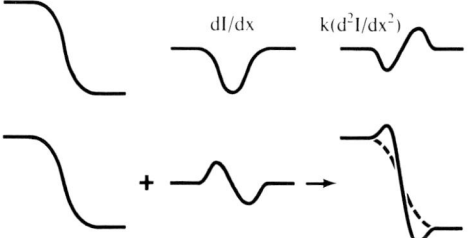

FIGURE 10.4. Emphasis of boundaries can be accomplished by mixing with any brightness signal some of its own negative second derivative. In a scanning system one must pass through each point in both perpendicular directions to make emphasis independent of edge orientation. The visual system normally does some of this or an approximate equivalent.

no loss of information. "Soft scenes" as in soft-tissue radiography or tomography might generally be enhanced to uncover possibly unnoticed detail.

As mentioned, more information on all these matters is to be found in Jacobson and Mackay (1958).

10.3. EDGE EMPHASIS

Changes in adjacent materials are seen by noticing the boundaries or lines between structures. Indeed it has been suggested that the only thing sensed by the visual system is the edges or boundaries of objects, rather than the objects themselves, with the visual system filling in the appearance of uniform material in between changes (Walls, 1954); this would at least explain the existence of and necessity for physiological nystagmus. Whether this interesting argument is true or not, it is certainly true that the emphasis of edges in an image can make structure more noticeable. Gradual changes are especially difficult to notice.

An example of the emphasis of edges is shown in Figure 10.4. Scanning along a line across an image, there appears a drop in intensity. The rate of change of intensity there can be calculated. From this the second derivative can also be calculated, and it appears as shown. On the second line is shown the effect of adding some fraction of the negative second derivative of a signal to the signal itself. The transition from light to dark is made more rapid, and the light side is made a little extra light, while the dark side is locally made extra dark. Attention will be drawn to this transition and it will become more noticeable, though if too much derivative is added it can appear as if extra structure were present. Combinations of other derivatives can alter the detailed form of this emphasis of an edge.

Such a process is easy to implement in a scanning system such as a television system. As the scan progresses, the rate of change of the signal is readily indicated by a simple capacitor circuit, which can be followed by a second one to indicate the second derivative. It is simple to electronically subtract this voltage from the original signal so that the combination can be displayed. With a simple scan, an edge extending across the direction of scan would be emphasized, while one parallel to the scan direction would not be emphasized. Thus it is necessary to produce an isotropic scan that passes both up-and-down and side-to-side at each point on the image.

These capacitor differentiating circuits are equivalent to high-pass filters, in that they pass high frequencies more readily than lower ones. Thus the emphasis of edges is equivalent to the emphasis of rapid changes or fine structure. This also tends to emphasize undesirable noise, and this places a limit on such processes.

The human eye normally seems to produce such a modification of the pattern resulting from observing a scene. Thus observing a dimly lit roof of a distant house shadowed against the sky after sunset will yield the appearance of the sky being somewhat brighter around the edge of the roof, and the roof being darker at its edge than elsewhere. These parallel stripes of nonuniform intensity are called Mach bands, and a better example is given in Jacobson and Mackay (1958). The probable mechanism of their generation is the "lateral inhibition" between adjacent receptors in the retina of the eye. If every receptor diminishes the signal delivered by adjacent receptors then the bright signal will darken a nearby darker signal more than the dark signal will reduce the adjacent brighter region, which gives the above form of pattern change. Microelectrode studies show this mechanism to be active in the eyes of other animals as well.

With the practicality of the large-scale integration of electronic circuits, such lateral inhibitions can be built into a detector mosaic to any desired degree. Mimicking the eye's lateral inhibition would be another example of parallel processing. Other processes can also yield such an emphasis of edges. For example, the development of a Xerox image involves the application of dark powder to charged regions. If the powder is blown before the charged surface then a little of it destined to hit a slightly charged region will be diverted into an adjacent more charged region, resulting in bright being extra bright and dark being extra dark at a transition boundary. It has already been mentioned that lack of agitation of the developer of a photographic film will result in a region of high density somewhat depleting the developer so that the adjacent or dim region will be even dimmer due to less development, producing a similar effect at boundaries.

The use of a computer allows great flexibility in comparing a point with surrounding points, and modifying what is displayed at the central point to suit the occasion. One of the possibilities is the above mentioned sharpening action in which any changes or derivatives of intensity accompanying a change in position are augmented to become more rapid. Some of the alternatives for computer processing of images is indicated in Figure 10.5. The brightness of each point in an image is recorded or stored in the form of a proportionate number in one electronic circuit, there being as many such circuits as there are points in the length and breadth of the image. There is also a second identical set of counters that can similarly store the image. Any point on the image can have its number combined or averaged with that of adjacent points and this combination number can be transferred to the other set of storage counters as one corresponding point of the new modified image. At the top of the figure is shown the averaging of the value at

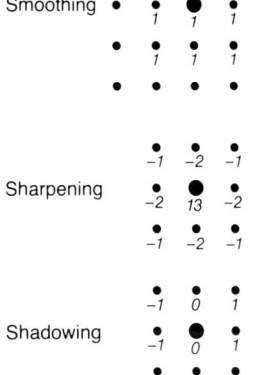

FIGURE 10.5. In a computer the intensity at each point can be combined with the intensity at nearby points to achieve different effects. Weighting all points in each group equally before display tends to smooth out changes in the image. Sharpening or edge emphasis can also be achieved as indicated, as can shadowing. Each point is combined with its neighbors, and the value stored, before the next point is taken as the new center for a computation with what are now its neighbors. An entire image takes a few seconds to process, even when more distant points than the nearest neighbors are included.

a point with the value corresponding to each of the adjacent eight points, all values being weighted equally. By uniformly averaging each point in with its neighbors the action is a smoothing or blurring as if some scanning process were being done with a larger spot. After computation of the indicated point and transferral of the value produced into the second matrix of points, the process is repeated for the point one step to the right with its neighbors, and so on. An entire image can be thus modified in a few seconds for display.

If the neighboring points are weighted more nearly as shown at the center of the figure then a sharpening or crispening action results. Some points adjacent to the one momentarily under consideration are given a negative weight, and this is also what is involved in some of the deblurring reconstruction methods for computed tomography in Chapter 4. More points around the momentarily central one can be included in the averaging process in order to extend the regrouping process to any desired relationship. The weighting for shadowing, as if the illumination of the scene were coming from the side, is shown at the bottom of the figure (and an example is later given in Fig. 12.7).

What happens between dots in the sharpening case is the same sort of thing happening in the eye due to lateral (mutual) inhibition between nearby receptors. Here the weighting coefficients add to unity, which tends to preserve the overall average brightness due to an effective overall gain of 1. Any combination of only positive weighting numbers will merely spread or smear the image.

An example of the sharpening of an image, which can be interpreted as the mixing in of some rate of change between a point and its neighbors in all directions, is seen by comparing Figure 10.6 with Figure 10.7. The original CT image (Fig. 10.6) shows the two kidneys and the overlying "hot dog" shape of the pancreas, along with air in the gut appearing black. This image was then processed in several different ways, to be seen in subsequent sections, using a 5×5 element, that is, more points around the central one than those just adjacent to it

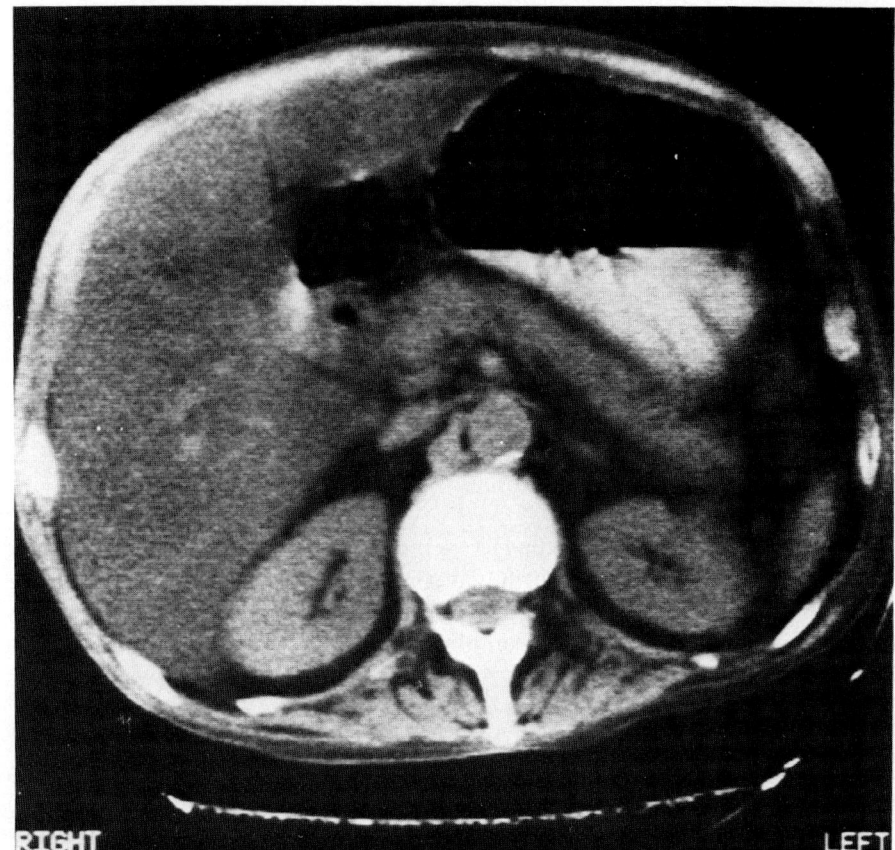

FIGURE 10.6. An X-ray CT image which will be seen processed in various ways. The large black areas are gas in the gut.

were involved. All processing was done on the image data, though it could have been done on the original CT raw data which was then processed by the original algorithm to form the cross sectional image. The effect of "sharpening" is seen in Figure 10.7, after using the following rows of weighting numbers: 0, −2, −4, −2, 0; −2, 2, 4, 2, −2; −4, 4, 12, 4, −4; −2, 2, 4, 2, −2; 0, −2, −4, −2, 0. The appearance is somewhat modified, and noise is somewhat more noticeable. However, some fine detail that could not be seen in the original picture is also now noticeable.

It might be said that the emphasis of edges is the opposite of what one does in removing a spot from clothing. Even if not much of the mass of dirt is removed, spreading the edges to make them less distinct can make the spot less noticeable.

The emphasis of edges has been described as being implemented by time-dependent electronic circuits in scanning systems or by computers in a point-by-

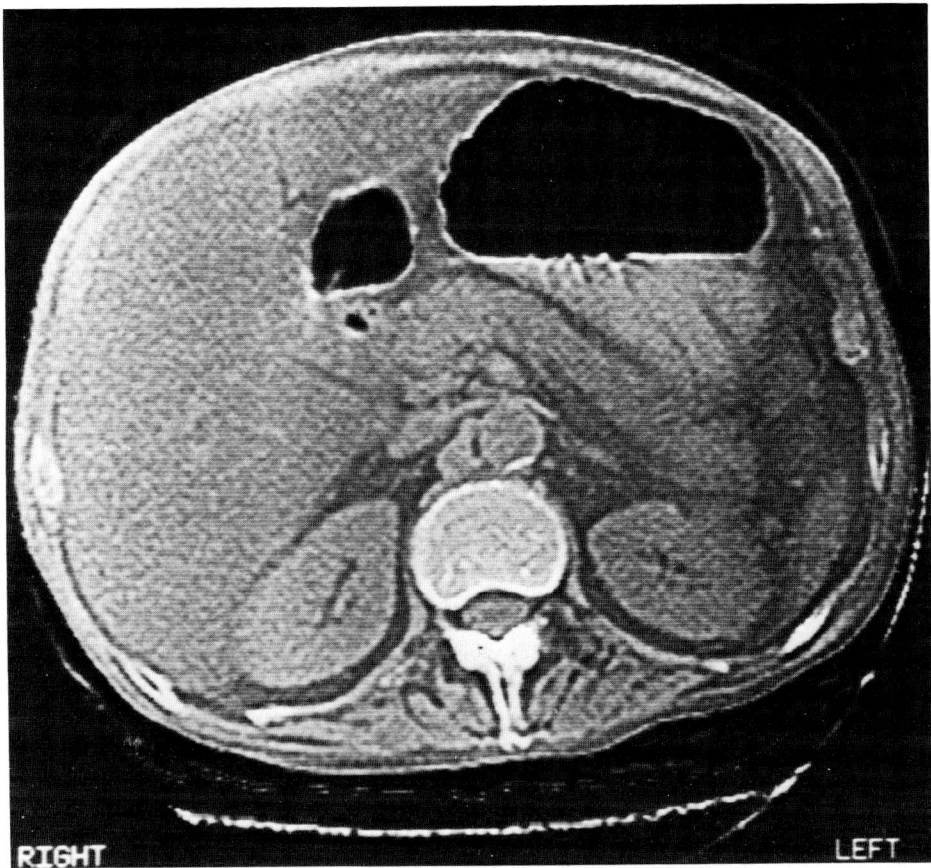

FIGURE 10.7. Computer processing of the image in the previous figure. Edges and noise are both emphasized.

point procedure. Many of these same processes can be carried out extremely rapidly by optical systems, and these are arbitrarily placed in the next section.

10.4. DEBLURRING

If it is known in detail how each point in a degraded image is smeared out or blurred, then the brightness around every point can be regrouped back to the central point in order to undo this blurring process without violating any of the laws of entropy or information theory. In many cases emphasis of high spatial frequencies is involved. The process can not be perfect because noise inevitably interferes, and, in practice, a high frequency emphasis can also emphasize photon and grain noise. If blurring has deteriorated high frequency information rather

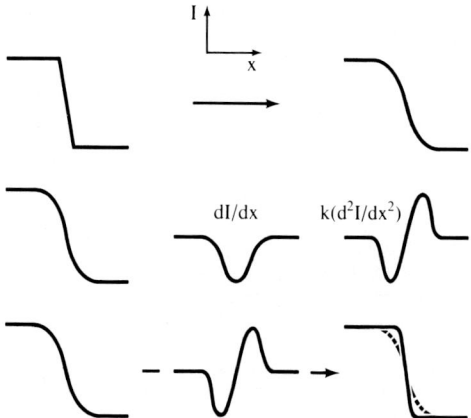

FIGURE 10.8. Blurring generally takes some intensity from a more intense region and puts it into a region of less intensity. Some restoration can be accomplished by mixing with the signal some of its own negative second derivative, though the addition of too much will emphasize edges or possibly add artifacts. A similar result is accomplished by suitable computer averaging among the intensities of adjacent points.

than totally obliterating it, one can in practice often reconstruct the image if the way energy from a known point became spread out is also recorded.

A feeling for what is involved might be communicated by a modification of Figure 10.4, as shown in Figure 10.8. Many forms of image degradation or blurring tend to take light from brighter regions and put it into darker regions, whether the cause is lens aberrations, limited frequency response of amplifiers, or even light scattering in the eye itself. In Figure 10.8 a light distribution in an image is depicted at a boundary as actually dropping abruptly while the blurred perception of it is somewhat rounded off or degraded. What one has to work with is the degraded version, and, again, one could combine with the pattern some of its own negative second derivative. If the amount chosen is correct then sharpness or crispness can be restored as in the bottom line. Again, scanning would have to be in both directions through each point, or a computer would have to make comparisons in both directions around a central point. Information about how light was spread from an object known to be a small point would aid in judging the amount of derivative to incorporate, assuming this was the same over the whole field. The image of a point source can be recorded separately to yield a "point-spread function," or the recording of a known point on the film to be processed can supply this information.

Modifying an image at every point in the same way can also be done by purely optical systems. In Figure 10.9 is shown a point source of rather monochromatic light being rendered parallel by a lens to traverse a subject on whose far side another lens brings the rays to a focus at the position of a variable opacity mask. The subject is considered as being made up of alternate dark and light regions having a variety of periodicities, one frequency being shown here acting like a

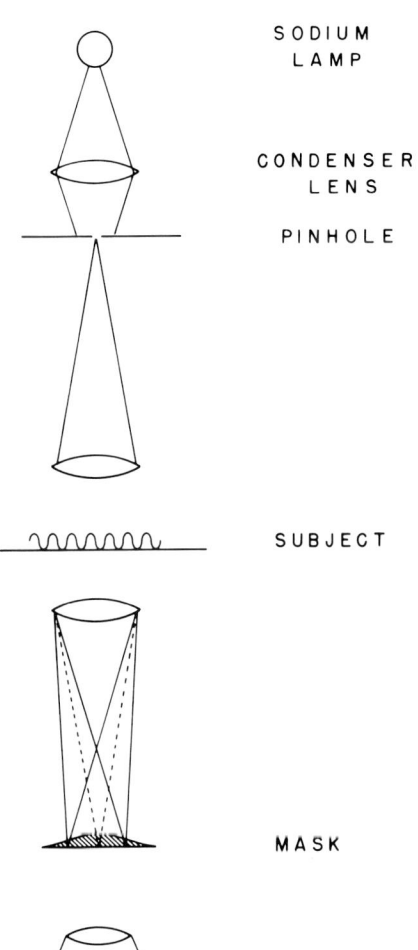

FIGURE 10.9. Deblurring or edge emphasis by an optical system that emphasizes the higher spatial frequencies in the subject. This was tried before lasers were available, and in some cases the use of incoherent light is less troublesome. The camera is focused to photograph the subject while the mask selectively blocks some periodicities more than others in the image. (From Mackay, 1962.)

diffraction grating to deviate light at the mask to both sides. The mask is shown as darker at the center than at the edges, thus passing more of the high frequencies than of light changing more slowly across the subject. This light then goes on to the camera which can be focused to photograph the subject with, however, slow variations in the subject now being deemphasized with respect to the higher frequency spatial components. In an airplane image of a city, for example, cobblestones in the street would be seen more distinctly than rooftops. This was tried before the coherent light of a laser was available and the effect was there. It was suggested, for example, that this might be used to sharpen an X-ray image made with a broad focal spot (Mackay, 1962). This has since been carried out by Krusos et al. (1970). In principle, the best resolution-enhancing filter for this case is a radiogram made of a pinhole aperture (point source) Fourier transformed by a lens onto film with photographic processing giving an overall gamma of 2, and used in contact with a hologram made at the same transform plane by including a reference wave; contrast also will be increased, and to some extent deblurring of specific planes is possible thus providing three-dimensional information from one two-dimensional photograph. This type of processing is what was suggested for removing the blur of back projection in computed tomography in Figure 4.5.

The spreading of each point in an image into a point-spread function overlapping its neighbors as a result of the imaging process is often described in terms of the spatial frequency components of the result. The Fourier transform of the point-spread function is called the optical transfer function in lens systems, for example, and the Fourier transform of an image is multiplied by this to obtain the Fourier representation of the blurred image after passage through the lens. Higher frequency coefficients often are attenuated, which accounts for lack of sharpness, and some Fourier coefficients may have become zero to yield irretrievable loss of information. Multiplying the Fourier transform of the blurred image by the reciprocal of the transfer function of the lens system should give the frequency representation of the unblurred image, which can be converted from frequency components back to an image by a Fourier transformation. Noise will be especially troublesome at spatial frequencies where the transfer function approaches zero, since there the contribution from the original scene may be less than from the noise. A blurred image will have some frequency components missing (the zeros mentioned above), and, with normal scenes, the spacing of these missing frequencies gives information about the nature or degree of defocus and blur. General methods for determining the nature of blur from a degraded image are being investigated by a number of groups. At the present time these complex general methods are interesting, but most workers in practice find some "filter" (weighting numbers in Fig. 10.5) that is useful for them and routinely apply it to those images that would seem likely to be improved by processing.

Resolution enhancement generally increases with increasing signal-to-noise ratio, no matter what imaging modality is involved. A specific example with ultrasound is indicated by Vollmann (1982). Degradation of this ratio can reduce performance by an observer and, in more drastic cases, can render a signal useless.

One may have an image with a lower contrast than is desired, because the average illumination level is large compared to the variations in illumination level. Reducing the average illumination level can increase the contrast of the image. In Figure 10.9 this can be done by placing an opaque spot, or partially transparent spot, on the optical axis at the position of the mask. This will remove all or part of the "dc component" or average illumination. Thus contrast enhancement can be done by such an optical system by reducing the average component of the signal. Similarly, either an opaque or a transparent annulus can be placed at the mask position to either pass or reject any spatial frequency or periodicity in an image.

Such methods are extremely rapid. Parallel processing of the entire field is involved, and the required time is only some cycles at the frequency of light.

Spatial filtering methods can be used to undo the effects of repeatable sources of blur and to emphasize edges. Most recent books on optical processing have a chapter on spatial filtering since the availability of lasers facilitates the methods (while introducing some problems due to interference effects).

If a laser is to be used, there are a few practical points to be observed. Between the laser and the collimator lens can be placed a condenser lens focused on a pinhole aperture that effectively filters out irregularities in light due to dust and so on. Illumination is generally not uniform across a laser, but one can use the center 10% of a laser beam for uniformity to 10%. Sufficiently accurate lenses are expensive, and it is generally cheaper to use just the center of a larger lens for a sufficiently good figure.

One point should be emphasized with regard to Figure 10.9. The brightness-changes across the subject having one frequency are shown as continuous and unbroken like a diffraction grating, but this is not required. A single pair of points will deflect light appropriate to their separation and another pair elsewhere with the same separation will similarly deflect light independent of their position with respect to the first pair. Thus the presence of both pairs can be revealed or concealed by placing one dot or hole in the mask (actually two holes or dots, or an annulus for cylindrical geometry, because a spectrum appears to both sides of center). Indeed this was suggested as a practical method for measuring the average diameter of a random grouping of small objects such as hairs or red blood cells (McNicholas and Curtis, 1931) based on a similar suggestion by Thomas Young in 1824, who had applied it to measuring particles in blood, milk, dust, and barley smut. (Obviously, a single red corpuscle does not give a very bright pattern when its size is being measured.) Such ideas have now evolved to the automatic screening of exfoliated cervical cytological samples (Pap smears) where abnormal cases are infrequent and observer fatigue an important factor (Pernick et al., 1978).

Considering a lens as producing a Fourier transform was mentioned in Chapter 3. In this writing an attempt has been made to use the term frequency analysis to avoid distracting those less accustomed to thinking in terms of Fourier transformations, but for many it is a useful concept (e.g., Bracewell, 1978). Computers can calculate a Fourier transform, and there are special procedures with certain limitations that are able to do this rapidly (the "fast Fourier transform"). This can be used, for example, to analyze the free induction decay signal coming from the receiving coil in a NMR observation. It can be considered as one dimensional since it is a sequence in time. But in the present image considerations, frequency components both up-and-down and side-to-side are to be determined, and some rejected in order to emphasize edges or to reject large overall brightness variations that are of low frequency. This requires a two-dimensional Fourier transform followed by some attenuation of certain frequencies, followed by conversion from the remaining combination of frequencies back to an image. The lenses in Figure 10.9 effectively could do this extremely rapidly, but the accuracy of computer computation is sometimes desired. A compromise method for rapidly producing a one-dimensional or a two-dimensional Fourier transform is indicated in Figure 10.10. The electrical signal carrying information is superimposed on a carrier

214   IMAGE PROCESSING

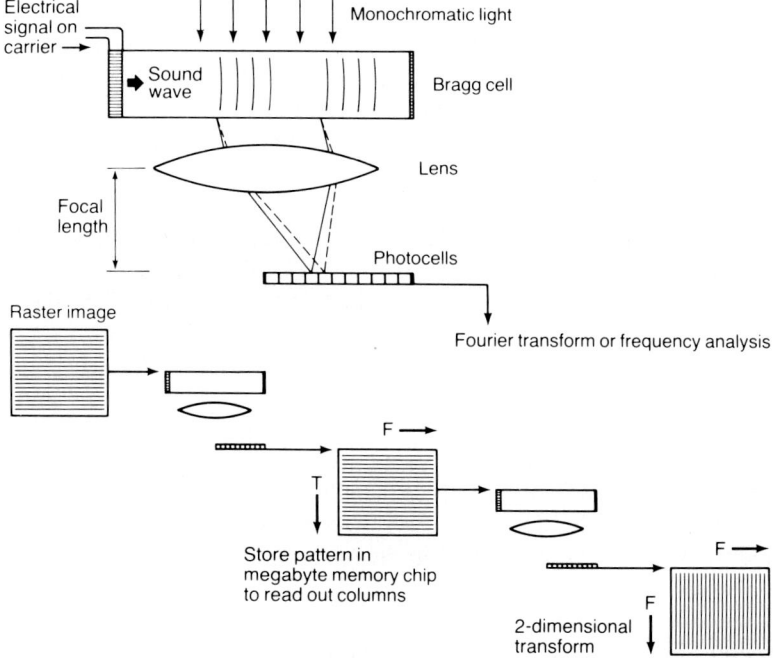

FIGURE 10.10. Conversion of an electrical signal into a sound pattern controlling a pattern of light allows the rapid production of a Fourier transform in the form of a signal from a row of photoelectric cells. The signal can be read out from these either in sequence or in parallel. Two such units separated by semiconductor memory provide for the rapid taking of a two-dimensional Fourier transform.

wave of high frequency. This is applied to an ultrasonic transducer to set up sound waves in a transparent tube. The compressions and rarefactions of the sound will act like a diffraction grating for light shone through the tube, and thus two different frequencies of signal will be diverted to two different positions on a row of photoelectric detectors (perhaps a charge-coupled device). The pattern of waves need not be moving along the tube but it can be. The photocells thus deliver a frequency analysis or Fourier transform of the electrical signal. Signals can be read out of the photodetector in parallel or in sequence.

A difficult problem has been how to deal with a two-dimensional situation. At the lower part of the figure an approach is indicated that retains the extreme rapidity of the basic frequency analysis produced by the method at the top of the figure. The image is assumed available in electrical form, probably by scanning in a sequential raster pattern. Applying this signal to the acoustic cell (Bragg cell) indicates the frequency components present at successive instants of time. This electrical signal can then be applied for momentary storage to one of the large-scale integrated memory chips that has recently become available. This is sequentially read out in the perpendicular direction into another acoustic cell in order to obtain the frequency components in the up-and-down direction. The use of the integrated circuit allows one to "turn the corner" and produce a two-dimensional Fourier transform as indicated. This signal can be further worked on optically or by computer. Extension to three dimensions can be visualized by considering columns running into a series of planes parallel to this one. This scheme has many applications for the rapid analysis of data produced in computed tomography or in some of the nuclear magnetic resonance imaging methods.

With regard to the objectives of this section, Figure 10.7 can be considered as the approximately deblurred version of Figure 10.6 since part of the smearing of limited geometrical resolution is corrected. As mentioned, most workers determine some filter that is helpful and routinely apply it to all cases rather than making a complex determination for each image of what might be ideal.

10.5. SMOOTHING

At the top of Figure 10.5 was indicated a process by which the indication at each point was averaged equally with its neighbors to yield an image that could have been produced by scanning the subject with a larger spot. This obviously reduces the fineness of detail that can be seen. But it is useful for finding defects in a homogeneous gray organ, for example, in finding big lesions in the liver. Such averaging gets rid of all kinds of noise, and one trades spatial resolution for contrast resolution. Each pixel represents more photons without having to increase the dose to the patient.

In Figure 10.11 the result of applying such processing to the image of Figure 10.6 is seen. As expected, the effect on fine detail is the opposite of Figure 10.7. This is also the type of processing carried out in the final part of Figure 3.4. In that case, however, the reduction of displayed photon noise was unable to bring back the adrenal gland because of its small size.

Another application of these blurred images is to seeing detail in and near dense bone (e.g., the petrus bone in the head) which often appears as a white blur. One can easily store the sharp image, store a blurred version, and then subtract the latter from the former for display. A display of finer detail then sometimes is possible, as in the unsharp mask method done photographically.

10.6. CHANGES, MOVEMENT, AND DIFFERENCES

A sheet of film can be exposed in contact with an X-ray film or other photographic image in order to make a positive from a negative. If it is then developed to a gamma of about unity, its changes will be just the reverse of the original, and if held in contact with the original, the two will about cancel to give a uniform opacity. This positive could be placed in contact with a negative made earlier, and if any changes had taken place these would be noticeable by not canceling. Differences between the two images become noticeable whether they are due to changes, movement, or intrinsic differences between two objects to be compared.

216 **IMAGE PROCESSING**

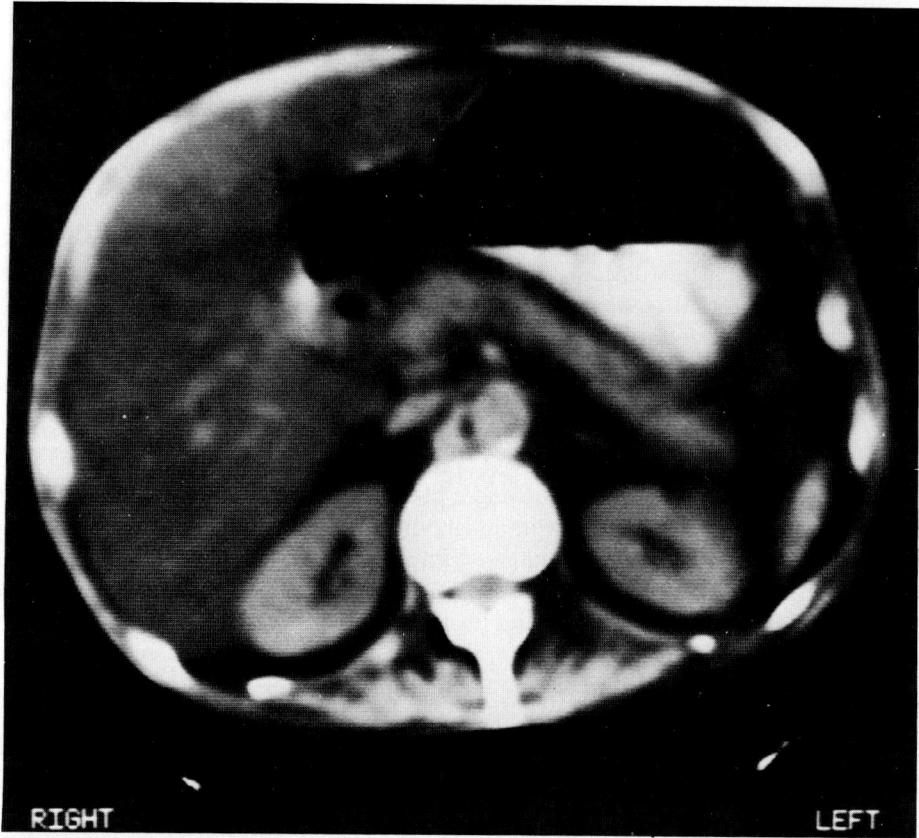

FIGURE 10.11. Weighting the points in computer processing of an image for smoothing gives this effect from the image of Figure 10.6. Noise is reduced and nonuniformities in the liver become more noticeable.

The methods are generally not accurate enough for the purposes of Chapter 8, where the image-changes in going from one wavelength to another are small, in which case low noise and excellent stability or accuracy are required for comparability. The same processes can be done electronically, and this will be discussed in Chapter 11.

Good resolution in space can be achieved by photographic procedures, and an example of this is seen in Figure 10.12. In this case what were to be compared were the original image of a person's chest (Fig. 10.2) with an image of the same person after injection of iodinated contrast to outline the various arteries. A negative of one view can then be printed in contact with a positive of the later view to show only the distribution of the iodine. The result is seen in Figure 10.12. The human aorta is again seen with its three branches, but the paths of many other vessels also are noticeable. Even the ribs cancel well, though the registration of the two films is not absolutely exact everywhere. (The small group of streaks at

CHANGES, MOVEMENT, AND DIFFERENCES 217

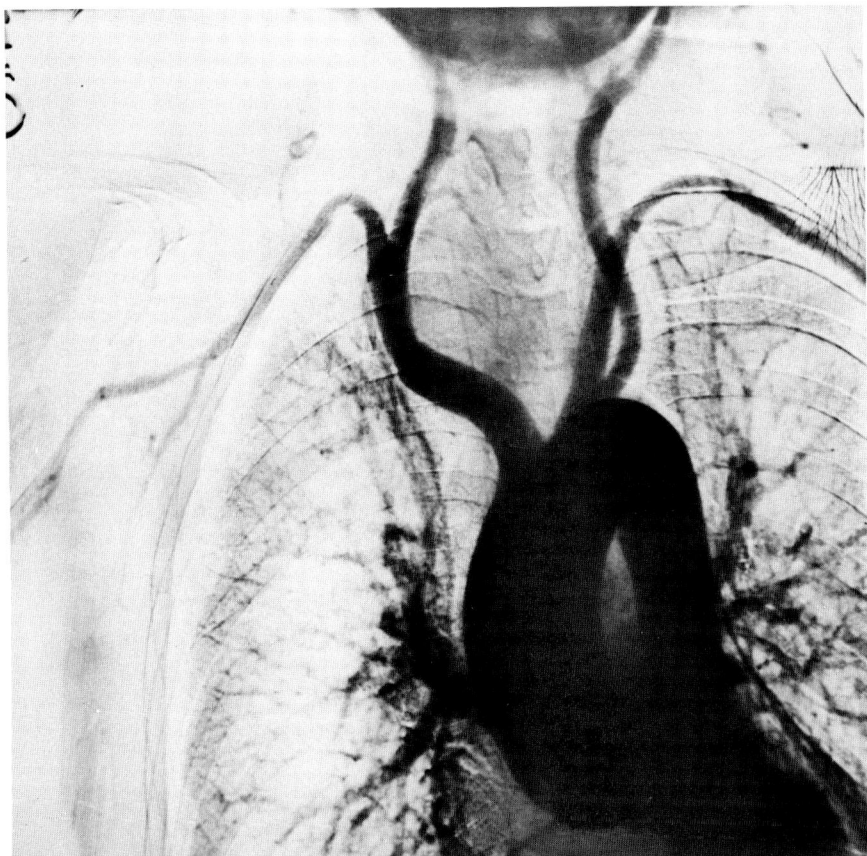

FIGURE 10.12. The image after injection has subtracted from it the image of Figure 10.2 in order to make changes (injected vessels) more noticeable. The artifact at the upper right is an electrical discharge associated with moving the film.

the upper right is the effect of static electricity producing a pattern as films are moved.) The result of an equivalent electronic process will be given in Chapter 11, where film storage is replaced by digital electronic storage and then subtraction.

Simple photographic methods have also proven valuable in following the progress of changes as they evolved. In Figure 10.13 are seen some further details of the ultrasonic equipment for examination of the guinea pig legs such as were seen in Figure 6.22. The initial sweep quickly produced a Polaroid print that could be inserted into a holder built to fit on the front of the display oscilloscope. A sheet of glass was positioned as a partial reflector so that this reference photo would appear superimposed upon the scene on the cathode ray tube being viewed through the glass. Changes in this complex scene became immediately noticeable (Rubissow and Mackay, 1971). Another way of observing changes in this case was to record the average reflecting properties of the whole cross section of the leg for

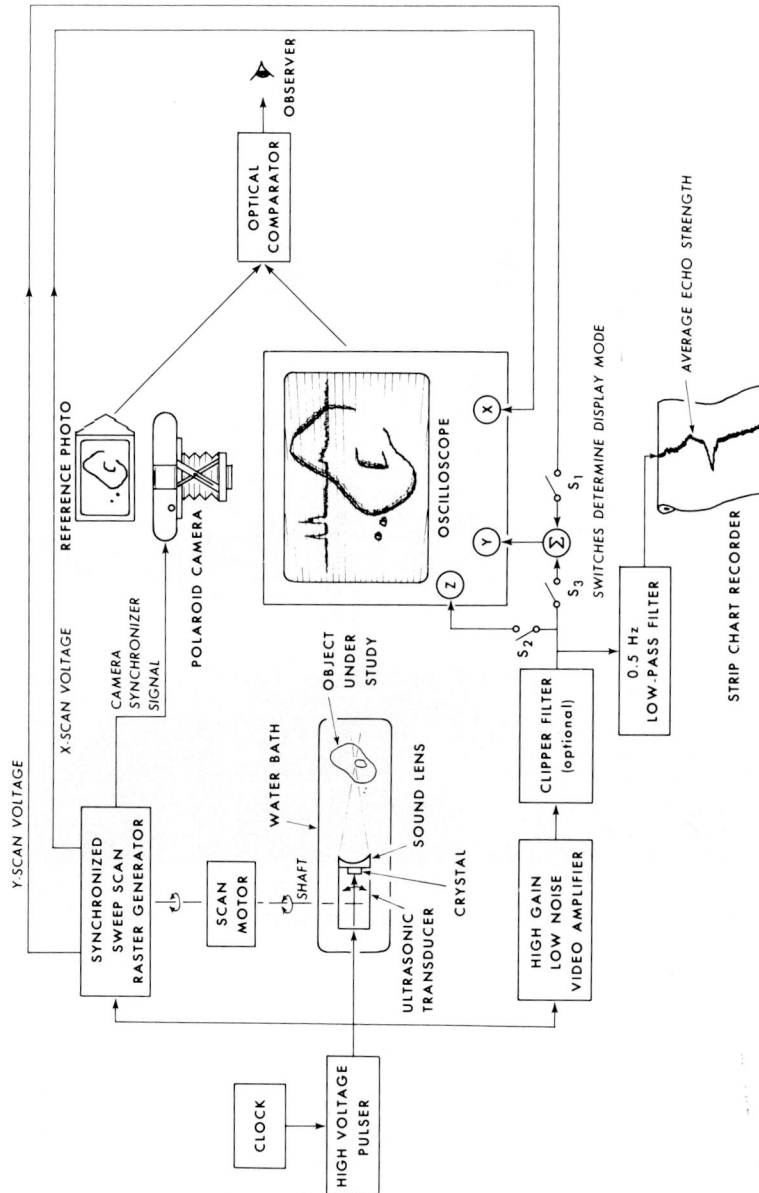

FIGURE 10.13. The ultrasonic system used to produce some of the previously seen images of the formation of tiny bubbles during decompression. Rather than complicated subtraction and storage circuitry, it was sufficient at that time to use a reference photo inserted into the display oscilloscope to make noticeable the appearance of any new echoes. Both A- and B-scans were available, as well as an averaging recorder to indicate overall changes or comparisons. (From Rubissow and Mackay, 1971.)

a complete scan, graphed as a slowly changing line on a pen recorder. Whatever the position of the line, an increase in the number of echoes resulted in an increased deflection of the trace, thus making noticeable any changes. These two simple recording techniques were invaluable at a time when electronic storage and display were much more difficult, and under some circumstances they can still be convenient.

10.7. QUANTITATIVE ANALYSIS

Some images can be interpreted quantitatively, and this can be helpful in evaluating conditions. There are several aspects to this.

Dimensions are often read off from ultrasonic images for various purposes. With the help of simple computer displays one can read out heart chamber diameters or wall thicknesses, and derived calculations can indicate left heart volume or cardiac output based on shape assumptions; abnormal patterns of motion can be quantitated by indicating slopes in M-mode displays. Estimates of fetal age are quickly determined from fetal dimensions such as head diameter or crown-rump length. Such examples can be implemented with nothing more than a ruler if the magnification is known.

The spectral methods of Chapter 8 allowed quantitative evaluation of the amount present of a particular element, as in Figure 8.4. The electronic signal that produced this figure could be measured directly or, after suitable calibration, the information could be extracted from the printed image by measuring the darkening at each point with a microphotometer. Later it became routine to measure attenuation at a point rather than the average along a column through the subject by making observations from several directions (as in computed tomography or the method of Oldendorf mentioned in Chapter 4). Again the equipment had to be stabilized against the effects of any drifts. When this was done the idea of Hounsfield numbers became a reality. These matters have been discussed in the previous chapters so that more need not be said here.

Certain types of displays lend themselves to relatively accurate intercomparisons between parts of an image to the extent that they may be used quantitatively, if desired. In Section 12.2 the conversion of a black and white image into various colors will be mentioned, with each hue being made to correspond to a given gray or brightness level. It is essentially impossible to compare two separated gray regions without making measurements upon them but the brain can, so to speak, remember different colors. Thus two regions that are red have the same degree of some property. The more complicated color images of Chapter 8 have similar properties, and their effectiveness can be predicted from a chromaticity diagram. There is a type of deflection display that has this same property of allowing for relatively accurate intercomparisons between regions rather than the mere noting of a change, and this will be mentioned presently.

In Figure 10.14 is seen the display on an oscilloscope of the television signal being recorded in an X-ray motion picture process (Mackay, 1961 or 1959). Across

220   IMAGE PROCESSING

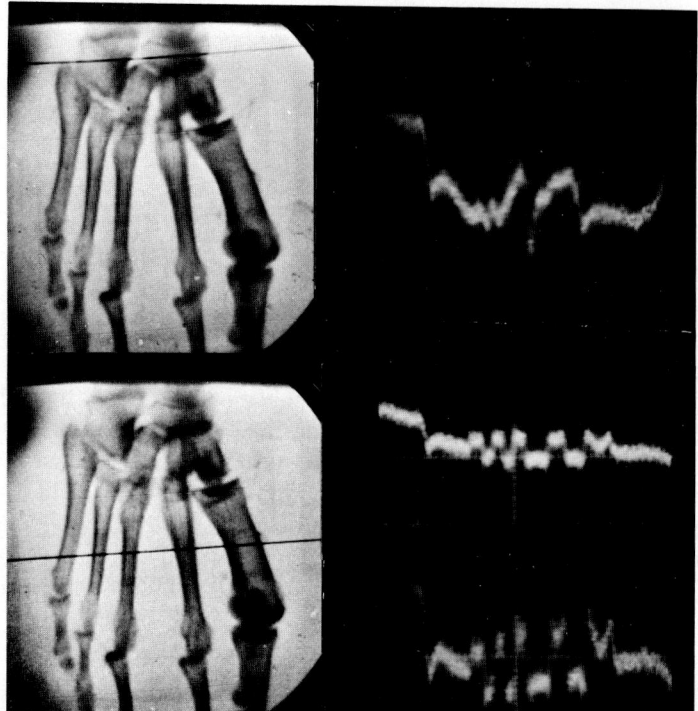

FIGURE 10.14. On a television display of an usual X-ray image it is simple to make appear upon an oscilloscope the intensity versus position along any chosen line. This aids in choosing X-ray intensity and amplifier gain, and also provides for the quantitative comparison of detail with small contrast. (From Mackay, 1961.)

the image appears a single dark line on the oscilloscope, and to the right appears the pattern of intensity along this line. Increasing darkness is upward. In its original use, the brightness pattern was properly matched to the exposure range of the photographic film by adjusting the amplification and the average value so that the trace (or the particular part of it of interest) fell just between two lines on the oscilloscope. However, different brightness levels can readily be evaluated, and this was specifically suggested as a way of making radiology more quantitative (Mackay, 1962) though overlap of structures does not allow this to be done as neatly as with a CT image. With this method one can emphasize a particular structure in a systematic way, or alternatively be sure of including all structures of interest. Also, if X-ray intensity is separately set so that the darkest region of interest matches a predetermined intensity then structure there will be statistically adequately defined, as it will be in all brighter regions. The intensity recording is somewhat analogous to an A-mode scan in ultrasound, and the effect of displaying many such lines under each other to produce a different sort of quantitative image will be discussed in Section 12.3. From an image in this form one can make quantitative comparisons between regions that are adjacent or not adjacent.

An entirely different type of quantitation of an image involves analysis of the statistics of its components in order to measure how properties distribute themselves. This can be done with any image and can be easier than might be expected. The field of stereology deals with exploration of three-dimensional space when only two-dimensional sections through solid bodies or their projections on a surface are available. Thus far the greatest application has been to analyses of three-dimensional structures by reference to two-dimensional polished surfaces (geology and metallurgy) or thin sections (microscopy) or projections (astronomy). In 1847, for example, the geologist Delesse showed that the volume fraction of ore in rock could be estimated from the area fraction of ore on a random cut surface of the rock. It should be obvious that the fraction of the liver occupied by metastases as seen in a CT slice image could similarly be considered. Aspects other than the volumes of organs can similarly be considered, for example, the true vascular length of opacified vessels in an angiogram can be accurately estimated from a single ordinary X-ray image (radiographic projection).

A good general treatment of the field is the book by Underwood (1970). The comments here are based on an excellent, and potentially important, two-part article by Reid (1982a) which treats the applications of these methods to radiology; both CT or ultrasound and traditional projection radiography are included. An abbreviated treatment of these matters in a clinical setting has also been given (Reid, 1982b).

The character of the approach can be seen from Figure 10.15. The area of an irregular figure can be approximated by laying a grid pattern over it, and counting the number of squares (of known area) falling within the pattern. If less than half a

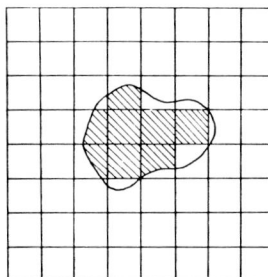

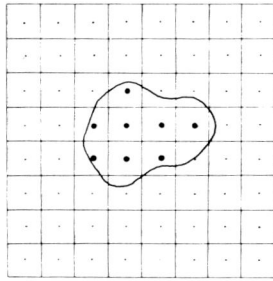

FIGURE 10.15. Counting squares to estimate an area is equivalent to counting the number of dots falling in that area, the latter being considerably easier. An application of the analysis of X-ray images is seen in the next figure.

square falls within the pattern it is not counted, while if somewhat more than half a square is within the pattern then it is counted, as at the top part of the figure. If the squares are reasonably small, the template of squares can be laid down at random and give a fair approximation to the area, but the method is somewhat tedious. Below is seen a much easier equivalent method in which the template is made up not of squares but of dots spaced so that they have the position of the centers of the squares that otherwise would be involved. It is then merely necessary to count the number of dots falling within the boundary of the area to be estimated.

One merely has a piece of transparent film, upon which are inked dots equally spaced in both directions, and this template is laid over an image to be analyzed, as in the bottom half of the figure. Here the equivalent squares are shown, but they would not appear on the template. The template could be laid upon the image several times and the dots counted to be certain that there was nothing peculiar about the orientation of the first placement. Successive values should agree closely if the pattern is uniform in the sense of being isotropic. Actually, the points need not be regularly spaced but can be random, and the resulting template calibrated on a known area, though the variance in measurement will then be larger.

Counting dots is the easiest and fastest method of area measurement in many cases. It can be applied to CT sections that are thinner than the object to be observed, and these sections can be spaced either uniformly or at random. The number of dots falling in the liver, for example, is a measure of the area of the liver in that section, and from several sections it is possible to compute the total volume of the liver. Similarly, the number of dots in metastases can be counted to evaluate the degree of invasion, or the fraction involved. In a single representative section, for example, Figure 10.16, if there are 11 points intersecting liver and 3 points intersecting metastases then the fractional volume of liver replaced by metastases is 3/11 = 27%. Similarly, the fraction of lung replaced by bullae can be evaluated in emphysema. From such observations it should be possible to quantitate the effect of chemotherapy on cancer.

If absolute volumes rather than ratios are of importance in evaluating CT images then it is sufficient to once count the number of points on the scan circle of known area (Figs. 4.7 and 4.15), and to compare all subsequent counts to this number as a reference. It might be questioned if automatic methods were not easier than counting dots. It is perfectly possible to have appear on the television display of a CT image a brightened point whose position can be moved about the image by the operator. If this point is guided around the periphery of a structure, the computation of the enclosed area is easily made automatic, but the counting of dots is generally faster and easier than outlining an edge. Similarly, the use of a planimeter is not as easy. It is also possible to automatically add all the areas in which the CT number (Hounsfield number) is the same, but this is not always expected to be good enough since it is often necessary for the observer to inform the computer which region is which by outlining shape when absorptions are similar.

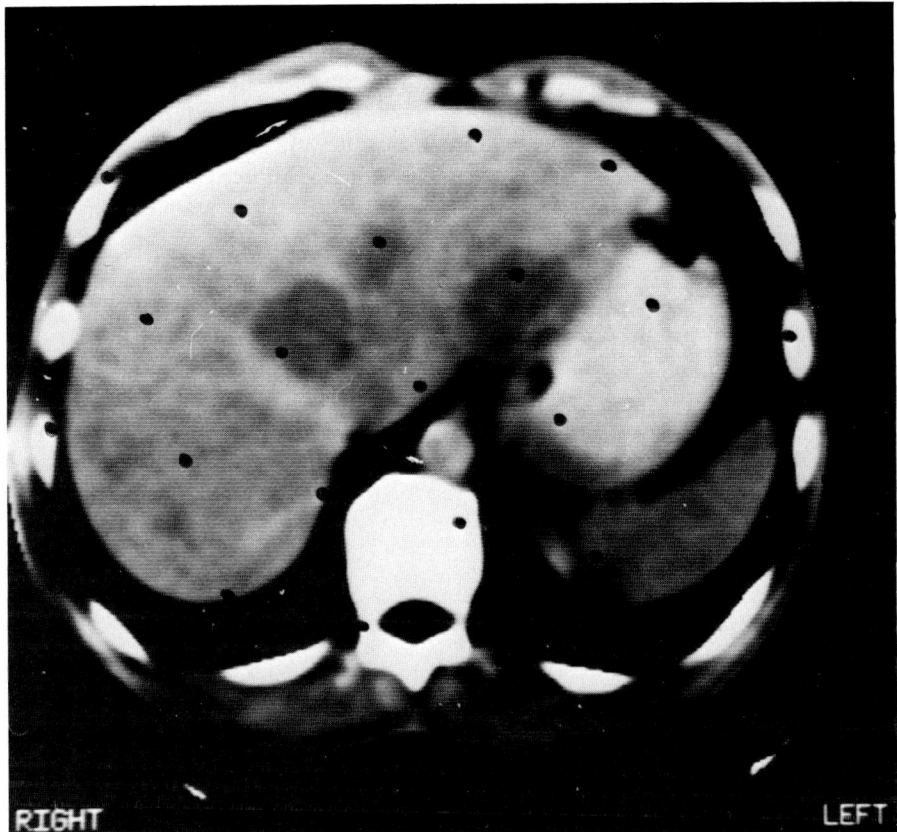

FIGURE 10.16. Random placement of a template of dots over a CT image here places 11 dots over the liver (the structure at the right of the figure being the stomach) and 3 dots overlying large low density metastases that appear dark, indicating that in this region about 3/11 is metastatic. In this image the portal vein appears light while the vena cava is still dark following intravenous iodine. Finer dot spacing would give an improved estimate.

From such techniques one can estimate the volume of liver, kidney, or brain, and also determine fluid and air volumes. It is possible to estimate the volume of thoracic gas without a body box.

Counting dots falling on an image of an organ in a classical transmission radiograph allows estimation of the projected area. For a convex object (no dents) the surface area is four times the average area of several projections in different directions. This is easy to remember from the special case of a sphere whose area is $4\pi r^2$ and every projection is a circle of area πr^2.

Similar methods will allow evaluation of average number, average diameter, average area, and average curvature. The last would have implications for the passage of therapeutic agents into a lesion, where the ratio of surface to volume of a region is important. It becomes a simple matter to estimate the length of tumor vessels, for a given diameter-size range, per unit volume of tumor compared to normal tissue. Traditional projection images (as opposed to CT) are useful in analyzing linear features such as occur in angiography, bone trabeculae, bronchovascular markings, and so on. Templates with other than dot patterns may be helpful in studying the differences between normal and tumor vascular distributions (perhaps anisotropic vs. isotropic); every angiographer recognizes vascular "irregularity" as one possible sign of malignancy, and it may be possible to develop an even more precise definition of this factor.

Such methods will not replace or even compete with the detection or differential diagnosis of pathology demonstrated radiographically, but they should help in quantitating the often qualitative observations of diagnostic radiology.

11

DIGITAL METHODS

Digital radiography, as with other modalities, is an electronic way to store images using more efficient electronic detectors. In the X-ray case, there is generally less spatial resolution, more brightness resolution, more linearity of response, slightly larger equipment, freedom from development time, greater meaning to "negative light," a convenient form for processing by computer, no positive prints required for subtracting, and reduced film costs. The ideas are not difficult to understand.

In most electrical circuits a steadily increasing voltage is accompanied by a smoothly increasing current. In a digital circuit there generally somewhere is positive or regenerative feedback from the output back to the input, to the extent that a small input change produces not a smooth continuous response at the output but rather a sudden discontinuous rise or step. As the feedback is increased further, the return down of the output does not take place at the same value of input as the upward switching, and there is hysteresis or memory. That is, switching up and down at different values of an applied signal means that the present condition in the circuit depends on what happened last, and this is a two-state memory. The feedback in digital circuits can also result in a current that decreases over some range for a steadily increasing applied voltage. This response can be construed as due to an electrical "negative resistance," such as allows living nerves to generate sudden impulses, and many electronic devices are able to show this property (Mackay, 1958c). The storage of numbers in calculators involves such circuits. The suddenly changing voltages can be used to control the storage of charge in capacitors or the orientation of a recording electromagnet, or to switch other currents, but their presence is characteristic.

Such switching circuits allow the subdivision of a voltage range into a series of

226 DIGITAL METHODS

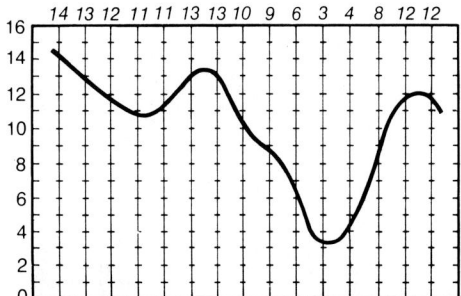

FIGURE 11.1. A changing voltage being represented by a succession of numbers (across the top) at successive instants of time. Fifteen steps of voltage plus zero are the possible digital values in this case, indicated at the left. In a typical calculator this requires 4 binary counters to store 4 bits.

steps that can be electronically counted in order to store the magnitude of an applied momentary input value. Storage can be done with arbitrarily high accuracy (with a sufficient number of steps) and for any duration. Once the input is converted into a number it is difficult for noise to affect indications since only certain values of voltage are allowed in these circuits, that is, when the signal is converted to a number that does not change, it is less likely that subsequent processes introduce noise.

In a set of counters can be stored the value corresponding to brightness at each point in a television image, and recent developments in integrated circuits readily provide many components in a small space so that this can be done cheaply and as rapidly as television pictures are normally transmitted. Various things can then be done with the stored image, including either modification or simple display. Though fast and convenient, overall performance is not necessarily better than with film storage of the same image. Combinations of counting circuits, or what radiation physicists call scaling circuits, plus switching circuits, yield digital computers.

In Figure 11.1 the use of counters to represent a voltage is indicated. The wavy line depicts a changing voltage that is to be represented digitally. As an example, it might be the signal coming from a television camera. At the first instant, counters rapidly count up through increasing voltages until at 14 they arrive at the momentary value of the input as indicated at the left. The count 14 is then stored as at the top to record the value of voltage at that instant. An instant later the count is 13, and then shortly thereafter 12, and so on. By sampling the voltage at these regularly spaced intervals its time course can be recorded, with the succession of numbers across the top of the figure describing this pattern during this interval. The pattern shown all falls within 15 steps plus the number zero representing nothing at the input. This number of possibilities can be obtained by cascading four bistable circuits (circuits which can be in either of two states, depending upon their previous input). Each time another bistable circuit would be added, the number of possible steps would be doubled. Thus the addition of one more circuit of this simple sort would allow subdivisions that are twice as fine, or equivalently allow recording twice as large a voltage excursion. With a sufficient number of circuits, any desired fineness of resolution can be achieved, including the accurate recording of noise. The numbers across the top can be read out after the passage of any amount of time to reproduce this voltage waveform.

In a typical calculator each of the numbers in this example would require 4 binary counters for storage. In the example shown, 15 groups of such counters were required to encompass the entire duration from left to right. If the waveform were reproduced from these numbers it would appear slightly different. Thus the value indicated as 10 would be made exactly 10, and the value 3 would similarly be lowered to exactly 3, while the first 11 would be raised and the second lowered slightly. This is characteristic of digital systems where a continuously changing input is quantized, and the effect can be made as small as desired by sufficiently fine subdivision in both time and magnitude. If the vertical scale is in volts, then to achieve an indication of the closest integer as shown, the switch point between zero and one would be at a half, and successive ones at 1.5, 2.5, 3.5, and so on. That is, any input between 1.5 and 2.5 is indicated as 2.0 on the recording and so on.

With a sufficiently large number of transistors, a long detailed waveform could be recorded, corresponding to the recording of an image with as much fineness of detail as desired. Several successive images would require correspondingly more storage capacity, whether the images were produced rapidly or slowly. Some electronic detectors of radiation have considerably higher detective quantum efficiencies than does film, and the recording of their electrical outputs in electrical form is appropriate. Since each input step corresponds to the same increment in voltage, such recording is quite linear even when the range is large. (Familiarity with counting thousands of dollars to the nearest penny tends to divert one from the fact that many systems do not show linearity in dealing with fixed small increments, such as is implied here.)

To store a television image having 500 lines, with the same resolution across, will involve 500×500 distinguishable points or pixels, and, at each of these quarter million, one may have to store any of 16 levels of brightness for a reasonable image, requiring $4 \times 500 \times 500$ or a million binary counters. This number is needed for each single frame or image, which must be recorded in $\frac{1}{30}$ sec (at the American power frequency), thus requiring fast counters that are able to respond to high frequencies.

There are a few practical points if the voltage being recorded is actually coming from a television camera, rather than from a set of separate detectors as in an X-ray computed tomography system. The uniformity of a television camera can be off 50% if set up poorly, due to shading and so on. Uniformity to 10% can be fairly easily obtained. A charge-coupled-device camera can be uniform to 1%, but spatial resolution is slightly limited. At the time of this writing there are units giving an image with 380×480 picture elements. The scanning spot on a vidicon is typically not very small, and so in practice they are not much better in this respect than a charged-coupled device.

Counters can count either upward or downward, that is, one can either add counts or subtract counts. This makes it relatively easy to subtract a later image from an earlier one since the counter representing one point in the image is first

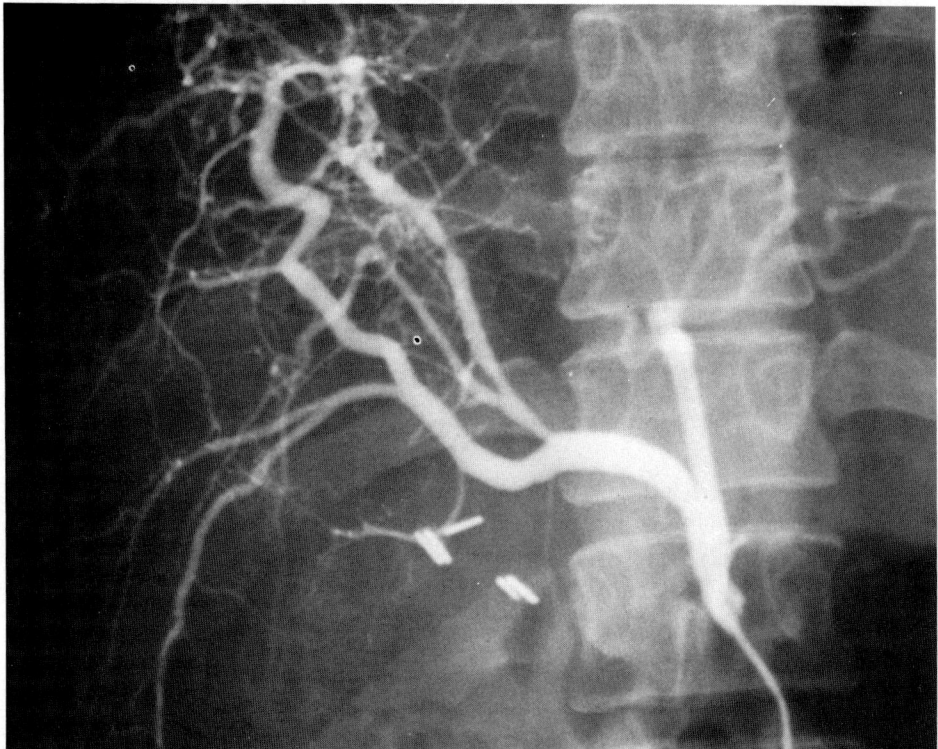

FIGURE 11.2. Image recorded before an injection of iodine, the recording being done on film.

counted upward proportional to brightness and then later counted downward proportional to the subsequent brightness. One application of this is to remove distracting background after an injection of iodine contrast medium by subtracting away an earlier image made before the injection. In that case the second image will average somewhat darker due to the presence of iodine, and intensity regulating feedback circuits within the image intensifier or brightness feedback to the X-ray source itself will produce undesirable intensity changes. Thus such circuits must be disabled in such an application.

Some radiology television sets employ a bandwidth of 15 MHz. For digital work this may be purposely reduced to 5 MHz to get rid of noise, at the cost of reduced spatial resolution. A typical angiogram may involve 3 or 4 frames per second with film, and for heart work (for coronary arteries with a movie camera on an intensifier) 30 or more frames per second. For comparison, some commercial digital units store 30 images/sec with a spatial resolution of 512 × 512 lines (essentially 262,000 points), with an 8 bit "deep" gray scale at each point.

In Figure 11.2 is seen an example of a type of image that would be subjected to such processing. It is the film of an angiogram. The catheter is through the celiac artery with the tip in the hepatic artery. Mostly hepatic arteries and liver are seen

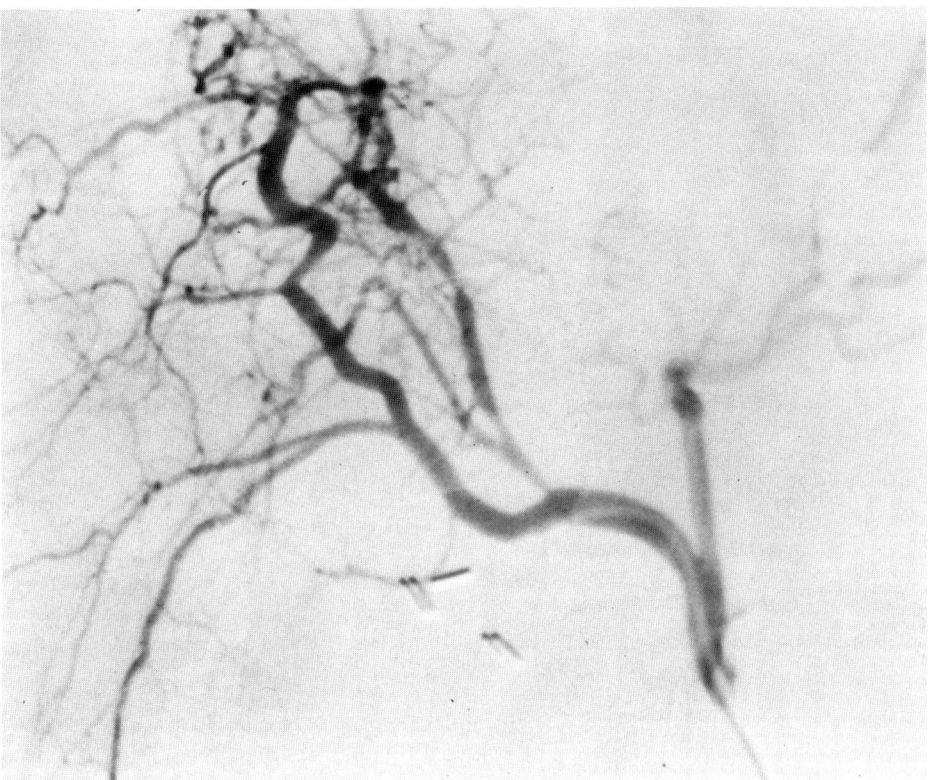

FIGURE 11.3. Photographic subtraction of the image in the previous figure from the image after injection, produced by printing a positive of one with a negative of the other.

to the left. In Figure 11.3 is seen a film subtraction from the pre-injection mask of the film in the previous figure. Printing together a positive and a negative is seen to produce an effective result, though some steps in the darkroom are required.

In Figure 11.4 is seen the digital subtraction angiogram. The same subject and sitting were involved as in the prior two figures. The injection for film was 15 cc/sec with a total of 35 cc of standard agent (76% iodine). In the digital case, the amount of iodine was reduced to one-quarter.

These subtraction techniques allow for image contrast increases by an order of magnitude, and it has been hoped that this would allow the injection of a contrast medium into a peripheral vein rather than requiring cardiac catheterization for ventriculography. Such a procedure is "relatively noninvasive," but the dilution of the contrast agent from the point of entry until left heart structures become opacified is significant. Injection into a jugular vein, followed by flushing, has been used to detect human left-to-right shunts, and to calculate left ventricular volume and ejection fraction, and to evaluate wall motion abnormalities (Bogren, 1982). The convenience of contrast enhancement by these methods is made available for

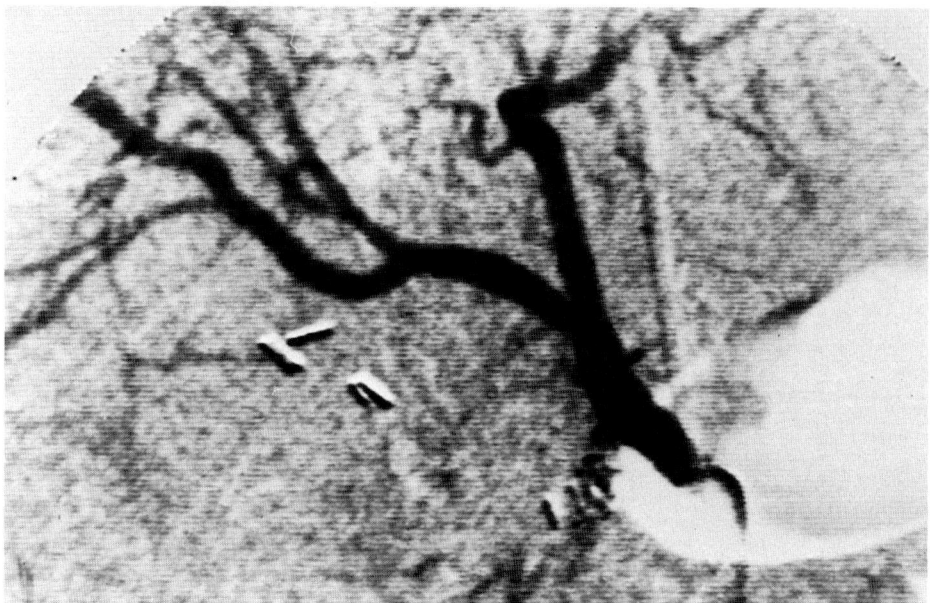

FIGURE 11.4. Digital subtraction image of essentially the same situation as the previous two figures.

rapidly changing objects by comparing successive television fields so that one acts as a mask for another. A history of some of this has been given by Mistretta and Crummy (1981). Others continue to study the possibilities of using a smaller amount of iodine in arterial injections (Miller et al., 1983). A reduced dose of iodine has the practical effect of disturbing some patients less, thereby minimizing movement which can disturb the adequacy of a prior image to act as a mask for the post-injection image.

If dichromography is done with two X-ray spectra both above the K edge of iodine (each beam having a spread of energy) then either bone or soft tissue can be made to not appear, but they cannot both be made to cancel perfectly (Chap. 8); energy subtraction followed by subtraction of an early mask from a final image is one way to improve cancellation of everything but the iodine. Some work on this has been reported by Foley et al. (1983).

It was originally hoped that injection into a peripheral vein would work. But it presently appears that the catheter must be at least in the vena cava with a large dose (50 cc) in order to visualize the pulmonary vessels, carotid arteries, femoral arteries, and aorta. If one needs to observe the fine structure of the liver and other abdominal organs, the distal extremities, and the brain, then an arterial injection is presently desirable even with digital storage of the image.

It should be noted that Figure 3.3 is, of course, digital, as are most CT and NMR images, including Figure 2.2, and many modern ultrasonic images including Figure 6.12. A certain glamor has come to surround the term digital when applied to radiology, but two of the main real advantages are the conveniently provided contrast adjustment and the convenient possibility of masking one image with another by subtraction. (Previous television systems had, of course, the option of contrast adjustment while observing, and also the acceptance of signals from more efficient detectors that would allow lower patient dose.) Attention in many cases has centered upon the masking function. Some adjustment of the position of the mask can be done automatically by the computer, and several groups are working on the possibility of slightly deforming an earlier masking image so that it will better match a subsequent image. Some local stretching or deformation hopefully could be used for compensation of involuntary patient movements. Combining dichromography with this sequential form of masking to remove masking motion and residual bone effects will have to be compared for relative convenience and effectiveness with any computer methods that evolve. Presently what is sometimes done to help eliminate misregistration artifacts associated with patient movement is to subtract two subtraction images, both formed with some original mask which now cancels out, so that the first of the two more recent images serves as a new mask for the second. Even though a mask can be moved, the subject must breath-hold during the procedure.

Though the equipment is somewhat expensive, some economies of both time and money accrue from its use. Digital methods are fast and convenient, and there is little film expense in a complete run of perhaps 50 exposures (20 films on each of 2 or 3 projections or injections).

An aspect of the difference between optical and digital computation should be mentioned. It is often stated that one can not have negative light, which is not totally true since light on a carrier (a reference beam, as in a hologram) can act as a negative quantity with respect to other parts of the scene by having a reversed phase. But what is difficult to accomplish, except in computer storage, is to take away some signal that was laid down at an earlier time. This complicates filtered back projection (modifying projections before combining them) by optical methods (Chap. 4).

Storing images in electrical form for convenient processing (e.g., contrast alteration or differencing) was another of our original reasons for introducing video magnetic tape recorders into radiology, and for some purposes they are still quite adequate for storing many successive frames or versions of an image. However, modern magnetic disks are in many cases an excellent substitute, and they can record large quantities of digital information quite cheaply. Video tape recorders normally record signals as a continuous variation in degree of magnetization, though they can be modified to record digitally. However, the mass market for computer components has so reduced size and cost, while increasing performance, that there is decreasing incentive to work without them. At the time of this writing there are standard relatively inexpensive $5\frac{1}{4}$ inch hard disk units that will store 140 million bytes (each consisting of 8 binary bits) with an average access time of 30 msec, and that can magnetically record over 5 million bits per second. For comparison, a single 12 inch diameter laser optical disk such as is used for entertainment may contain 2.6 billion bytes, though present ones are not erasable for rerecording as are magnetic disks.

12
DISPLAYS

An object is explored by interacting with it some penetrating modality, and the resulting distribution of this probably invisible modality is collected for examination. Primary processes select this modality to be as effective as possible by choice of wavelength or other parameters, and then secondary processes, including edge emphasis, are used to modify the distribution in order to make it as accessible as possible to easy observation. It is then converted into a visible display for observation. There are a number of possible displays, and they are not all equally effective in every application. A hint of some of this diversity can be indicated by displaying a single image in several different ways. The example chosen is an ultrasonic echo image of the cross section of the leg of a guinea pig using a mechanical sector scan that was sufficiently rapid to permit real-time viewing (Chap. 6). A dozen alternatives are seen in Figure 12.1. Since the formation of decompression bubbles was being monitored, a time indicator was included at the right in each image.

For comparison, the last image labeled l is the usual variable brightness display of a B-scan. Ultrasound came from the left and the echoes returned to the left through the surface of the leg which is the long curved line. Scan lines are generated at a 2.5 kHz rate here, which spaces them close enough almost to fuse together. The video signal is clipped to eliminate negative peaks, but not filtered. In this case the image of a point is distorted into a small vertical line pair whose vertical length corresponds to the azimuthal (angular) resolution of 1.5 mm, while the horizontal width of a line pair corresponds approximately to the range resolution of about 1.5 cycles at 7.5 MHz, or 0.3 mm. All the other pictures shown use a video signal that is pulse-stretched by clipping and filtering with a 2 MHz low-pass

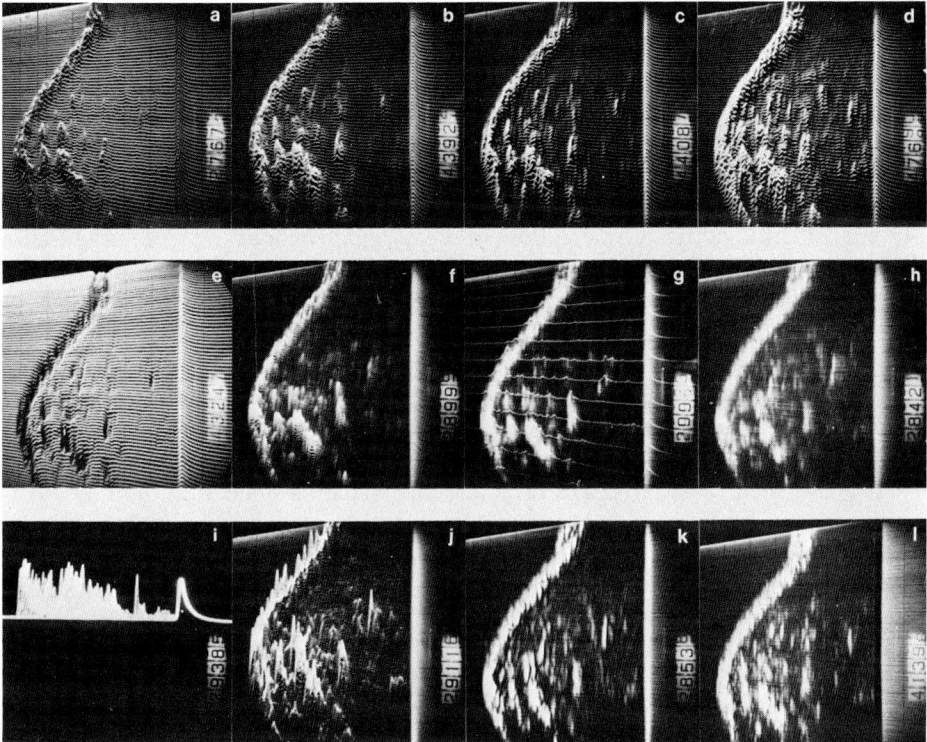

FIGURE 12.1. Various displays of the ultrasonic image of a cross section of the leg of a guinea pig. (See text for details.) A combination of brightening plus a vertical deflection proportional to returning echo amplitude was found to be most useful in this application. (From Rubissow and Mackay, 1971.)

filter. An effect of this is seen at h where the line pairs become single lines of about the same length and width as the line-pair combinations, and thus no appreciable resolution is lost. (The difference in echoes between the two images is due to a difference in time of 413 minus 284 minutes.)

The effect of mixing in some first derivative of the video signal with the modulation is seen at k, where an increase in local contrast of echoes is apparent. Also the fronts (left sides) of echoes tend to be brightened and the backs darkened, giving the illusion of shading as if illumination had been from the left, thus casting shadows. Although this is not direction-independent edging, this display helps resolve closely adjacent echoes and makes small echoes more noticeable.

If quantitative data on the magnitude of echoes are required, as is necessary in evaluating the size of bubbles, then intensity modulation is not convenient. (Film exposure, development, and storage must all be controlled, and photometric measurements of brightness made and interpreted as a nonlinear function of magnitude.) Deflection modulation in which the scan line is deflected vertically in proportion to the video echo signal gives a characteristic display with mountains, ridges, and valleys, as shown at a in the figure. It is necessary that the lines be spaced somewhat far apart or the peaks will all fuse into each other, as they do at f. Adding some intensity modulation to the deflection-modulated scan yields the extremely effective display of part b, which can be further enhanced for some applications by mixing in some first derivative signal to the intensity modulation, with the result shown at Figure 12.1c. The display of part b has brightening to attract attention to a region where something is happening, plus a deflection which allows the taking of quantitative data everywhere in the image. The image at c permits slightly better visualization of very small echoes, at the expense of difficulty in measuring larger echoes whose backsides are lost in the shadow. The bubble studies discussed in Section 6.7 generally used the display at b.

By increasing the gain setting of the video amplifier, the display at b can be made to show more detail, as at d. However, it becomes more difficult to discriminate small echoes because of the increased background information.

Large deflection modulation, with some brightness modulation, and with scan lines spaced closely enough to fuse, results in the appearance seen at j.

At e the scan raster is made brighter, and negative deflection modulation is used, together with negative brightness modulation, including a small amount of negative first derivative. The result is a "mesa and canyon" display, in which the information appears as obvious depressions rather than peaks. The valleys are darker, and the leading (left) edges are also shaded darker, to aid the eye in more easily interpreting the data.

An image may have an isolated peak, and it may be desired to measure only this one peak amplitude during an experiment. By allowing the ultrasonic transducer to continue to scan, but not applying the scan voltage to the oscilloscope, a modified A-scan is obtained as at i of Figure 12.1, where all the scan lines are superimposed on the same base line. The peak at the right is clearly distinguished, and more easily measured than it is, for example, at j, which shows the same image and modulation but with the scanning raster.

The display shown at g combines brightness and deflection-modulated images in such a way that advantage can be taken of the better localization and detail of the former, and better presentation of echo amplitudes of the latter. Here, a very widely spaced deflection-modulated image is overlaid on top of the conventional brightness-modulated image. Most information on echo location and strength is assimilated by the eye in the brightness display. However, where quantitative data are required, a reasonably good estimate can be obtained from the deflection-modulated scans. Such a display can be made to appear on an oscilloscope, or a double-exposure photograph can be made to include the two scan types.

Different returning amplitudes can be converted into a change of hue in a color display, as will be discussed later. Experiments also were conducted to determine if pairs of ultrasonically imaged photos of a target seen by the scanner from slightly different angles would convey information more readily by separate viewing with the two eyes. The effect here is purely artificial since the pair of images being compared by the eye do not correspond directly to the two views at slightly offset angles that the eye is accustomed to interpret as three dimensional; the stereo effect is recorded from one plane in the cross section being imaged and viewed from another. Slightly rotating the subject between views yielded pictures that sometimes showed peaks or ridges in depth when viewed in a stereo viewer. This may have application where it is exceedingly difficult to see the desired signal in the background. In the previously mentioned studies on bubbles, the display at b was the most useful. If deflection modulation is kept small so that peaks do not rise more than several linewidths, the brightness modulation still conveys much information with relatively small distortion in the actual profiles. A magnifier can still yield quantitative data on echo height, even with very low echoes that remain apparent because of their increased brightness. More information on all this is to be found in Rubissow and Mackay (1971) and Rubissow (1973).

With the overview of these brief comparisons completed, a few more details are relevant.

12.1. GRAY SCALE

We are accustomed to photographs containing black and white, and all shades of gray in between. X-ray images are similar in containing varying degrees of darkening corresponding to differing degrees of opacity. Most of the early ultrasonic images were recorded as a variable oscilloscope brightness falling on a photographic film, and so these also had a scale of gray in their displays. It is true that ultrasonic echo images were realized as tending to outline structures by returning echoes largely from boundaries, and thought was given to filling in the enclosed regions by the integral of an image, but this seemed not to improve performance. Thus a gray scale is the norm, and, to some extent, the more distinguishable steps of brightness or darkness, the finer the discriminations that can be made. An example of the effect of limiting the number of discernible steps on the evaluation of an image has been displayed in Figure 3.5.

A rather interesting event took place in one segment of the medical profession in the early years of the proliferation of ultrasonic equipment. Especially with abdominal scanners, during the scanning process of mechanical movement of the probe the image was collected on a storage tube, from which it could be later read out in its entirety as an electrical signal. In many cases this was printed out on a sheet of paper as "hard copy" for examination and permanent record with the patient's other records. Some of these storage tubes could retain only two conditions corresponding to fully bright and fully dark, and in addition, some of the paper recorders themselves had a very limited range of intermediates between black and white. Thus some practitioners came to regard this as the norm. With the development of storage devices that could record and display a range of brightness values at each spot, including the present extremely simple integrated semiconductor units, advertisements began to appear emphasizing gray scale. A surprisingly common question asked at lectures on medical ultrasonics in 1973 and 1974 was "Is gray scale really better?" It was only one short span in history when gray scale was not the norm, at least to some extent. The advances in technology have rendered the question somewhat irrelevant, but the answer remains—of course it is generally better.

Variations in structure actually contribute to different degrees of darkening, and this information is lost if gradations are not retained and reproduced. In some B-scan images the background texture is associated with the microstructure of tissue parenchyma, which can be considered as a signal and used as a basis for diagnosis (e.g., Kossoff et al., 1976; Nicholas et al., 1980). Structural information is discarded if gradations in darkening are removed.

Comparison steps of darkening in the form of a gray step wedge are often included to the side of an image to facilitate comparison or evaluation within the image independent of film processing, as in Figures 3.4, 4.12, and 9.8.

12.2. COLOR

It is possible to perceive more changes of hue (differences in color) than differences in brightness or shades of gray. There are advantages, other than aesthetic, to color vision. Several pieces of information can be gathered and compared by conversion to color, as was mentioned in Chapter 8. Two early examples were given in the frontispiece. However, conversion to color can also be a secondary process in which the distribution of a single variable quantity is displayed not as shades of gray but as a spectrum of colors. Rather than combining three black and white images, a single black and white image has its several ranges of darkness made to correspond to several different colors. One can visualize doing this by printing a positive dyed one color against an identical negative dyed a different color. Ever since color television sets have been readily available, this process has become extremely convenient to implement and adjust electronically.

It should be emphasized that the conversion to color of a single image accomplishes a different result and is for a different purpose than the combining of three independent pieces of information by conversion to color. The latter can bring in spectral information relating not merely to questions of absorption or other interactions, but also how this quantity varies with a change in conditions such as a shift in wavelengths. Three colors are not essential, and two differing images can similarly be combined. Two of the radiographs that were involved in making the top image of the frontispiece were the ones that were combined by projection through two projectors, one being with white light and one not, in order to give a "full color" image of the sort often associated with Land and mentioned in Section 2.6 (Walls, 1960). Color shifts in this unfamiliar scene were indeed observed, and the effect was not equivalent to placing a positive and negative of a single image into the two projectors. The fact that color was seen in this observation is itself instructive about the visual system and simultaneous color contrast (see Walls, 1960).

Combinations other than a simple change in one variable can be displayed as changes in color. Thus, variations in texture can be given different colors. Or one can modulate color by low frequency or large scale structure, and brightness by high frequency spatial variations to make inconspicuous structure more noticeable. When using a computer to process images it becomes easy at each point on the image to take the ratio of red to green to form a new red input that emphasizes contrast between features because the weak colors will become weaker and the strong stronger. In all cases a triangular chromaticity diagram can describe the result of the process, and it is known what magnitude of shift on such a diagram is required to perceive a change in color (Judd, 1935). The amplification of the signal can be increased so that a very small input change will result in any desired shift in color, though this is obviously considerably easier to do with a television system than photographically. As mentioned in Chapter 8, so many images are now routinely converted to color that the whole matter is taken for granted.

It should be noted that the excellent sharp discrimination of color differences by the eye is achieved by comparing three rather unsharp or broad overall responses to stimuli of different frequencies (the tri-stimulus curves of the cones of the retina). Similarly, images made in three rather broad X-ray or ultraviolet distributions can result in more subtle discriminations than might be expected, though absorptions in the X-ray region do not change as dramatically as in the visible region. To further emphasize this point, there is an example of the comparison of the signals from three detectors in quite a different field. Thus, Carter Collins has constructed an "artificial nose" (1980) that differentiates smells by their different effect on tiny thermistors, coated with a lipid (lanolin), a protein (albumin), and a long-chain polymer (polyvinyl chloride). The three outputs are compared to appear as a point on a Maxwell diagram (Judd, 1935), and these various smells could be displayed as various colors. The small changes in temperature are sufficiently different to group certain smells in different regions of the diagram. Some odorants showed a negative absorption (desorption), which Collins chose to depict as a different color on a black and white triangular diagram, thus allowing emphasis or inclusion of a fourth piece of information. This is analogous to the earlier suggestion of including in an ultrasonic echo image reflections of reversed phase depicted in a different color to denote a decrease rather than increase in acoustic impedance. Again is seen fine discrimination without highly specific receptors.

The colors seen in the frontispiece are not necessarily what would be seen if the mouse or hand were cut open for viewing by visible light. As was mentioned in Chapter 8, there were predominantly red mice and predominantly blue mice, depending on which dye was used on the low voltage and which on the high voltage radiographs. There is no necessity to make short X-ray wavelengths translate to short visible wavelengths, and the reverse may sometimes be found more satisfying. The term pseudo-color has sometimes been recently applied to color translation processes in general.

Too much emphasis of edges (Sec. 10.3) can introduce lines that may be interpreted as extra structure. Experience has proven that there is a strong tendency for this to happen with a black and white image that has been converted to color since any boundary with a simple change in brightness can be converted into a region bounded by a rainbow; a shift from blue to red, with all intermediates, may be regarded as parallel separate structures. Thus caution needs to be used in these processes.

Conversion of a black and white image to color does allow absolute comparisons between separated regions in an image. Thus, two similarly yellow regions must be equal in the property being displayed, while two shades of gray could not be well compared simply by looking. Even adjacent regions are somewhat difficult to compare in black and white if the line of demarcation between is not relatively abrupt.

Eight percent of males have defective color vision, and some account should be taken of their reduced capacity to distinguish color if they are among the user population. Again, a television display can easily be switched between different modes to help with this.

12.3. PSEUDO 3D

Structure was indicated in Figure 12.1a,e by an appearance of depth. This not only provided for the display of very small signals (low contrast) by a deviation in a line, but provided for the reproduction of quantitative information everywhere in the figure at once. The same method can be applied to other modalities, including visible light. Variations in density of a structure are usually displayed in X-ray images as a variable darkening, but they can be displayed as deflections in a line, somewhat in the fashion of an A-scan in ultrasound. A succession of such lines arrayed one under the other produces the appearance of changing elevation, and indeed, for a uniform body with one surface flat, this is actually a display of differences in height or thickness along the X-ray beam. In general, however, the display is pseudo three dimensional, but convenient in many cases.

If the image is in the form of a television signal, then it can be applied to the circuit of Figure 12.2 in order to modify it for presentation in this form. The three signals of brightness plus horizontal and vertical position of a point are converted to a vertical and horizontal output for display on a television screen. Turning a knob rotates the entire image by changing the angle indicated as Θ in the drawing. This can be done electronically with the multiplication indicated, but one can instead employ a potentiometer with windings spaced so that resistance is proportional to the sine and cosine of the angle to which the shaft is turned. Applying a voltage such as the vertical signal to such a variable resistance device automatically delivers a voltage proportional to the input voltage multiplied by the sine of the angle and also the cosine of the angle of the shaft. One segment of this voltage has a fraction taken out to produce the effect of tilting the entire flat display onto its side to give the appearance of mountains; it is the effect of flying in a helicopter from over mountains down into a gully beside them, with the degree of tilt of the display being adjustable. At whatever degree of tilt is desired, the apparent amplitude of the hills can be adjusted. Because of the use of such circuits to produce

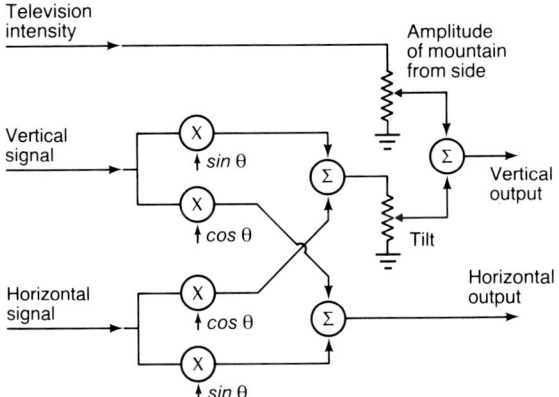

FIGURE 12.2. A standard television signal representing an image can be modified by suitable circuits to give the appearance of variations in elevation viewed from different directions. Examples of some different effects are seen in the next three figures.

special effects in commercial television broadcasts, some of these circuit configurations can be purchased already assembled.

The effect is seen in Figure 12.3. An X-ray film of a human shoulder was viewed by an ordinary television camera and the display on the television screen was as at the upper left. Neither the film nor the camera was moved in producing the modifications seen in the other parts of the figure. Some of the overall effect of turning knobs is shown, but better perception of the fine detail can be obtained from Figure 12.4 as the image is turned while the subject is fixed. In Figure 12.5 different degrees of tilt are seen, again compared with an ordinary view, while in Figure 12.6 rotation of a highly tilted image is seen.

Common experience by engineers observing oscilloscopes and fundamental studies by specialists in vision both indicate that very tiny deviations or bumps in a straight line are quite noticeable. Thus these displays are able to make accessible small changes that otherwise would require careful contrast enhancement for observation. There is the further advantage that the circuits can be stabilized to yield quantitative information about absolute values. Thus there are obvious applications of these displays to some forms of video densitometry, for example.

At the bottom of Figure 10.5 was indicated a way of computer processing an image such that a shadowing effect was produced as if illumination came from the side. (It is, of course, well known that the mountains on the moon are much better observed when the light of the sun comes from the side, rather than during a full moon.) The computed tomography sectional image shown in Figure 10.6 has been processed in this general way, yielding the result seen in Figure 12.7. The appearance of depth is obvious and can be advantageous in some cases.

Two other false impressions of depth have already been mentioned. One is to display to the two eyes a pair of images that are to be compared. An effect of

240   **DISPLAYS**

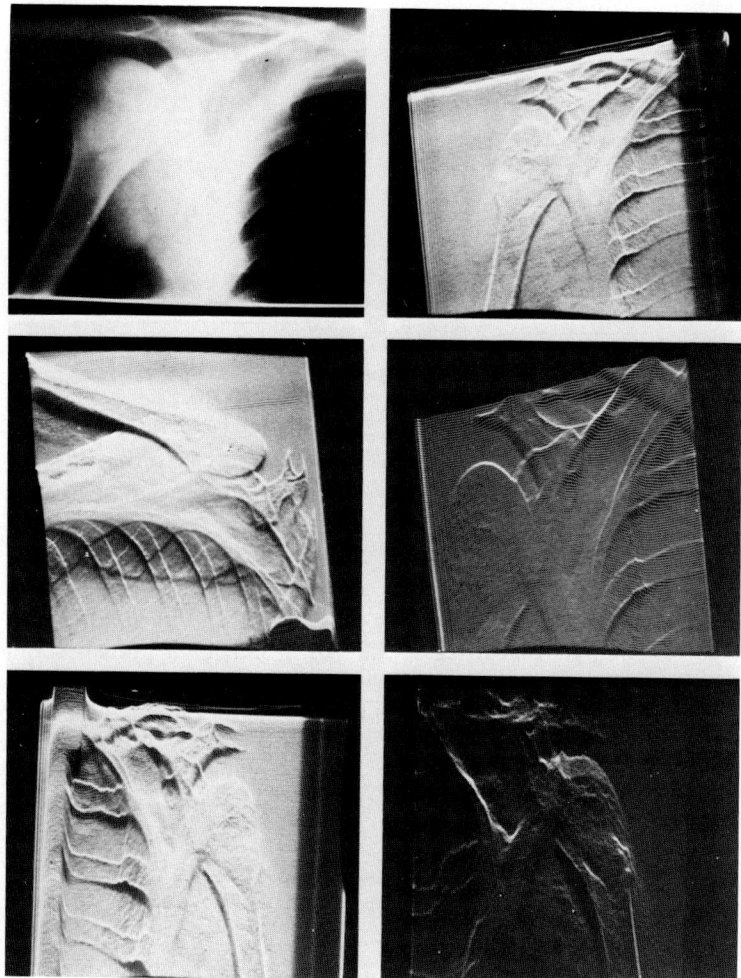

FIGURE 12.3. Rather than steadily irradiating a subject, the television camera was aimed at a radiograph of a shoulder, as seen at the upper left. Some possible modifications of the image without moving the radiograph are seen in other parts of the figure. An effect of depth is achieved, and deflections in a line have both some quantitative significance and the potential for making noticeable very small changes in intensity.

depth is achieved and can be quantitated by well known methods, wherever the two images differ. Thus the velocity pattern of a flowing fluid containing particles might be determined from two successive pictures (or rather, the projection of this pattern on the plane perpendicular to the direction of view), or a piece of counterfeit money might be identified by comparison with a bill known to be authentic. Also was mentioned the production of ultrasonic cross sectional images from two slightly different directions in the scan plane, and their eventual viewing by the two eyes in a direction perpendicular to the scan plane.

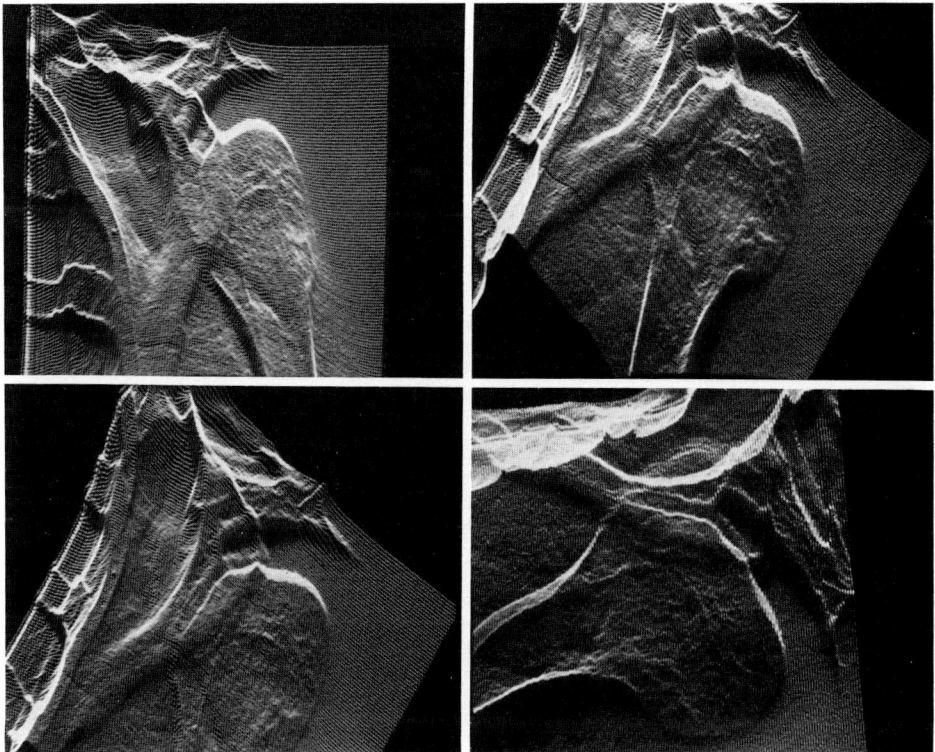

FIGURE 12.4. Other deflection displays derived from a radiograph of a human shoulder with a circuit of the sort depicted in Figure 12.2.

12.4. TRUE 3D

Many cues indicate distance and depth to an observer, but the one of interest here is the fact that the two eyes view an object from slightly different directions, thus making the images on the two retinas slightly different. This is interpreted as depth.

The effect can readily be demonstrated as in Figure 12.8. Two lights are shone upon a flat surface. If any object such as a chair is held between the lights and the surface then two shadows of the chair will appear. If one of these shadows can be viewed with one eye, and the other shadow with the other eye, then one will perceive a dark chair with depth, that is a three-dimensional shadow. In the present case it is suggested that the light from the lamps be polarized so that an observer wearing glasses consisting of crossed polarizers will see the light of one lamp in one eye and the other lamp in the other eye. In that case the flat surface should be a metal sheet so as not to depolarize the light. The shadows of the chair are views from two directions (along the beams of light) just as normal vision with

242 **DISPLAYS**

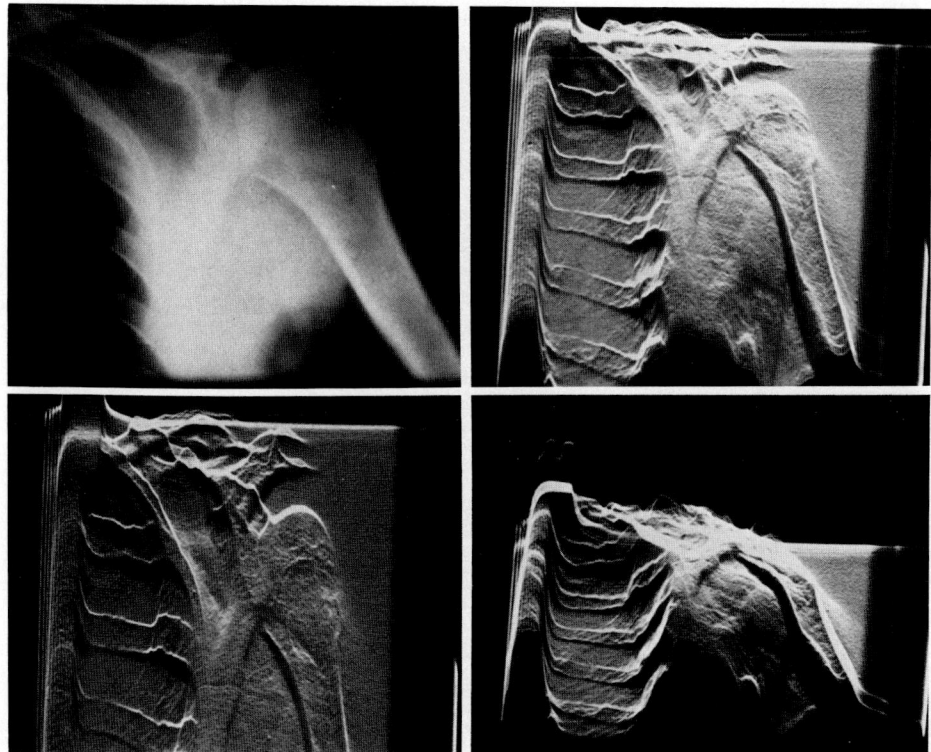

FIGURE 12.5. The radiograph of the shoulder and some possible modifications of the display at similar size.

two eyes is a view of the chair from two slightly different directions. Thus the shadows are interpreted as having depth.

The traditional way for making a three-dimensional X-ray image is similarly to produce a pair of radiographs taken with two separate X-ray sources at a small angle to each other. The projection of shadows then yields two images that can be viewed in any number of ways by the two eyes. Such an X-ray image pair of the human head is seen in Figure 12.9, while Figure 12.10 is a stereo pair of another biological object, a large shell. Suggestions for viewing are given in connection with the next two figures.

The same can be done with modalities other than X-rays. The nuclear resonance method of Figure 7.3, for example, gives an image of everything out of the page projected onto the plane of the page, and tilting the gradients shown along the page to a plane slightly different from the page (which can be done by energizing another set of coils rather than by a mechanical shift) can give an image tilted with respect to the page, to form a stereo pair. Often gradients out of the page are used to isolate a thin section parallel to the page and yield a slice image (e.g., Fig. 2.2), just as ultrasound also produces an image of a thin section, and a series of

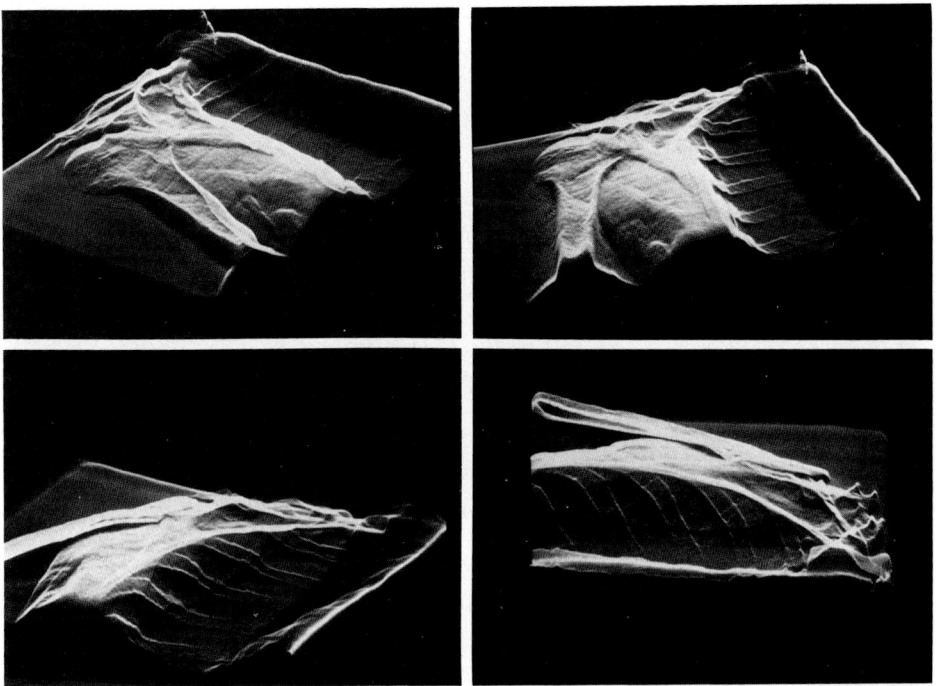

FIGURE 12.6. Apparent rotation of the deflection display when set to give the illusion of rather extreme viewing from the edge.

such slices can be combined into a stereo pair by methods described next. In the X-ray case this allows limiting the depth included to just the region of interest, and permits one to work at otherwise inaccessible angles.

There are various ways of collecting and displaying three-dimensional information. One possibility is shown in Figure 12.11. At the upper left a series of computed tomography sections is depicted as having been arranged in a stack in the same sequence as the corresponding sections were arranged in the subject. The right and left eyes of the observer are shown slightly crossed looking into the stack. This would not work at all well since various sections would get in the way of each other and the stack would tend to be quite dark. However, one can make prints of these sections as shown at the right. For the print to be viewed by the right eye, each section is printed in succession with a slight displacement to the right, while on a print to be viewed by the left eye the same sections are printed in succession with a displacement to the left. The effect is not perfect but viewing the two prints separately by the two eyes does give an impression of depth. Obviously only a finite number of sections can be included, and the fact that each is distinct seems to permit the incorporation of the information rather than just producing a massive blur at every place.

Two such images are reproduced here. In order that the reader be able to view

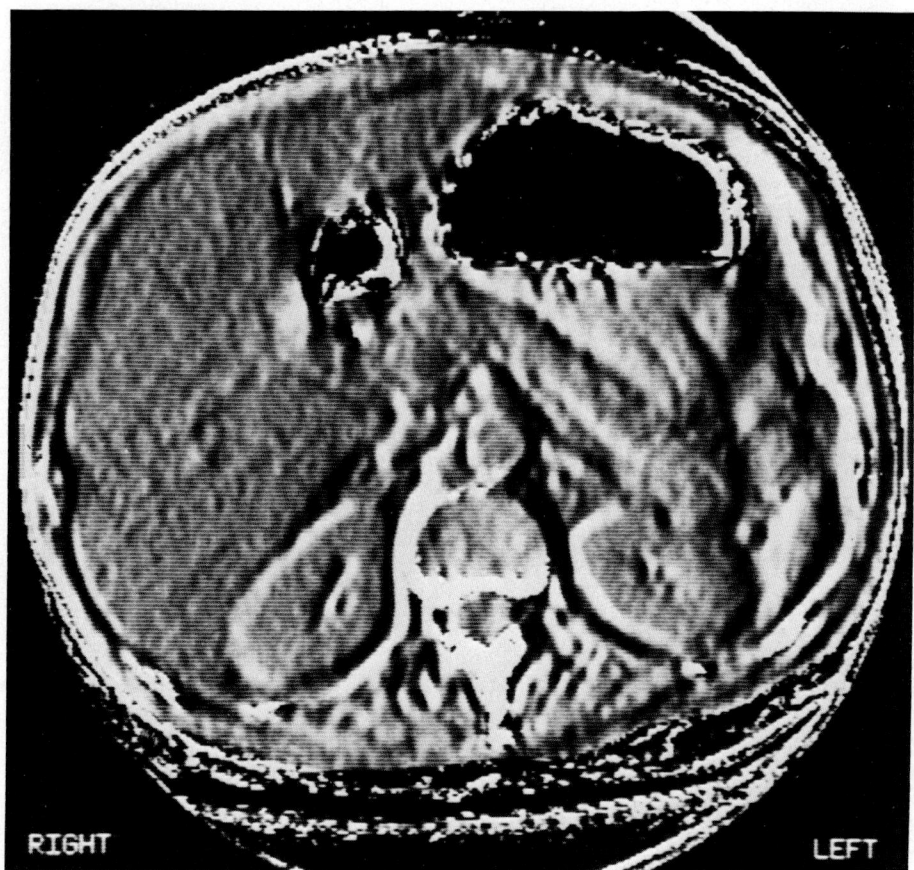

FIGURE 12.7. The impression of depth also can be given by creating the illusion of illumination from the side. Here is seen the image of Figure 10.6 as modified by computer processing in the fashion depicted at the bottom of Figure 10.5.

them without a special stereoscope, it is necessary to look at the left image with the right eye and the right image with the left eye, as is suggested in the lower part of the figure. Some readers will have mastered this technique, which is useful for viewing larger photographs as well without the help of lenses or prisms. Practice in viewing can be had by holding Figure 12.12 about a foot from the face. The index finger of one hand is then held about halfway in between and, with the right eye closed, the finger is adjusted so that it appears under the right pair of lines, as observed with the left eye. Without moving the finger the left eye is closed and the right eye should then observe the finger to be under the left set of lines. Alternately opening the eyes allows adjustment of the finger until this is achieved. When both eyes are then opened to look at the fingertip, three images will be perceived. Attention is then concentrated on the center one which will appear in depth due to the overlap of the other two images. A little practice will allow this to

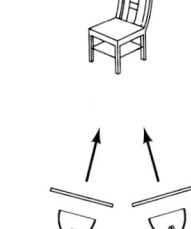

FIGURE 12.8. It is possible to cast three-dimensional shadows on a flat wall. Illumination along two directions casts two shadows, and one image is here shown being channeled to one eye while the other goes to the other eye by means of polarized light and a metallic screen. The analogy with the production of a stereo pair of X-ray images through motion of the X-ray tube between two exposures should be clear.

be done rapidly with any appropriate stereo pair. The result is the desired one indicated at the bottom of Figure 12.11.

Eight CT slices were combined in the manner of Figure 12.11 to produce Figure 12.13. Each section was 2 mm thick and they were spaced every 6 mm (4 mm between edges). The original films, on which the width of the thyroid cartilage was 40 mm, were each shifted $\frac{1}{3}$ mm. When viewed as has been suggested, one obtains the effect of looking down along a human larynx. Similarly, from cross sectional images, one is able to get the effect of looking down the mid-lumbar human spine in Figure 12.14. The spine is seen to have a complex fracture of the vertebral body with a bone fragment in the canal.

These two images are printed on glossy paper because it will reproduce a larger brightness range (Sec. 10.1). Each CT image itself has a certain range of brightness, and when these are added on top of each other the total range to be encompassed can extend beyond what can be reproduced. In the present case, glossy paper allows incorporation of more CT sections, with sufficient range in each, before running into saturation of full bright or full dark.

The CT apparatus can be made to so display a set of sections in this fashion directly on an oscilloscope. The brightness range that can then be encompassed is considerably greater. The effect of this can be demonstrated for four slices with eight slide projectors whose beams are arranged to slightly not overlap. Not only is the effect superior because of an increased dynamic range, but combination is by addition rather than subtraction. For example, with 20 projectors one can add lights, and an opaque region in one section would not obscure the rest (an additive process) while if one places slides in front of each other (a subtractive process) an opaque region obscures everything beyond.

In the above prints, the maximum black in any one line of sight is the sum of grays, and therefore there is not much contrast in any one. Each section of the eight was given about 1/8 the normal exposure onto the final print so that it is possible to see through a dark region in any one.

Some people will find it easy, instead, to look at the right image with the right

246 **DISPLAYS**

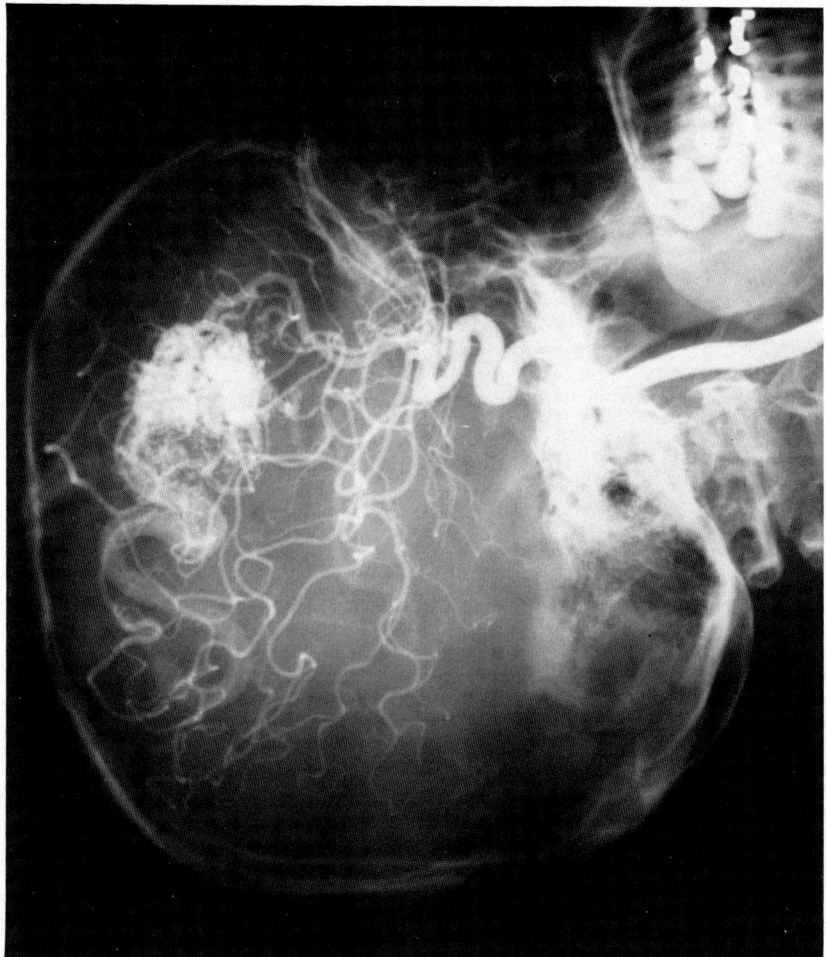

FIGURE 12.9. Stereo pair of a large arteriovenous malformation in a human head. Injection was into the left common carotid artery.

eye and left with left by relaxing their gaze, even though convergence of the eyes is coupled to accommodation. Observation will then seem to be from the other direction. The cross-eyed technique not only helps with focus of the eyes by their converging on a near point, but it is a useful technique for looking at two large objects nearby such as a pair of full sized radiographs. Projected slides of these figures can be observed similarly.

Placing a Fresnel prism, or other prism, before one eye can superimpose the two images without effort, in which case they need not be side-by-side but can be in any relative position, including one above the other, which has some advantages. A suitable prism power would be about 15 prism-diopters, one prism diopter giving a deflection of 1 cm at 1 meter.

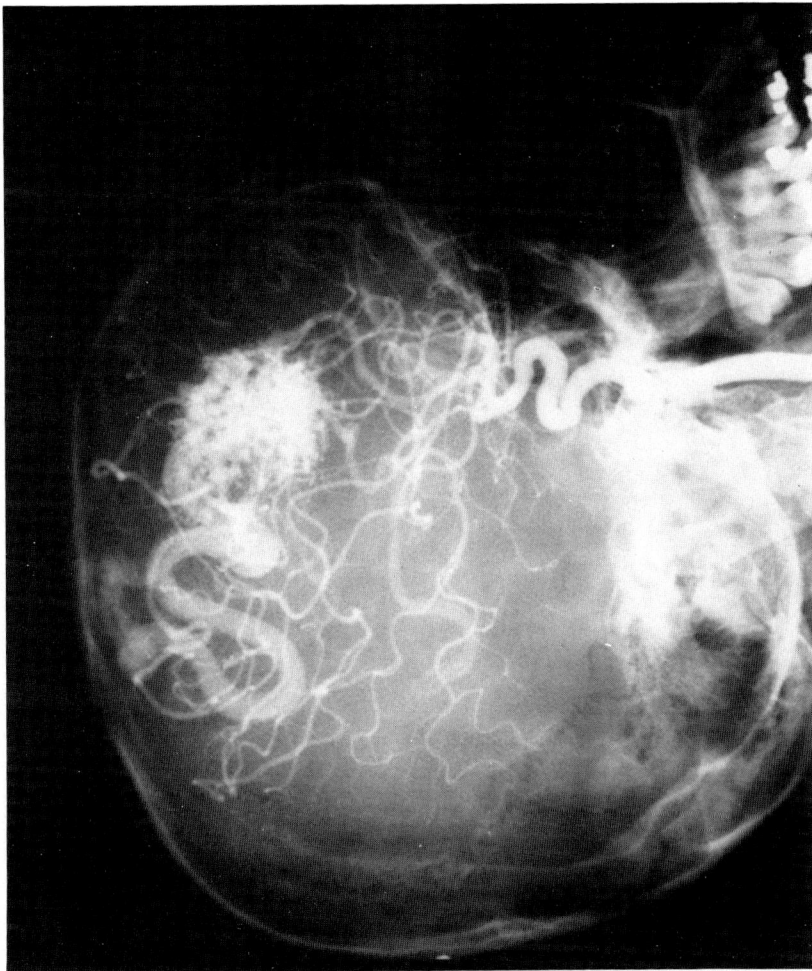

FIGURE 12.9. (*Continued*)

The two images of Figure 12.10, for example, can be viewed in different ways to give various effects, including looking through the shell from the top or the bottom. The reader might wish to predict the effect of turning over both images, turning tops down (viewing from back), turning right sides to left (only), and doing each of these three with the right and left images interchanged. Inverting the individual pictures rather than the whole book gives the effect of looking through the lip of the shell rather than into it, which is what will be seen by those observing without eyes crossed.

It has been mentioned that if sections can be viewed in sequence as fast as about 5 per second then the mind can put together a full three-dimensional impression. Modern computer displays of CT information can be sequenced about this

FIGURE 12.10. Stereo X-ray view of a Polynesian trumpet formed by removing the tip from a large triton shell. Note surface features as well as internal structure. The X-ray tube was shifted by approximately 10% of its distance from the subject between exposures (6 inches at 6 feet). The tube voltage was 80 kV and the exposure 70 milliampere seconds, which is about what would be used for medium bones in the body.

FIGURE 12.10. (*Continued*)

250   DISPLAYS

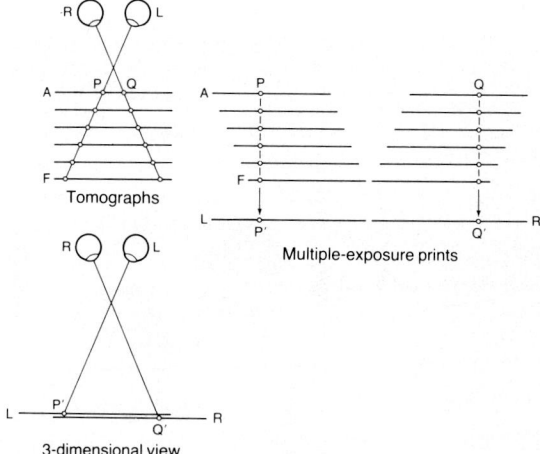

FIGURE 12.11. It is essentially impossible to look down through a stack of slice images to obtain a three-dimensional impression of the body, but a stereo pair can be formed by successively shifting the sections relative to each other in opposite directions for the two images. One can then look at these two images without any equipment by slightly crossing the eyes so the right eye views the left image and the left eye views the right image. An exercise in this technique is given in the next figure.

rapidly, allowing an observer to apparently "run back and forth" through a subject, or as was mentioned, to apparently rotate a subject about some line or axis in the body (Chap. 4).

With information stored as in Figure 4.11, the same shift of view can be achieved from data put in about a distribution of any other property, including patterns of elasticity and density from ultrasound, or degree of staining in actual slices observed by visible light. A computer can also draw lines between points of equal magnitude both in the plane and between planes to give a three-dimensional sketch of the structures that can then be rotated or viewed from any angle.

The question of filling a volume in space with a display is of interest. In the case of a microscope one can vibrate a sheet of frosted glass up and down above the

FIGURE 12.12. Looking at the fingertip, as indicated in the text, held between this figure and the eyes will allow fusing the two drawings into a single image. Does the short vertical line then appear at the same distance as the taller line? This method of viewing takes a little practice at first, but it is useful for viewing pairs of large radiographs or even suitable slides projected on the wall. If needed, cocking the head slightly to one side to make the level of the horizontal lines the same can be helpful in achieving fusion. Once the technique is learned, it is simple without the help of a finger.

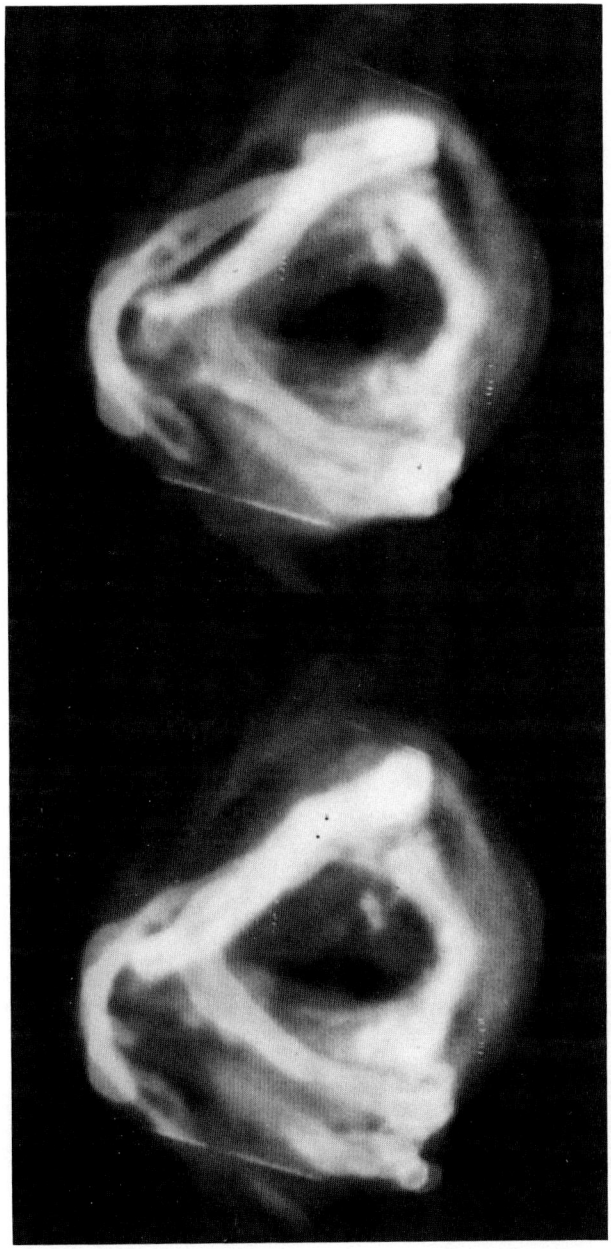

FIGURE 12.13. View along a human larynx formed from eight X-ray computed tomography sections that were combined to produce a three-dimensional view in a direction perpendicular to the plane of the sections. Viewing is by slightly crossing the eyes, as has been indicated. At the top of the picture is the hyoid bone, the thyroid cartilage is light at right and left, and the cricoid cartilage is light across the bottom. The two spots just above the ends of the cricoid cartilage are the arytenoid cartilage. The vocal cords are at the upper center at the margins of the dark region, but not emphasized in this display. Outside are muscles.

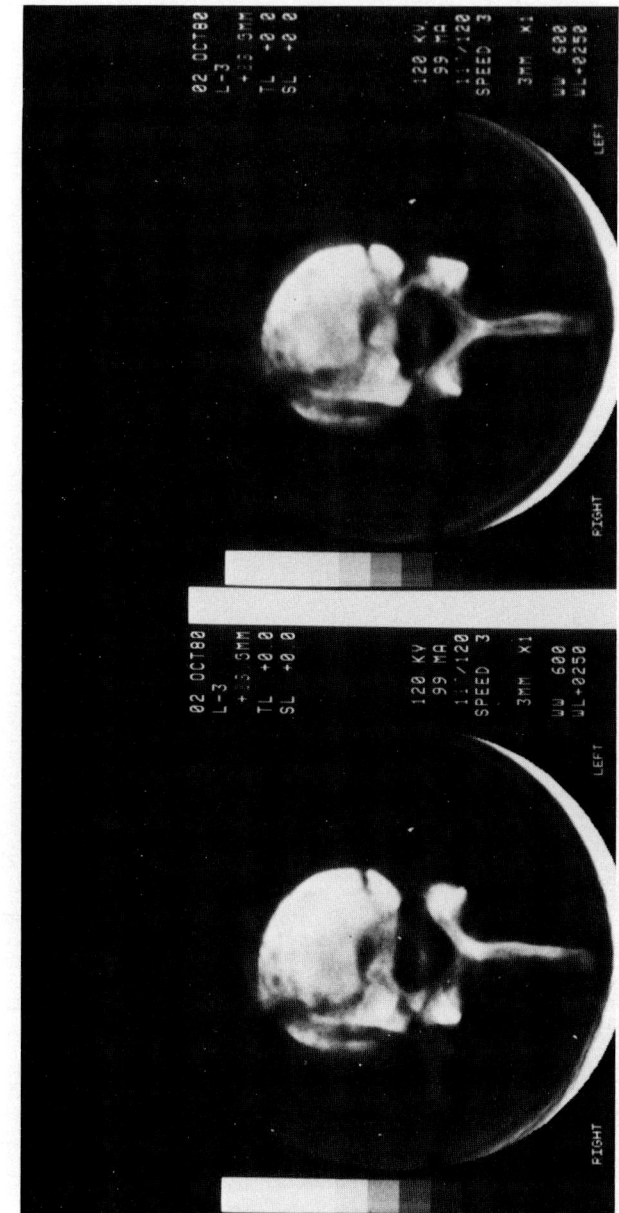

FIGURE 12.14. The mid-lumbar spine (L3) after an automobile fell on the subject. The stereo pair was formed from a series of X-ray CT sections in the way that has been indicated.

eyepiece to spread the projected image out through a region of space, while the specimen itself is vibrated in synchronism up and down below the objective lens. Similarly, if one had information stored in electrical form about all points in the volume of a subject (as in the first part of Fig. 4.11), then this information might be read out on the face of a cathode ray tube that was being vibrated toward and away from the observer. Shaking a cathode ray tube for a 3D display is not a new idea (Peters and Isbister, 1952), but vibrating an oscilloscope rapidly does not yield a very robust configuration. Instead it would be preferable to vibrate or rotate an array of light-emitting diodes. Also preferable would be to vibrate through a volume of space the end of a fiber optics bundle terminating at the other end on a cathode ray tube. One could also aim a cathode ray tube up and, with a lens and mirror, project onto a spinning translucent screen that rotates at the end of an arm in order to cyclically approach the viewer as successive planes are displayed.

Electron beams crossing in a gas to produce a glow have been suggested as a means of producing a three-dimensional indication (Howell, 1952). Nonlinear properties in the light-distributing medium are required, and thus light beams crossing in a smoke-filled box would be less effective. Various ways can be considered for making a glow appear at a desired distance from a cathode and they range from mechanical to electronic while providing for the three-dimensional shift of a spot of light.

When a sequence of flat pictures or slides of sections can be viewed in succession, then these need only be staggered in distance for appropriate viewing by a periodically pulsed light (a strobe unit) in order to give a 3D view. The principle is the same whether the pictures are on an eccentric cylinder, the blades of a fan, holes in a wheel, and so on. The only requirement is that each successive picture be momentarily glimpsed at its correct distance from the observer in rapid sequence (e.g., Szilard, 1974; Kopans, 1980).

Stereo pairs made at different times can have an unusual appearance if lighting changes. Metals in part look metallic or lustrous because of a difference in intensity of corresponding points on the retinal images of the two eyes, that is, in unfamiliar subjects, metallic luster is a binocular phenomenon that can appear in stereo pairs where it is inappropriate.

It is also possible to combine a series of section images into a single hologram. All slices can be viewed together in the relative positions they had when the hologram was produced (Baum and Stroke, 1975) to give a three-dimensional effect. Surgeons, because of their customary view looking down into an incision, might be helped by a 3D image in which upper layers are printed lighter.

A moving picture of a rotating object is an interesting special case. Successive frames view the subject from slightly different angles, and thus if two prints are run through two projectors, with one shifted a few frames, one can view the result separately with the two eyes in order to see an image in depth of a rotating subject (Walls, 1942). There are, of course, many methods for shunting the two images separately to the eyes. X-ray motion pictures have been made this way with the effect of depth (Weinberg et al., 1954).

There is a way of seeing true depth in a motion picture of a rotating object made with a one-eyed camera while projecting a single film through a one-eyed projector, or over a single-channel television system (Mackay, 1954a). The time delay between the viewing of a scene and the arrival of the impression in the brain seems to depend on the brightness of the scene, the time being a decreasing function of brightness. Thus, if a person views a pendulum from a direction perpendicular to the plane of swing and places a neutral filter before one eye, the bob will appear to swing in an ellipse, since the filtered eye is supplying an image of a previous position which fuses with the other or present image to give a resultant position in front of or behind the plane of swing. Such effects are usually associated with the name Pulfrich (1922), who had only one eye and thus never saw his effect. Similarly, if a motion picture of a rotating subject is viewed with a neutral filter before one eye then one eye will effectively perceive the present image while the other eye will perceive a previous position at a slightly different angle. The effect is not quite as large as would be desired for practical purposes, but it can be seen. In those studies the effect was also demonstrated using several drops of 4% homatropine in one eye to dilate the pupil. The pharmaceutical effect was not seen by everyone, but was as predicted with most. (This datum was not in the original article because a referee objected to possible danger, presumably from glaucoma.) A better drug would be neosynephrine which less affects focus. It is interesting that a three-dimensional effect is possible without special positioning or glasses by putting a drop in one eye.

12.5. MOTION PICTURES

If a pair of lights is rapidly cycled on and off in such a way that they alternated being on, then there is a range of frequencies in which one perceives a light hopping back and forth between the two positions rather than merely flickering. This so called "phi" motion would seem to be the basis for moving pictures, though flicker fusion contributes by tending to remove some of the fluctuating aspect when the frequency is sufficiently high. Some animals that show no interest in television may not perceive the motion effect. A motion picture can give an impression of the time course of an event, and its components, much more effectively than any collection of individual still pictures. Thus Figure 6.22 indicates what is happening, and can be subjected to analysis, but the single frames are less effective than an ultrasonic motion picture from which they can be derived.

There were X-ray movies almost as soon as there were motion pictures. The first moving picture projector was demonstrated within months of the discovery of X-rays in 1895, and within a year John MacIntyre, a Glasgow physician, showed blurred roentgen movies of the leg of a frog before the Glasgow Philosophical Society. The present 16 mm sound film rate is 24 frames per second which means completing 24 X-ray images in each second, making dose of radiation a significant consideration. The use of television in connection with X-rays to reduce the dose

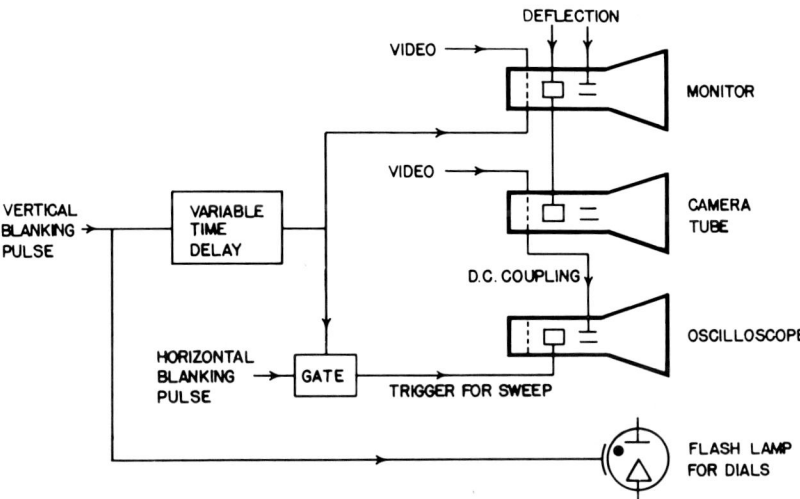

FIGURE 12.15. One method for recording a television image on motion picture film is to have the film run continuously in a shutterless camera past an oscilloscope carrying the image but with no vertical deflection. Objects beside the oscilloscope can be included in the image by the periodic flash of a strobe lamp. The brightness pattern along any one line can be displayed on a separate oscilloscope for the purposes that have been mentioned previously. This results in minimum lost information or excess dose, and sound can still be recorded though the frame rate for motion pictures and television images is different. (From Mackay, 1961.)

to patient and physician was patented by the Frenchman Deauvillier in 1915 and demonstrated in 1928; it has applicability here.

Various combinations of a motion picture camera (24 frames per second) and a television system (30 frames per second, two field per frame interlace) had been tried and they typically gave a dose 2.5 times that necessary to produce the given image quality, with a resolution only approximately half that potentially present. The former was due to the camera shutter being closed for three-fifths of the time during film advance, and the latter because successive fields do not appear on the same or on successive movie frames. Because of storage in the detector, dose need not be reduced if the X-ray beam is turned off whenever the shutter is closed, but this can produce a four-to-one exposure difference across each frame. One solution to the problem (Mackay, 1961) was to build a camera in which the film ran continuously while viewing a cathode ray tube upon which appeared a single horizontal line due to lack of any vertical deflection, as in Figure 12.15. For speech studies, words printed on a disk also were exposed in each frame by a strobe light. This arrangement is also the one for producing the quantitative display seen earlier in Figure 10.14 that allowed matching the X-ray intensity to the subject and the television brightness to the film latitude. (For display at a particular vertical position, one could instead of using a delay circuit simply count lines, or even point to the screen edge with a small photoelectric cell and display the pattern when the cell "sees" a line starting at that level.) The continuously running film need lose the effect of no interacting quantum and, as was there pointed out, this was also a property of the newly developed magnetic video tape recorders. The display of the intensity pattern across one line of the image proved quite valuable, and its comparison with the pseudo 3D images of Section 12.3 also is helpful.

An important aspect of the recording of motion pictures is that a small section can be reviewed repeatedly to reveal details not noticed in the first viewing. Thus a subject need not be retained and repeatedly dosed for an extended examination. The convenience of a tape recorder in immediate viewing has been mentioned, and, along with its other advantages, this reduces the necessity for taking extra precautionary footage in case some aspect was missed during the first exposure.

The same considerations apply to any other modality as well. The motion pictures from which sequences such as in Figure 6.22 can be derived were made with a camera having a shutter normally open until triggered to suddenly advance the film (a 35 mm Flight Research Corporation, model 4). This was a convenient way for recording a sector scan of an unusual frequency. Electronic storage of the signal, point by point, would have allowed it to be read out in the usual television raster pattern suitable for recording on a video tape recorder. The helical scan feature of such recorders allows the later study of single frames, which can be important in diagnosis. Such scan converter equipment for coupling into a video recorder is now much simplified by the semiconductor large-scale integrated chips that are available. The real-time phased array or spinning transducer ultrasonic units are specifically made to work at the usual television frame rates, and can thus be readily coupled into magnetic recorders.

12.6. HOLOGRAMS

A hologram is a diffraction pattern formed when coherent light from a laser returns from a subject to interfere on a piece of photographic film with a reference beam coming directly from the laser. This pattern contains all the information about the shape of the subject, and suitable viewing allows a true three-dimensional image to be seen. The concept was a product of the pioneering thinking of Gabor (1949). An advantage was expected to be that images would form without the distortion of lenses in applications ranging from ultrasonics to electron microscopy.

The diffraction pattern has little obvious relationship to the subject that generated it. As an example of the pattern resulting from a very simple flat subject, for those inexperienced with holograms, an enlargement of one is shown in Figure 12.16. The original is about 3 cm wide. When this transparency is held near the eye and a distant spot of light observed, the letters STU appear above and below the light, the lower set being inverted. If the transparency is torn into many pieces, each piece will display the same effect. Holograms come in many forms and can supply great detail.

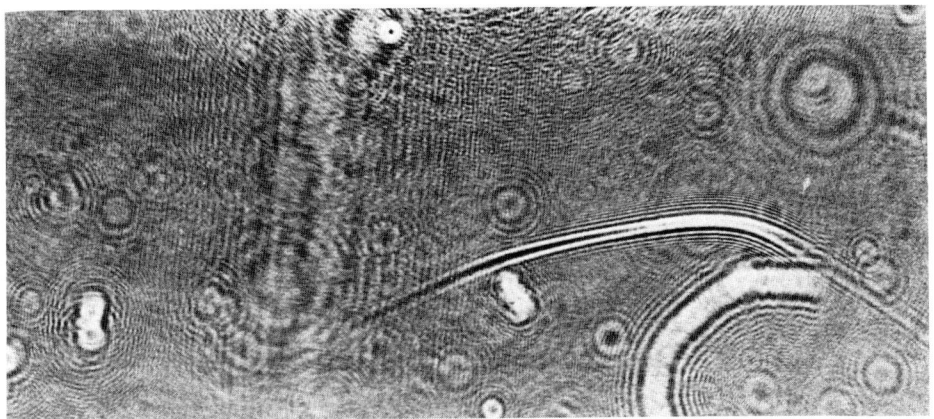

FIGURE 12.16. The hologram of even a simple flat subject bears little resemblance to the object. This is the somewhat enlarged pattern that produces the block letters STU. When white light is shown through the transparency the letters appear fringed with color, the image being more distinct if viewing is by light in a restricted range of color.

One way of regarding them is to assume that the light from any point on the object combines with the reference beam at the film to form a group of concentric circles that constitutes a zone plate. When viewed, this zone plate will then yield a point of light (Chap. 3). Thus the group of all points in the subject becomes a group of similar points in the image. A hologram contains a full three-dimensional image. Thus a hologram prepared of an object above which is a small magnifying glass can be viewed from one direction to see the magnified image of the object but if viewed from further to the side one can look around the magnifying glass and see the unmagnified image of the object. It is even possible to prepare a hologram in which the virtual image of a lens appears in front of various objects, and this lens can act in combination with other real lenses placed before the hologram to view the objects (Rugheimer and Kirkpatrick, 1977). In some cases lightweight holograms can substitute for mirrors and lenses, though a "hologram of a telescope" cannot provide angular modification or a complete system (Close, 1975). Every technical library contains books on holograms, and some writings have been devoted exclusively to their applications to medicine and biology (e.g., Feleppa, 1972). Movement can be recorded, or other changes. For microholography of living organisms it is desirable to use X-rays tuned to a resonance of nitrogen near 0.3 nanometers (Solem and Baldwin, 1982). In light microscopy a single brightfield hologram can give subsequent observation by phase contrast, dark-field interference, or motion-induced darkening, and dry weight of a sample can be measured. Once a hologram is recorded, the three-dimensional structure of the object can then be repeatedly explored at leisure.

There are magnification changes that accompany a change in wavelength between the recording and the viewing radiation. This has proven a problem in some cases, though one way of dealing with the problem in the case of X-rays was mentioned in Chapter 3. Other modalities can yield holograms. Both radio and ultrasound come from continuously oscillating sources, and they are coherent like a laser. One method of ultrasonic holography involves immersing the subject in a liquid bath within which is placed the transducer. Undulations of the surface of the liquid then constitute a hologram which can be viewed with a laser source (Smith and Brenden, 1969; Holt and Coldrick, 1969; Anderson and Curtin, 1973). In general, ultrasonic holography has been disappointing, though certainly there are some applications.

The use of a hologram not for the display of an image but for the improvement of images was mentioned in connection with Figure 4.5, where functioning as a matched filter could give a form of deblurring. However, that application is a correlation process that goes far beyond deblurring. A distribution of objects or patterns is simultaneously each correlated with a sample object, and where there is a match a spot of light is produced. Thus a radiograph of a box of starfish imaged through such a system should display a spot wherever one of these was located that had the orientation of the sample, and other objects should give little signal.

As mentioned in Section 12.4, a series of slice images can be combined for simultaneous display by sequentially exposing them to form a single hologram. Though a computer is able to display such stored information as if from any direction, holography still can be an important way to couple such information into an observer.

REFERENCES

Abragam, A. (1961). *Principles of Nuclear Magnetism*. Clarendon Press, Oxford, pp. 71–73.

Akutagawa, W., G. Huth, R. Lewis, G. Drianis, and R. Davis (1980). Increased tissue differentiation using color display of multiple-energy CT scans. *Radiology* **134**, 739–756.

Albright, R. J., J. Harris, and N. Zinne (1969). Ultrasonic diagnosis of micturition. In *Proceedings of the 8th International Conference on Medical and Biological Engineering*. IEEE, New York, p. 32-1.

Alvarez, L., J. Anderson, F. El Bedwei, J. Burkhard, A. Fakhry, A. Girgis, A. Goneid, F. Hassan, D. Iverson, G. Lynch, Z. Miligy, A. Moussa, M. Sharkawi, and L. Yazolino (1970). Search for hidden chambers in the pyramids. *Science* **167**, 832–839.

Alvarez, R., and A. Macovski (1976). Energy-selective reconstructions in X-ray computerized tomography. *Phys. Med. Biol.* **21**, 733–744.

Anderson, R., and H. Curtin (1973). Ultrasonic holography. *Radiology* **109**, 417–421.

Andrew, E. R. (1956). *Nuclear Magnetic Resonance*. University Press, Cambridge.

Anger, H. (1958). Scintillation camera. *Rev. Sci. Instrum.* **29**, 27–30.

Anger, H. (1967). In *Instrumentation in Nuclear Medicine,* G. Hine (ed.), Vol. 1. Academic, New York, pp. 485–522.

Arnold, J. R., J. Testo, P. Friedman, and G. Kambic (1983). Computed tomographic analysis of meteorite inclusions. *Science* **219**, 383–384.

Auphan, M., and H. Dormont (1977). Pulsed acoustic radiation of plane damped transducers. *Ultrasonics* **15**, 159–168.

Baker, M., and G. Dalyrymple (1978). Biological effects of diagnostic ultrasound: A review. *Radiology* **126**, 479–483.

Baker, R., E. Ramberg, and J. Hillier (1942). Photographic action of electrons in the range 40–212 kV. *J. Appl. Phys.* **13**, 450–456.

Bamber, J., and J. Phelps (1977). Effective directivity characteristic of a pulsed transducer. *Ultrasonics* **15**, 169–174.

Bangert, V., and P. Mansfield (1982). Field gradient coils. *J. Phys. E: Sci. Instrum.* **15**, 235–239.

Barrett, A. H., and P. C. Myers (1975). Subcutaneous temperature: A method of noninvasive sensing. *Science* **190**, 669–671.

Barrett, H., W. Stoner, D. Wilson, and G. de Meester (1974). Coded apertures from zone plate. *Opt. Eng.* **13**, 539–545.

Barth, D., W. Sutherling, J. Engel, and J. Beatty (1982). Neuromagnetic localization of epileptiform spike activity in the human brain. *Science* **218**, 891–894.

Bates, R., and T. Peters (1971). Improvements in tomography. *New Zealand J. Sci.* **14**, 883–896.

Battocletti, J., R. Halbach, A. Sances, S. Larson, R. Bowman, and V. Kudracev (1979). Flat crossed-coil detector for blood flow measurement using NMR. *Med. Biol. Eng. Comput.* **17**, 183–191.

Baum, G., and G. Stroke (1975). Optical holographic 3D ultrasensography. *Science* **189**, 994–995.

Beaver, W., D. Hameron, and A. Macovski (1977). Ultrasonic imaging with an acoustic lens. *IEEE Trans. Sonics Ultrason.* **24**, 235–243.

Beischer, D., and J. Knepton (1964). Influence of strong magnetic fields on the ECG of squirrel monkeys. *Aerosp. Med.* **35**, 939–944.

Béné, J., B. Borcard, E. Hiltbrand, P. Magnin, and R. Sechehaye (1977). Magnétographic nucléaire en champ foible. *C.R. Acad. Sci. Ser. B* **284**, 141–143.

Berliner, L. (ed.) (1976, 1979). *Spin Labeling Theory and Applications,* Vols. I, II. Academic, New York.

Birzis, L., and S. Tachibana (1964). Local cerebral impedance and blood flow during sleep and arousal. *Exp. Neurol.* **9**, 269–285.

Biberman, L. (ed.) (1973). *Perception of Displayed Information.* Plenum, New York.

Bleha, W., L. Lipton, E. Wiener-Avnear, J. Grinberg, P. Reif, D. Casasent, H. Brown, and B. Markevitch (1978). Application of liquid crystal light values to real-time optical data processing. *Opt. Eng.* **17**, 371–384.

Bloom, A., and D. Mansir (1954). Measurement of nuclear induction relaxation times in weak magnetic fields. *Phys. Rev.* **93**, 941.

Bogren, H. (1982). Experimental and early clinical applications of digital angiocardiography. In *Advances in Medical Imaging 1982,* M. Reid (ed.). Sacramento Radiology Research and Education Foundation, Sacramento, Calif., pp. 117–122.

Bom, N. (1972). *New Concepts in Echocardiography.* Kroese Publishers, Leiden.

Borcard, B., E. Hiltbrand, P. Magnin, and G. Béne (1979). Identification de fluides physiologiques humains, *in situ,* par relaxation protonique dans le champ terrestre. *C.R. Acad. Sci. Ser. B* **288**, 41–42.

Bottomley, P. A., and A. R. Andrew (1978). RF Magnetic field penetration, phase shift and power dissipation in biological tissue: Implications for NMR imaging. *Phys. Med. Biol.* **24**, 630–643.

Bovee, W., K. Getreuer, J. Smidt, and J. Lindeman (1978). NMR detection of human breast tumors. *J. Natl. Cancer Inst.* **61**, 53–55.

Bowman, R., and V. Kudravcev (1959). Blood flowmeter utilizing NMR. *IRE Trans. Med. Electron.* **6**, 267–269.

Boyd, D., C. Cann, J. Couch, D. Faul, R. Guld, and K. Peschmann (1982). Future advanced CT technology. *J. Comput. Asst. Tomog.* **6**, 202.

Bracewell, R. N. (1956). Strip integration in radioastronomy. *Austr. J. Phys.* **9**, 198–217.

Bracewell, R. N. (1978). *The Fourier Transform and Its Applications,* 2nd ed. McGraw-Hill, New York.

Bracewell, R. N., and A. Riddle (1967). Inversion of fan-beam scans in radio astronomy. *Astrophys. J.* **150**, 427–434.

Brasch, R. C. (1981). Major contrast media toxicity. In *Diagnostic Radiology,* Margulis and Gooding (eds.). Masson, New York, pp. 275–281.

Brody, W., D. Cassel, G. Sommer, L. Lehmann, A. Macovski, R. Alvarez, N. Pelc, S. Riederer, and A. Hall (1981). Dual-energy projection radiography: Initial clinical experience. *Am. J. Roentgenol.* **137**, 201–205.

Brooks, R. A., and G. DiChiro (1975). Theory of image reconstruction in computed tomography. *Radiology* **117**, 561–572.

Brooks, R., G. Glover, A. Talbert, R. Eisner, and F. DiBianca (1979). Aliasing: A source of streaks in CT. *J. Comput. Asst. Tomog.* **3**, 511–518.

Brown, P., and R. Galloway (1976). Ultrasonic image converter tube. *Ultrasonics* **14**, 273–277.

Brownell, G. L., and W. H. Sweet (1953). Positron emitter localization. *Nucleonics* **11**, 40–44.

Budinger, T. F. (1981). Potential medical effects and hazards of human NMR studies. In *NMR Imaging in Medicine,* Kaufman, Crooks, and Margulis (eds.). Igaku-Shoin Ltd., Tokyo and New York, pp. 207–231.

Burchardt, C., P. Grandchamp, and H. Hoffman (1974). An experimental 2 MHz synthetic aperture sonar system. *IEEE Trans. Sonics Ultrason.* **21**, 1–6.

Burgener, F. A., and D. Hamlin (1981). Contrast enhancement in abdominal CT: Bolus vs. infusion. *Am. J. Roentgenol.* **137**, 351–358.

Calderon, C., D. Vilkomerson, R. Mezrich, K. Etzold, B. Kingsley, and M. Haskin (1976). Differences in attenuation of ultrasound by normal, benign and malignant breast tissue. *J. Clin. Ultrasound* **4**, 252–256.

Campbell, F. W., and L. Maffei (1974). Contrast and spatial frequency. *Sci. Am.* **231** (No. 5), 106–114.

Carden, E., B. Chir, and W. Doll (1970). Doppler flowmeter for detecting emboli, with interference screening. *Anesthesiology* **34**, 551–552.

Chance, B., Y. Nakase, M. Bond, J. Leigh, and D. McDonald (1978). Surface coils. *Proc. Natl. Acad. Sci. USA* **75**, 4925–4929.

Chiervitz, O., and G. Hevesy (1935). Radioactive indicators in the study of phosphorus metabolism in rats. *Nature* **136**, 754–755.

Cho, Z., J. Chen, E. Hall, R. Kruger, and D. McCaughey (1975). A comparative study of 3D image reconstruction algorithms with reference to number of projections and noise filtration. *IEEE Trans. Nucl. Sci.* **22**, 344–350.

Clark, L. C., and F. Gollan (1966). Liquid breathing. *Science* **152**, 1755–1756.

Close, D. H. (1975). Holographic optical elements. *Opt. Eng.* **14**, 408–419.

Cohen, J., and W. Gibson (1962). Vector model for color sensations. *J. Opt. Soc. Am.* **52**, 692–697.

Cohen, Z., S. Seltzer, M. Davis, and R. Hanson (1981). Iodinated starch particles for CT of liver. *J. Comput. Asst. Tomog.* **5**, 843–846.

Collins, C. (1980). Apparatus for analyzing and identifying odorants. U.S. Patent Application 220001.

Collins, C. C., and J. M. Madey (1974). Tactile sensory replacement. *Proc. San Diego Biomed. Symp.* **13**, 15–26.

Cormack, A. M. (1963). Representation of a function by its line integrals, with some radiological applications. *Int. J. Appl. Phys.* **34**, 2722–2730.

Cormack, A. M. (1980). Early two-dimensional reconstruction and recent topics stemming from it. *Science* **209**, 1482–1486.

Craig, D. R. (1955). *Photogramm. Eng.* **21**, 556–560.

Crooks, L. (1980). Selective irradiation line scan techniques for NMR imaging. *IEEE Trans. Nucl. Sci.* **27**, 1239–1244.

Crooks, L., T. Grover, L. Kaufman, and J. Singer (1978). Tomographic imaging with nuclear magnetic resonance. *Invest. Radiol.* **13**, 63–66.

Crosby, B., and R. S. Mackay (1978). Some effects of time *post-mortem* on ultrasonic transmission through tissue under different modes of handling. *IEEE Trans. Biomed. Eng.* **25**, 91–92.

REFERENCES

Cuffin, N. (1981). Effects of torsogeometry on magnetocardiographic lead fields. *IEEE Trans. Biomed. Eng.* **28**, 742–749.

Damadian, R. (1971). Tumor detection by NMR. *Science* **171**, 1151–1153.

Damadian, R., L. Minkoff, M. Goldsmith, M. Stanford, and J. Koutcher (1976). Tumor imaging in a live animal by field focusing NMR. *Physiol. Chem. Phys.* **8**, 61–65.

D'Arcy, F. J., and N. A. Porter (1962). Detection of cosmic ray μ mesons by the human eye. *Nature* **196**, 1013.

Deatherage, B. H., L. A. Jeffres, and H. C. Blodgett (1954). A note on the audibility of intense ultrasonic sound. *J. Acoust. Soc. Am.* **25**, 582.

De Layre, J., J. Ingewall, and C. Malloy (1981). Gated sodium-23 NMR images of an isolated perfused working rat heart. *Science* **212**, 935–936.

Delesse, A. (1848). *Ann. Mines* **13**, 379–381.

De Rosier, D. J., and A. Klug (1968). Reconstruction of 3D structures from electron micrographs. *Nature* **217**, 130–134.

Di Chiro, G., R. Brooks, R. Kessler, G. Johnston, E. Jones, J. Herdt, and W. Sheridan (1979). Tissue signatures with dual-energy computed tomography. *Radiology* **131**, 521–523.

Diegel, J. G., and M. Pintar (1975). Accentuation of T_1 differences at reduced fields. *J. Natl. Cancer Inst.* **55**, 725–726.

Dillon, S., and M. Mored (1981). A new laser scanning system for measuring action potential propagation in the heart. *Science* **214**, 453–456.

Doerner, E. (1962). Wiener-spectrum analysis of photographic granularity. *J. Opt. Soc. Am.* **52**, 669–672.

Donovan, G. E. (1951). Radiography in colour. *Lancet* No. 6659, 832–833.

Doyle, F., J. Gore, J. Pennock, G. Bydder, J. Orr, R. Steiner, I. Young, M. Burl, H. Clow, D. Gilderdale, D. Bailes, and P. Walters (1981). Brain imaging by NMR. *Lancet* **2**, No. 8237, 53–57.

Drabek, C. M. (1977). Seal hearts and aortae. In *Functional Anatomy of Marine Mammals*, R. J. Harrison (ed.), Vol. 3. Academic, New York, pp. 226–231.

Drayer, B., D. Gur, S. Wolfson, and E. Cook (1980). Experimental xenon enhancement with CT cerebral imaging. *Am. J. Roentgenol.* **134**, 39–44.

Drost, C., and J. Milanowski (1980). Self-reciprocity calibration of arbitrarily terminated ultrasonic transducers. *IEEE Trans. Sonics Ultrason.* **27**, 65–71.

Dunn, W., H. Wahner, and L. Riggs (1980). Measurement of bone mineral content in human vertebrae and hip by dual photon absorptiometry. *Radiology* **136**, 485–487.

Dussik, K. T. (1942). Über die möglichkeit lochfrequente mechanische schwingungen ab disgnostiches hilfsmittel zu verwierten. *Z. Neurol. Psychiat.* **174**, 153–168.

Edelstein, W. A., J. Hutchison, G. Johnson, and T. Redpath (1980). Spin warp NMR imaging and applications to human whole-body imaging. *Phys. Med. Biol.* **25**, 751–756.

Eden, J. A. (1955). *Photogramm. Rec.* **1**, 5–10.

Eller, A., and H. Flynn (1969). Generation of subharmonics or order one-half by bubbles in a sound field. *J. Acoust. Soc. Am.* **46**, 722–727.

Elachi, C. (1982). Radar images of the earth from space. *Sci. Am.* **247** (No. 6), 54–61.

Engstrom, A. (1946). Quantitative micro and histochemical elementary analysis by Roentgen absorption spectrography. *Acta Radiol.*, Supp. LXIII.

Enzmann, D., and S. Young (1979). Perfluorinated compounds as contrast agents in CT. *J. Comput. Asst. Tomog.* **3**, 622–626.

Erdmann, W., F. Fry, K. Johnston, and N. Sanghir (1982). Ultrasonic transkull visualization. *IEEE Trans. Sonics Ultrason.* **29**, 5–11.

Evans, R. D. (1968). X ray and X ray interactions. In *Radiation Dosimetry*, Attix and Roesch (eds.), Vol I, Chap. 3. Academic, New York.

Faul, D., J. Couch, C. Cann, D. Boyd, and H. Genant (1982). Composition-selective reconstruction for mineral content in axial and appendicular skeleton. *J. Comput. Asst. Tomog.* **6**, 202–204.

Feleppa, E. (1972). Holography and medicine. *IEEE Trans. Biomed. Eng.* **19**, 194–205.

Foley, D., G. Keyes, D. Smith, B. Belanger, L. Sieb, T. Lawson, K. Thorsen, and E. Stewart (1983). Temporal energy hybrid subtraction in intravenous digital subtraction angiography. *Radiology* **148**, 265–271.

Fox, F., S. Curley, and G. Larson (1955). Phase velocity and absorption measurement in water containing bubbles. *J. Acoust. Soc. Am.* **27**, 534.

Frank, G. (1940). Verfahren zur Herstellung von Körperschnittbildern mittels Röentgenstrahlen. German Patent 693374.

Franklin, D., W. Schlagel, and R. Rushmer (1961). Blood flow measured by Doppler frequency shift of backscattered ultrasound. *Science* **134**, 564–565.

French, L. A., J. Wild, and D. Neal (1950). Detection of cerebral tumors by ultrasonic pulses. *Cancer* **3**, 705–708.

Fry, W. J. (1958). Intense ultrasound in investigations of the central nervous system. In *Advances in Biological and Medical Physics*, Tobias and Lawrence (eds.), Vol 6. Academic, New York, pp. 282–347.

Fry W., and F. Dunn (1956). *J. Acoust. Soc. Am.* **28**, 129–133.

Gabor, D. (1949). Microscopy by reconstructed wavefronts. *Proc. Roy. Soc.* **A197**, 454–487.

Galvin, J. M., and B. Bjarngard (1975). A defence of Johann Radon. *Phys. Med. Biol.* **20**, 839–843.

Gammell, P., and D. Le Croissette (1978). Wideband transducer for tissue characterization. *Ultrasonics* **16**, 233–234.

Garroway, A., P. Grannell, and P. Mansfield (1974). Image formation in NMR by selective irradiation. *J. Phys. C: Solid State Phys.* **7**, 457–462.

Geddes, L., and L. Baker (1975). *Applied Biomedical Instrumentation*, 2nd ed. Wiley, New York.

Geselowitz, D. B. (1979). Magnetocardiography: An overview. *IEEE Trans. Biomed. Eng.* **26**, 497–504.

Goodman, R., A. Bassett, and A. Henderson (1983). Pulsing electromagnetic fields induce cellular transcription. *Science* **220**, 1283–1285.

Gordon, R., R. Bender, and G. Herman (1970). Algebraic reconstruction techniques (ART) for three-dimensional electron microscopy and X-ray photography. *J. Theor. Biol.* **29**, 471–481.

Gordon, R., P. Hanley, D. Shaw, D. Gadian, G. Radda, P. Styles, P. Bore, and L. Chan (1980). Localization of metabolites in animals using ^{31}P topical magnetic resonance. *Nature* **287**, 736–738.

Graham, M. (1965). Guard ring use in physiological measurements. *IEEE Trans. Biomed. Electron.* **12**, 197–198.

Gramiak, R., P. Shah, and D. Kramer (1969). Ultrasound cardiography: Contrast studies in anatomy and function. *Radiology* **92**, 939–948.

Griffin, D. R., R. Hubbard, and G. Wald (1947). The sensitivity of the human eye to infrared radiation. *J. Opt. Soc. Am.* **37**, 546–554.

Grover, T., and J. Singer (1971). NMR spin echo flow measurements. *J. Appl. Phys.* **42**, 938–940.

Guillemin, E. A. (1949). *Mathematics of Circuit Analysis*. Wiley, New York, pp. 485–496.

Gur, D., W. Good, S. Wolfson, H. Yomas, and L. Shabason (1981). Mapping cerebral blood flow by xenon CT. *Science* **215**, 1267–1268.

Hahn, E. L. (1950). Spin echoes. *Phys. Rev.* **80**, 580–594.

Hasegawa, B., R. Cacak, J. Mulvaney, and W. Hendee (1982). Problems with contrast-detail curves for CT performance evaluation. *Am. J. Roentgenol.* **138**, 135–138.

Haselgrove, J., H. Subramanian, J. Leigh, L. Gyulai, and B. Chance (1983). *In-vivo* one-dimensional imaging of phosphorus metabolites by phosphorus-31 NMR. *Science* **220**, 1170–1173.

Haughton, V., J. Donegan, P. Walsh, A. Syvertsen, A. Williams, C. Wilson, J. Cusick, and J. Schmidt

(1980). Clinical cerebral blood flow measurement with inhaled xenon and CT. *Am. J. Roentgenol.* **134**, 281–283.

Havron, A., S. Seltzer, M. Davis, and P. Shulkin (1981). Radiopaque liposomes for CT of spleen. *Radiology* **140**, 507–511.

Hayes, C., T. Case, D. Ailion, A. Morris, A. Cutillo, C. Blackburn, C. Durney, and S. Johnson (1982). Lung water quantitation by NMR imaging. *Science* **216**, 1313–1315.

Hevesy, G. (1923). Absorption and translocation of lead by plants: A contribution to the application of the method of radioactive indicators in investigation of the change of substance in plants. *Biochem. J.* **17**, 439–445.

Heyer, H., and B. Boone (1952). The present status of electrokymography. *Am. Heat J.* **43**, 206–210.

Hill, C. R. (1971). Ultrasonic thresholds for changes in cells and tissues. *J. Acoust. Soc. Am.* **52**, 667–672.

Hinshaw, W. S. (1976). Image formation by the NMR sensitive point method. *J. Appl. Phys.* **47**, 3709–3721.

Hinshaw, W., P. Bottomley, and G. Holland (1977). Radiographic thin-section of the human wrist by NMR. *Nature* **270**, 722–723.

Holland, G., and E. Heysmond (1979). Solid state high-power rf amplifier for pulsed NMR. *J. Phys. E: Sci. Instrum.* **12**, 480–483.

Holland, G., P. Bottomley, and W. Hinshaw (1977). Fluorine-19 magnetic resonance imaging. *J. Magn. Reson.* **28**, 133–136.

Holt, D., and J. Coldrick (1969). Acoustical holography and its applications. *Ultrasonics* **7**, 240–244.

Hospital Physicists' Association (1981). *Measurement of the Performance Characteristics of Diagnostic X-Ray Systems Used in Medicine.* Part III. Computed Tomographic X-Ray Scanners: Measurement and Use of the Associated Performance Parameters. The Hospital Physicists' Association, Topic Group Report—32. Hospital Physicists' Association, London.

Hoult, D., and R. Richards (1976). Noise in NMR. *J. Magn. Reson.* **24**, 71–85.

Hounsfield, G. (1973). Computerised transverse axial scanning. *Br. J. Radiol.* **46**, 1016–1022.

Hounsfield, G. (1980). Computed medical imaging. *Science* **210**, 22–28.

Howard-Flanders, P. (1981). Inducible repair of DNA. *Sci. Am.* **245** (No. 5), 72–80.

Howell, F. S. (1952). 3D indicator tube and circuit. U.S. Patent 2,604,607.

Howry, D. H. (1965). A brief atlas of diagnostic ultrasonic radiologic results. *Radiol. Clin. North Am.* **3**, 433–452.

Howry, D. H., and W. R. Bliss (1952). Ultrasonic visualization of soft tissue structures of body. *J. Lab. Clin. Med.* **40**, 579–592.

Hughes, W., G. Nussbaum, R. Connolly, B. Emamé, and P. Reilly (1979). Tissue perfusion rate determined from decay of oxygen-15 activity after photon activation *in situ. Science* **204**, 1215–1217.

Hulm, J., and B. Matthias (1980). High-field high-current superconductors. *Science* **208**, 881–887.

Hurter, F., and V. Driffield (1890). *J. Soc. Chem. Ind.* **9**, 455–460.

Hussey, M. (1975). *Diagnostic Ultrasound.* Blackie & Son, Glasgow.

Iskander, M. F., R. Maini, C. Durney, and D. Bragg (1981). A microwave method for measuring changes in lung water content. *IEEE Trans. Biomed. Eng.* **28**, 797–802.

Ikeya, M. and T. Miki (1980). Electron spin resonance dating of animal and human bones. *Science* **207**, 977–979.

Iyengar, K. S., and E. Richardson (1958). Measurement on the air nuclei in natural water which give rise to cavitation. *Br. J. Appl. Phys.* **9**, 154–158.

Jacobi, C. J. (1846). Über ein leichtes Verfahren die in der Theorie der Säculärstörungen vorkommenden Gleichungen numerisch aufzulösen. *Jour. für die reine und angewandte Mathematik* **30**, 51–94.

Jacobson, B. (1953). Dichromatic absorption radiography: Dichromography. *Acta Radiol.* **39**, 437–452.

Jacobson, B. (1958). Unpublished paper presented at First International Conference on Medical Electronics, Paris, June 28.

Jacobson, B., and R. S. Mackay (1958). Radiological contrast enhancing methods. In *Advances in Biological and Medical Physics,* Tobias and Lawrence (eds.), Vol. 6. Academic, New York, pp. 201–261.

Jenkins, F. A., and H. E. White (1937). *Fundamentals of Physical Optics.* McGraw-Hill, New York.

Jones, A. R. (1965). A γ dosimeter with a long memory. *Phys. Med. Biol.* **3**, 7 and 353.

Jones, R. C. (1968). How images are detected. *Sci. Am.* **219**, 110–117.

Judd, D. B. (1935). Maxwell triangle yielding uniform chromacity scales. *J. Res. Natl. Bur. Stand.* **14**, 41–57.

Kay, L. (1973). Synthesis of ultrasonic imaging. In *Research in Nondestructive Testing,* R. S. Sharpe (ed.) Vol. 2, Chap. 12. Pergamon, Oxford.

Kay, L. (1980). Air sonars with acoustical display of spatial information. In *Animal Sonar Systems,* Busnel and Fish (eds.). Plenum, New York, pp. 719–816.

Kay, L., J. Boys, G. Clark, and J. Mason (1977). Echocardiophone: A new means for observing spatial movement of the heart. *Ultrasonics* **15**, 136–141.

Kelly, E. (ed.) (1957). *Ultrasound in Biology and Medicine.* American Institute of Biological Sciences, Washington D.C.

Kirkpatrick, P. (1939). Theory and use of Ross filters. *Rev. Sci. Instrum.* **10**, 186–191; **15**, 223–227 (1944).

Koehler, A. M. (1968). Images with protons. *Science* **160**, 303–304.

Kopans, D. (1980). A strobe-sequenced device to facilitate the 3D viewing of cross-sectional images. *Radiology* **135**, 780–781.

Kossoff, G., D. Robinson, and W. Garrett (1966). Two-dimensional ultrasonography in obstetrics. In *Diagnostic Ultrasound,* C. C. Grossman et al. (eds.). Plenum Press, New York.

Kossoff, G., W. Garrett, D. Carpenter, J. Jellins, and M. Dadd (1976). Principles and classification of soft tissues by gray scale echography. *Ultrasound Med. Biol.* **2**, 89–105.

Kremkau, F., E. Carstinsen, and R. Gramiak (1969). Ultrasonic detection of cavitation at catheter tips. Elec. Eng. Tech. Rep. #GM09933-15. University of Rochester, N.Y.

Kristian, J., and M. Blouke (1982). Charge-coupled devices in astronomy. *Sci. Am.* **247** (No. 4), 67–74.

Krogness, K. (1977). The aqueduct echo method applied to brain stem tumors. *Ultrasonics* **15**, 142–143.

Krusos, G., S. Hilal, W. Scaman, and G. Myers (1970). Reduction of penumbra in X-ray images by optical spatial filtering. *Appl. Phys. Lett.* **16**, 27–40.

Kuhl, D. E., and R. Edwards (1963). Image separation radioisotope scanning. *Radiology* **80**, 653–661.

Kulmann, C. A. (1954). Study of electrically-adjustable continuously-variable delay lines using nonlinear reactances. Thesis: Master of Science in Electrical Engineering, University of California, Berkeley.

Kumar, A., D. Welti, and R. Ernst (1975). NMR Fourier zeugmatography. *J. Magn. Reson.* **18**, 69–83.

Lale, P. G. (1959). Examination of internal tissues rising gamma ray scatter. *Phys. Med. Biol.* **4**, 159–162.

Lam, F., and J. Szilard (1976). Pulse compression techniques. *Ultrasonics* **14**, 111–114.

Lampton, M. (1981). Microchannel image intensifier. *Sci. Am.* **245** (No. 5), 62–71.

Land, E. H. (1959). Experiments in color vision. *Sci. Am.* **200** (No. 5), 84–99.

Lantz, B. (1975). Relative flow measured by Roentgen videodensitometry in hydrodynamic model. *Acta Radiol.* **15**, 503–519.

REFERENCES

Lantz, B., B. Lindberg, and J. Huebel (1975). Three dimensional reconstruction of the human heart by video technique. *Acta Radiol.* **16**, 545–558.

Lauterbur, P. C. (1973). Image formation by induced local interactions: Examples employing nuclear magnetic resonance. *Nature* **242**, 190–191.

Lele, P., A. Mansfield, A. Murphy, J. Namery, and N. Senapati (1976). Tissue characterization by ultrasonic frequency dependent attenuation and scattering. National Bureau of Standards Special Publ. 453, p. 174.

Levi, C., J. Gray, E. McCullough, and R. Hattery (1982). The unreliability of CT numbers as absolute values. *Am. J. Radiol.* **139**, 443–447.

Lewitt, R. M., R. H. T. Bates, and T. M. Peters (1978). Image reconstruction from projections. *Optik* **50**, 19–33, 85–109, 189–204, 269–278.

Lichtenberg, G. C. (1777). *Novi. Comment. Gött* **8**, 168.

Liebeskind, D., R. Bases, F. Elequin, S. Neubort, R. Leifer, R. Goldberg, and M. Koenigsberg (1979). Diagnostic ultrasound: Effects on DNA and growth of animal cells. *Radiology* **131**, 177–184.

Light, H. (1969). Non-injurious ultrasonic technique for observing flow in the human aorta. *Nature* **224**, 1119–1121.

Longo, L., M. Delivoria-Papadopoulos, G. Power, E. Hill, and R. Forster (1970). Diffusion equilibrium of inert gases between placental capillaries. *Am. J. Physiol.* **219**, 561–569.

Lovell, W. V. (1946). Electromagnet. U.S. Patent 2,400,869.

Lovinger, A. (1983). Ferroelectric polymers. *Science* **220**, 1115–1121.

Lowenstam, H. A. (1962). Magnetite in denticle copping in recent chitons. *Bull. Geol. Soc. Am.* **73**, 435–438.

Lutsch, A. (1962). Solid mixtures with specified impedances and high attenuation for ultrasonic waves. *J. Acoust. Soc. Am.* **41**, 336–345.

Lynden-Bell, R., and R. Harris (1969). *Nuclear Magnetic Resonance Spectroscopy.* Nelson, London.

McCloud, J. (1954). Axicon: A new optical element. *J. Opt. Soc. Am.* **44**, 592–597.

McCormick, J., E. Wever, S. Ridgway, and J. Palin (1979). Sound reception in the porpoise. In *Animal Sonar Systems,* Busnel and Fish (eds.). Plenum, New York, pp. 449–467.

Mackay, R. S. (1949). In *Suggestions for Graduate Research in Electrical Engineering.* University of California, Berkeley.

Mackay, R. S. (1954a). 3D movies without special equipment. *Science* **119**, 905–906.

Mackay, R. S. (1954b). Color X-ray pictures. Convention Record of the IRE National Convention, Part 9. Institute of Radio Engineers (Now IEEE), New York, pp. 7–8.

Mackay, R. S. (1958a). Constant intensity image for minimum dose radiography. *Br. J. Radiol.* **31**, 642.

Mackay, R. S. (1958b). X-ray visualization and analysis of multicomponent subjects. *Science* **128**, 1622–1623.

Mackay, R. S. (1958c). Negative resistance. *Am. J. Phys.* **26**, 60–69.

Mackay, R. S. (1959). Display of only moving parts of a scene. *Science* **130**, 223–224.

Mackay, R. S. (1960). Two startling demonstrations with a magnet. *Am. J. Phys.* **28**, 678.

Mackay, R. S. (1961). Television cineradiography: Dose and contrast factors. *Am. J. Roentgenol. Radium Ther. Nucl. Med.* **85**, 342–351. [The same item was a book chapter in *Proceedings of the Second Institute of Medical Electronics Conference,* C. N. Smyth (ed.). Iliffe Publishers, London, 1959 pp. 525–531.]

Mackay, R. S. (1962). Electronics in clinical research. In anniversary issue: *Proc. Inst. Radio Eng.* **50**, 1177–1189.

Mackay, R. S. (1963a). Analysis and visualization using X-ray spectral information. In *Encyclopedia of X and Gamma Rays,* G. Clark (ed.). Rheinhold, New York, pp. 896–899.

Mackay, R. S. (1963b). In *Proceedings of the Second Symposium on Underwater Physiology.* National Academy Sciences/National Research Council Publ. 1181, Washington, D.C., p. 41.

Mackay, R. S. (1964). Application of physical transducers to intracavity pressure measurements. *Med. Electron. Biol. Eng.* **2,** 15.

Mackay, R. S. (ed.) (1965a). *Ultrasonic Imaging.* National Academy Sciences NRC:MAC: 2016, Washington, D.C.

Mackay, R. S. (1965b). Endoradiosondes. In *Biomedical Telemetry,* C. Caceres (ed.). Academic, New York, p. 225.

Mackay, R. S. (1966a). Experiments to conduct in order to obtain comparative results. In *Animal Sonar Systems,* R. G. Busnel (ed.), Vol. 2. Laboratoire de Physiologie Acoustique, Jouy-en-Josas, France, p. 1190.

Mackay, R. S. (1966b). Telemetering physiological information from within cetaceans, and the applicability of ultrasound to understanding *in vivo* structure and performance. In *Whales, Dolphins and Porpoises,* K. Norris (ed.). University of California Press, Berkeley, pp. 378–379, 445–470.

Mackay, R. S. (1970). *Bio-Medical Telemetry: Sensing and Transmitting Biological Information from Animals and Man,* 2nd ed. Wiley, New York.

Mackay, R. S. (1972). Non-invasive cardiac output measurement. *Microvasc. Res.* **4,** 438–452.

Mackay, R. S. (1980). Dolphin air sac motion measurements during vocalization by two noninvasive ultrasonic methods. In *Animal Sonar Systems,* R. Busnel and J. Fish (eds.). Plenum, New York, pp. 933–937.

Mackay, R. S. (1982). Dolphins and the bends. *Science* **216,** 650.

Mackay, R. S., and C. Collins (1957). Color X-ray images and enhanced contrast. *J. Biol. Photogr. Assoc.* **25,** 114–118.

Mackay, R. S., and S. Giovannoni (1977). Quantitative Doppler ultrasound measures marine mammal cardiac output noninvasively. In *Proceedings of the Second Conference on Biology of Marine Mammals.* NOSC, San Diego, p. 46.

Mackay, R. S., and S. E. Leeds (1953). Physiological effects of condenser discharges with application to tissue stimulation and ventricular defibrillation. *J. Appl. Physiol.* **6,** 67–75.

Mackay, R. S., and H. Liaw (1981). Dolphin vocalization mechanisms. *Science* **212,** 676–678.

Mackay, R. S., and G. Rubissow (1978). Decompression studies using ultrasonic imaging of bubbles *IEEE Trans. Biomed. Eng.* **25,** 537–544.

Mackay, R. S., and N. Seaton (1960). Deflection focusing of electron microscopes. *IEEE Trans. Med. Electron.* **7,** 87–94.

Mackay, R. S., K. Mooslin, and S. Leeds (1951). Effects of electric currents on the canine heart with particular reference to ventricular fibrillation. *Ann. Surg.* **134,** 173–185.

Mackay, R. S., G. Eilers, J. Horowitz, and E. Marg (1962). Ultrasonic echo imaging in eye research. In *Proceedings of the 15th Annual Conference on Engineering in Medicine and Biology,* Paper VII-9. IEEE, New York.

McKinnon, G. C., and R. H. T. Bates (1980). A limitation on ultrasonic transmission tomography. *Ultrasonic Imaging* **2,** 48–54.

McLeod, J. (1954). Axicon: A new optical element. *J. Opt. Soc. Am.* **44,** 592–597.

McNicholas, H., and H. Curtis (1931). Measurement of fiber diameters by the diffraction method. *Rev. Sci. Instrum.* **2,** 263–286.

Mackworth, N., and E. Thomas (1962). Head-mounted eye-marker camera. *J. Opt. Soc. Am.* **52,** 713–716.

Mansfield, P., and A. Maudsley (1977). Medical imaging by NMR. *Br. J. Radiol.* **50,** 188–194.

Massopust, L. C. (1952). *Infrared Photography in Medicine.* C. Thomas Co., Springfield.

Mastrangelo, C., J. Kendig, J. Trudell, and E. Cohen (1979). Nerve membrane lipid fluidity: Opposing effects of high pressure and ethanol. *Undersea Biomed. Res.* **6,** 47–53.

Meltzer, R. S., and J. Roelandt (eds.) (1982). *Contrast Echocardiography.* Martinus Nÿhoff, The Hague.

Merrill, E. H., and A. von Hipple (1939). Atomphysical interpretation of Lichtenberg figures and their application to study of gas discharge phenomena. *J. Appl. Phys.* **10**, 873–887.

Mertz, L., and N. Young (1961). Fresnel transformation of images. *Proc. Int. Conf. Opt. Instrum.*, 305–310.

Messino, D., D. Settle, and F. Wanderlingh (1965). On the stabilization of nuclei for ultrasonic cavitation in water. Fourth Meeting of International Commission on Acoustics, Liege.

Meyer, C. (1981). Preliminary results on a system for wideband reflection-mode ultrasonic attenuation imaging. *IEEE Trans. Sonics Ultrason.* **29**, 12–17.

Meyer, J., A. Hayman, M. Yamamoto, F. Sakai, and S. Nakajima (1980). Local cerebral blood flow with xenon and CT. *Am. J. Roentgenol.* **135**, 239–251.

Michenfelder, J., R. Miller, G. Gronert (1972). Evaluation of an ultrasonic device (Doppler) for diagnosis of venous air embolism. *Anesthesiology* **36**, 164–167.

Miller, F., E. Mineau, R. Koehler, J. Nelson, P. Luers, R. Sherry, P. Lawrence, R. Anderson, and R. Kruger (1983). Clinical intra-arterial digital subtraction imaging. *Radiology* **148**, 273–278.

Mistretta, C., and A. Crummy (1981). Diagnosis of cardiovascular disease by digital subtraction angiography. *Science* **214**, 761–765.

Møhl, B. (1967). Frequency discrimination in the common seal. In *Underwater Acoustics*, V. Albers (ed.), Vol. 2. Plenum, New York.

Moon, R. J. (1950). Amplifying and intensifying the fluoroscopic image by means of a scanning X-ray tube. *Science* **112**, 384–395.

Moon, R., and J. Richards (1973). Determination of intracellular pH by ^{31}P magnetic resonance. *J. Biol. Chem.* **248**, 7276–7278.

Moore, A. D. (1949). Fields from fluid flow mappers. *J. Appl. Phys.* **20**, 790–804.

Morrison, R., and R. Boyd (1973). *Organic Chemistry*, 3rd ed., Chap. 13. Allyn & Bacon, Boston.

Morse, O., and J. Singer (1970). Blood velocity measurement in intact subjects. *Science* **170**, 440–441.

Murray, P. W. (1981). Field applications in the head of a newborn infant and their application to the interpretation of transcephalic impedance measurements. *Med. Biol. Eng. Comput.* **19**, 539–546.

Myer, E., and E-G. Neumann (1972). *Physical and Applied Acoustics*. Academic, New York.

Nahring (1930). *Phys. Z.* **31**, 800–805.

Newberry, S. P. (1953). The X-ray microscope. In *Abstracts, Sixth Conference on Electronic Instrumentation and Nucleonics in Medicine*. IEEE, New York, p. 6.

Newhouse, V. and I. Amir (1983). Time dilation and inversion properties and the output spectrum of pulsed Doppler flowmeters. *IEEE Trans. Sonics Ultrason.* **30**, 174–179.

Nicholas, D., A. Barrett, J. Chu, D. Cosgrove, P. Garbutt, J. Green, S. Pussell, and C. Hill (1980). Computer analysis of gray scale tomograms. In *Acoustical Imaging*, Vol. 8. Plenum, New York, pp. 731–744.

Nishi, R. Y. (1972). Ultrasonic detection of bubbles with Doppler flow transducers. *Ultrasonics* **10**, 173–179.

Nyborg, W. (1973). In *Finite-Amplitude Wave Effects in Fluids*, L. Bjarna (ed.). International Publishing, London, p. 245.

O'Donnell, M. (1982). Proposed annular array imaging system. *IEEE Trans. Sonics Ultrason.* **29**, 331–338.

O'Donnell, M., J. Mimbs, B. Sobel, and J. Miller (1979). The relationship between collagen and ultrasonic attenuation in myocardial tissue. *J. Acoust. Soc. Am.* **65**, 512–515.

Oldendorf, W. H. (1961). Isolated flying-spot detection of radioactivity discontinuities; displaying the internal structure pattern of a complex object. *IEEE Trans. Bio.-Med. Electron.* **8**, 68–72.

Packard, M., and R. Varian (1954). Free nuclear induction in earth's field. *Phys. Rev.* **93**, 941.

Packer, K. J. (1969). The study of slow coherent molecular motion by pulsed NMR. *Mol. Phys.* **17**, 355–368.

Paul, F., S. Roath, and D. Melville (1978). Differential blood cell separation using a high gradient magnetic field. *Br. J. Haematol.* **38,** 273–280.

Paul, F., S. Roath, D. Melville, D. Warhurst, and J. Osisanya (1981). Separation of malaria-infected erythrocytes from whole blood: Use of a selective high-gradient magnetic separation technique. *Lancet* **2,** No. 8237, 70–71.

Pehek, J., H. Kyles, and D. Faust (1976). Image modulation in corona discharge photography. *Science* **194,** 263–270.

Pernick, B., R. Kopp, J. Lisa, J. Mendelsohn, R. Herold, H. Stone, and R. Wohlers (1978). Screening of cervical cytological samples using coherent optical processing. *Appl. Opt.* **17,** 21–51.

Peters, J., and E. Isbister (1952). Aircraft traffic control system. U.S. Patent 2,602,921.

Polson, M., A. Barker, and I. Freeston (1982). Stimulation of nerve trunks with time-varying magnetic fields. *Med. Biol. Eng. Comput.* **20,** 243–244.

Porter, R. A., and H. H. Miller (1978). Microwave radiometric detection and location of breast cancer. In Proceedings of the IEEE Electro/78 Session 30. Boston, May 24.

Pulfrich, C. (1922). *Naturwiss.* **10,** 533, 596, 751.

Radon, J. (1917). Über die bestimmung von funktionel durch ihre integralwerte länges gewisser mannigfaltigkeiten. *Bes. Verh. Sächs. Akad. Wiss.* **69,** 262–277.

Ramachandran, G., and A. Lakshminarayanan (1971). 3D reconstruction from radiographs and electron micrographs: Convolutions vs. Fourier transforms. *Proc. Natl. Acad. Sci. USA* **68,** 2236–2240.

Rankin, R. N. (1979). Gas formation after renal tumor embolization without abscess: A benign occurrence. *Radiology* **130,** 317–320.

Rao, P., K. Santosh, and E. Gregg (1980). Computed tomography with microwaves. *Radiology* **135,** 769–770.

Rayleigh, Lord (Strutt, J. W.) (1894). *Theory of Sound.* Dover, New York (1945 republication).

Reichmanis, M., A. Marino, and R. Becker (1975). Electrical correlates of acupuncture points. *IEEE Trans. Biomed. Eng.* **22,** 533–535.

Reid, J. M. (1965). Ultrasonic diagnostic methods in cardiology. Thesis, University Microfilms, Inc., Ann Arbor, Michigan.

Reid, J., and M. Spencer (1972). Ultrasonic Doppler technique for imaging blood vessels. *Science* **176,** 1235–1236.

Reid, M. H. (1982a). Quantitative stereology and radiologic image analysis. *Med. Phys.* **9,** 346–371.

Reid, M. H. (1982b). CT organ and lesion size measurements for diagnosis and therapy evaluation. In *Advances in Medical Imaging,* M. Reid (ed.). Sacramento Radiology Research and Education Foundation, Sacramento, Calif.

Reid, M., and A. Dublin (1984). QUAC, A modest proposal for optimum CT utilization: Multiple simultaneous patient examinations. *Am. J. Roentgenol.* **142** (4).

Reid, M., R. S. Mackay, and B. Lantz (1980a). Noninvasive blood flow measurements by Doppler ultrasound with applications to renal artery flow determination. *Invest. Radiol.* **15,** 323–331.

Reid, M., R. S. Mackay, and B. Lantz (1980b). Noninvasive measurement of fetal and neonatal blood flow. *Acta Radiol. Diagn.* **21,** 197–202.

Ritchings, R. T., and B. Pullan (1979). Technique for simultaneous dual energy scanning. *J. Comput. Asst. Tomog.* **3,** 842–846.

Ritman, E., J. Kinsey, R. Robb, L. Harris, and B. Gilbert (1980). Physics and technical considerations in design of the dynamic spatial reconstructor. *Am. J. Roentgenol.* **134,** 369–374.

Robb, R. A., J. Greenleaf, E. Ritman, S. Johnson, J. Sjostrand, G. Herman, and E. Wood (1974). Three-dimensional visualization of the intact thorax and contents. *Comput. Biomed. Res.* **7,** 395–405.

Robinson, A. L. (1983a). Laser light in the extreme ultraviolet. *Science* **220,** 1259–1261.

Robinson, A. L. (1983b). New magnets enhance synchrotron radiation. *Science* **219**, 1309–1311.

Robinson, L. C. (1973). *Physical Principles of Far-Infra Red Radiation.* Academic, New York.

Roentgen, W. C. (1895). On a new kind of rays. *Sitzungsberichte,* Physik.-Med. Ges. Würzburg, pp. 132–141; cxxxvii *Ann. Phys. Chem.* N.F.**64**, 1 (1898).

Rogers, A. R. (1973). *Techniques of Autoradiography.* Elsevier, Amsterdam.

Rogers, G. L. (1977). *Noncoherent Optical Processing.* Wiley, New York.

Romani, G., S. Williamson, and L. Kaufman (1982). Tonotopic organization of the human auditory cortex. *Science* **216**, 1339–1340.

Rose, A. (1957). In *Advances in Biological and Medical Physics,* Lawrence and Tobais (eds.), Vol. 5. Academic, New York, pp. 211–222.

Rowley, P. D. (1969). Quantitative interpretation of 3D weakly refractive phase objects using holographic interferometry. *J. Opt. Soc. Am.* **59**, 1469–1498.

Roy, O. Z., R. Beique, and G. D'Ombrain (1962). Method of iodine determination by characteristic X-ray absorption. *IRE Trans. Bio.-Med. Electron.* **9**, 50–53.

Rubissow, G. (1973). A Study of Decompression Sickness Using Ultrasonic Imaging of Bubbles. Ph.D. Thesis, University of California, Berkeley.

Rubissow, G., and R. S. Mackay (1971). Ultrasonic imaging of *in vivo* bubbles in decompression sickness. *Ultrasonics* **9**, 225–234.

Rubissow, G., and R. S. Mackay (1982). Ultrasonic studies during experimental decompression. In *Contrast Echocardiography,* Meltzer and Roelandt (eds.). Nijhoff, The Hague, pp. 30–43.

Rugheimer, M., and L. Kirkpatrick (1977). Demonstration holograph for comparing an image lens with a real lens. *Am. J. Phys.* **45**, 1027–1029.

Rutherford, R. A., B. Pullan, and I. Isherwood (1976). Measurement of effective atomic number and electron density using an EMI scanner.

Rutishauser, W., G. Noseda, W. Bussmann, and B. Preter (1970). Blood flow measurement through single coronary arteries by Roentgen densitometry. II. Right coronary artery flow in conscious man. *Am. J. Roentgenol.* **109**, 21–30.

Rutt, B., and A. Fenster (1980). Split-filter CT: A simple technique for dual energy scanning. *J. Comput. Asst. Tomog.* **4**, 501–509.

St. John, E. G., and D. R. Craig (1957). *Am. J. Roentgenol.* **78**, 124–128.

Satomura, S. (1959). Study of the flow patterns in peripheral arteries by ultrasonics. *J. Acoust. Soc. Jpn.* **15**, 151–158.

Shulman, R. (1983). NMR spectroscopy of living cells. *Sci. Am.* **248** (No. 1), 86–93.

Schulz, R., E. Olson, and K. Han (1977). A comparison of number of rays vs. number of views in reconstruction tomography. Joint Meeting of the Society of Photo-Optical Engineers and American Roentgen Ray Society, Sept. 25–29, Boston.

Schumacher, R. (1970). *Magnetic Resonance.* Benjamin, New York.

Schwan, H. P. (1965). In *Therapeutic Heat and Cold,* 2nd ed., S. Licht (ed.). Williams & Wilkins, New Haven, pp. 63–125.

Schwartz, B., and S. Foner (1977). Large scale applications of superconductivity. *Phys. Today* **30** (No. 7), 10–17.

Schwartz, M., R. Duara, J. Haxby, C. Grady, B. White, R. Kessler, A. Kay, N. Cutler, and S. Rapoport (1983). Down's syndrome in adults: Brain metabolism. *Science* **221**, 781–783.

Seltzer, S., D. Adams, M. Davis, S. Hessel, A. Hurlburt, A. Havron, N. Hollenberg, and H. Abrams (1979). Selective hepatic contrast agents for CT. *Invest. Radiol.* **14**, 356.

Sette, D. (1949). Ultrasonic lenses of plastic. *J. Acoust. Soc. Am.* **21**, 375–381.

Shepp, L. A., and J. A. Stein (1977). Simulated reconstruction artifacts in computerized X-ray tomography. In *Reconstruction Tomography in Diagnostic Radiology and Nuclear Medicine,* M. Ter-Pogossian (ed.). University Park Press, University Park, Md., pp. 33–48.

Siegel, S., J. Goddard, E. James, and E. Siegel (1979). Cellular attachment as indicator of effects of ultrasound. *Radiology* **133**, 175–179.

Sigel, B., J. Machi, J. Beitler, J. Justini, and J. Coelho (1982). Variable echogenicity in flowing blood. *Science* **218**, 1321–1323.

Simon, I. (1966). *Infrared Radiation*. Van Nostrand, Princeton, N.J.

Simpson, R., and H. Barrett (1980). Coded aperture imaging. In *Imaging in Diagnostic Medicine*, Nudelman (ed.). Plenum, New York, pp. 217–311.

Singer, J. (1959). Blood flow rates by nuclear magnetic resonance measurements. *Science* **130**, 1652–1653.

Singer, J. (1978). NMR diffusion and flow measurements and an introduction to spin phase graphing. *J. Phys. E: Sci. Instrum.* **11**, 281–291.

Singer, J. and S. Johnson (1959). Transistorized nuclear resonance magnetic field probe. *Rev. Sci. Instrum.* **30**, 92–93.

Skinner, L. A. (1970). Acoustically induced gas bubble growth. *J. Acoust. Soc. Am.* **51**, 378–382.

Smith, R., and B. Brenden (1969). Refinements and variations in liquid surface and scanned ultrasound holography. *Ultrasonics* **7**, 125–126.

Sokolov, S. (1936). Russian patent 49, Aug. 31, 1936; U.S. patent 2164125, 1939.

Solem, J., and G. Baldwin (1982). Microholography of living organisms. *Science* **218**, 229–235.

Souquet, J., P. Hanrath, L. Zitelli, P. Kremer, B. Langenstein, and M. Schlüter (1982). Transesophageal phased array for imaging the heart. *IEEE Trans. Biomed. Eng.* **29**, 707–712.

Spencer, M. P., and S. D. Campbell (1968). Development of bubbles in venous and arterial blood during hyperbaric decompression. *Bull. Mason Clinic* **22**, 26–32.

Stoner, W., J. Sage, M. Baum, D. Wilson, and H. Barrett (1976). Transmission imaging with a coded source. *Proc. ERDA X-Gamma Ray Symp. Ann Arbor*, 133–136.

Stong, C. L. (1968). Amateur scientist (quoting N. Wadsworth). *Sci. Am.* **218** (No. 2), 124–126.

Strandness, D., E. McCutcheon, and R. Rushmer (1966). Application of a transcutaneous Doppler flowmeter in evaluation of occlusive arterial disease. *Sur. Gyn. Obstet.* **122**, 1039–1045.

Strandness, D., R. Schultz, D. Summer, and R. Rushmer (1967). Ultrasonic flow detection: A useful technique in evaluation of peripheral vascular disease. *Am. J. Surg.* **113**, 311–320.

Suryan, G. (1951). *Proc. Indian Acad. Sci.* **A33**, 107.

Szilard, J. (1974). An improved 3D display system. *Ultrasonics* **12**, 273–276.

Szilard, J. and M. Kidger (1976). New ultrasonic lens. *Ultrasonics* **14**, 268–272.

Takahashi, S. (1957). *Rotation Radiography*. Japan Society for Promotion of Science, Tokyo.

Talbert, A., R. Brooks, and D. Morgenthaler (1980). Optimum energies for dual-energy CT. *Phys. Med. Biol.* **25**, 261–269.

Ter-Pogossian, M., M. Raichle, and B. Sobel (1980). *Sci. Am.* **243** (Oct.), 170–181.

Towe, B., and A. Jacobs (1981). X-ray Compton scatter imaging using a high speed flying spot X-ray tube. *IEEE Trans. Biomed. Eng.* **28**, 717–721.

Tretiak, O. J. (1978). Noise limitations in X-ray computed tomography. *J. Comput. Asst. Tomog.* **2**, 477–480.

Trokel, S., R. Srinivasan, and B. Braren (1983). Excimer laser surgery of the cornea. *Amer. Jour. Ophthal.* **96**, 710–715.

Tucker, D. G., V. Welsby, and R. Kandell (1958). Electronic sector scanning. *J. Br. Inst. Radio Eng.* **8**, 465–470.

Underwood, E. E. (1970). *Quantitative Stereology*. Addison-Wesley, Reading, Mass.

Underwood, J. H. (1978). X-ray optics. *Am. Sci.* **66**, 476–486.

Upton, A. C. (1982). Biological effects of low-level ionizing radiation. *Sci. Am.* **246** (No. 2), 41–49.

Valouch (1930). *J. Phys. Radium* **1**, 261–265.

Violante, M., P. Dean, H. Fisher, and J. Mahony (1980). Particulate contrast media for CT scanning liver. *Invest. Radiol.* **15,** 171–175.

Vollmann, W. (1982). Resolution enhancement of ultrasonic B-scan images by deconvolution. *IEEE Sonics Ultrason.* **29,** 78–83.

Walker, D., and R. Lumb (1964). Piezoelectric probes for immersion ultrasonic testing. *Appl. Mater. Res.,* 176–183.

Walls, G. L. (1942). 3D movies with a one-eye camera. *J. Opt. Soc. Am.* **32,** 693.

Walls, G. L. (1954). The filling-in process. *Am. J. Optom.* **31,** 329–340.

Walls, G. L. (1956). The G. Palmer story: Or, what it's like, sometimes, to be a scientist. *J. Hist. Med. Allied Sci.* **11,** 69–96.

Walls, G. L. (1960). Land! Land! *Surv. Ophthalmol.* **5,** 1–21 (reprinted from *Psychol. Bull.* **57,** Jan. 1960).

Webster, E. W. (1981). On the question of cancer induction by small X-ray doses. *Am. J. Roentgenol.* **137,** 647–666.

Weinberg, S., J. Watson, R. Gramiak, and G. Ramsey (1954). Stereo X-ray motion pictures. *J. Soc. Motion Pict. Television Eng.* **62,** 377–383.

Wells, P. N. T. (1969). *Ultrasonic Diagnosis.* Academic, New York.

Wells, P., M. Halliwell, R. Skidmore, A. Webb, and J. Woodcock (1977). Tumor detection by ultrasonic Doppler blood-flow signals. *Ultrasonics* **15,** 231–232.

Westheimer, G. (1963). Optical and motor factors in the formation of the retinal image. *J. Opt. Soc. Am.* **53,** 86–93.

Whittingham, T. (1976). Electronically switched array. *Ultrasonics* **14,** 29–33.

Wild, J., and J. Reid (1952). Ultrasonic ranging for cancer diagnosis. *Electronics* **25,** 136.

Wilson, R. R. (1941). A vacuum-tight sliding seal. *Rev. Sci. Instrum.* **12,** 91–93.

Wolff, S., L. Crooks, P. Brown, R. Howard, and R. Painter (1980). Tests for DNA and chromosomal damage induced by NMR imaging. *Radiology* **136,** 707–710.

Wyard, S. J. (1969). *Solid State Biophysics.* McGraw-Hill, New York.

Young, S., H. Muller, and B. Marincek (1981a). Contrast enhancement of malignant tumors after intravenous polyvinylpyrrolidone with metallic salts. *Radiology* **138,** 97–105.

Young, S., D. Enzmann, D. Long, and H. Muller (1981b). Perfluoroctylbromide contrast enhancement of malignant neoplasms. *Am. J. Roentgenol.* **137,** 141–146.

Zhernovi, A., and G. Latyshev (1965). *Nuclear Magnetic Resonance in a Flowing Liquid.* Translated from Russian by Consultants Bureau, New York.

INDEX

Absolute limits, 37-46
Absorption coefficient, 87-88
Acoustic impedance, 109
Acupuncture points, 7
Adiabatic fast passage, 154
A-mode, 191
A-scan, 105, 111, 220-234
Attenuation, radio signals, 17

Back projection, 46, 55, 66
Beam hardening, 78
Beer's law, 88
Bistable circuit, 226
Bits, 231
Black body, 10
Boltzmann's constant, 147
Bone scan, 19
Boxel, 75
Bragg cell, 214
Bremsstrahlung, 83
B-scan, 113, 189, 232, 236
Bubbles, 128, 129, 130-137
Bytes, 231

"Characteristic" X-rays, 84
Charge-coupled-device, 42, 227
Chirp, 107, 178, 190
Coded aperture, 48
Color images, 219, 236-238
Colors, infrared, 15
Color vision, eye properties, 166

Compound scan, 115
Compton radiography, 24
Compton scattering, 90, 176
Computed images, 46-49
Computed tomography (CT), 36, 42, 53-82, 93,
    98, 142, 159, 207-208, 222, 247
  digital, 62-82
  introduction, 53-62
Conical transducer, 103
Contrast agents:
  NMR, 163-165
  ultrasound, 127-129
  X-ray, 97-99
Contrast echocardiography, 137
Contrast enhancement, 44, 57, 200-205
Contrast transfer function, 51
Corona discharge photography, 26-27
Cosmic rays, 24
CT scan circle, 80
Cyclotron, 18, 24
Cylindrical lens, 54, 123

D curves, 198
Deblurring, 209-215
Detective quantum efficiency, 41, 65, 198
Dichromography, 168, 230
Digital computed tomography, 62-82
Digital methods, 225-231
Digital process, 45
Display orientation, computed tomography,
    76

273

Displays, 232-258
  color images, 236-238
  gray scale, 235-236
  holograms, 256-258
  motion pictures, 254-256
  pseudo three dimensional images, 238-240
  three dimensional images, 241-254
Doppler shift, 107, 134, 150, 180, 181, 183, 185, 190
Dynamic focus, ultrasound, 120
Dynamic spatial reconstructors, 80

Edge emphasis, 205-209
Electrons, 30
  interaction of X-rays, 86-91
Electromagnetic radiation, 3
Electrooculography, 15
Emanation and traversal, 29-30
Entopic perception, 34
Equipment evaluation, image formation, 49
Eye length, 2
Eye properties, 166, 206

Fan beam, 46, 58, 64
Far field, 27-29, 102
Fiber optics, 36, 58, 59-60
Filtered back projection, 66, 231
Flow measurement, NMR, 144
Fluorescence, 13, 16, 39
Fourier transform, 49, 66, 142, 145, 150, 151, 178, 212, 213
Fourth generation scanner, 64
Free induction decay, 145
Free precession, 146
Free radicals, 164
Frequency-dependent attenuation, 177
Fresnel lenses, 32

Gamma rays, 17, 19, 94, 198, 203, 215
Geiger counter, 19, 96
Gibbs phenomenon, 69, 93
Gray scale, 45, 160, 228, 235-236
Gyromagnetic ratio, 140, 148
Gyroscope precession, 139, 143

Hazard and dose, X-radiation, 94-97
Hazards of NMR, 161-163
Hazard of ultrasound, 125-127
H curves, 198
Helmholtz coils, 148
Holograms, 32, 48, 60, 124, 202, 212, 256
Hounsfield units, 47, 78, 89, 222

Image, speed of formation, 34-35

Image converter tubes (for ultrasound), 124
Image formation, 31-52
  absolute limits, 37-46
  computed images, 46-49
  equipment evaluation, 49-52
  scan vs. parallel processing, 31-37
Image intensifier, 41, 202
Image processing, 197-224
  changes, movement, and differences, 215-219
  contrast enhancement, 200-205
  deblurring, 209-215
  edge emphasis, 205-209
  photographic film, 197-200
  quantitative analysis, 219-224
  smoothing images, 215
Indocyanine green foam, 135-137
Infrared, 10-15, 30
Intensifier image, 36
Intensity of sound, 126
Interactions of sound, 107-110
Interactions of X-rays, 86-91
Inversion recovery, 154, 155, 160
Iodine, 174, 228
Ionization chamber, 65, 96
Ionizing radiation, 17, 161

K edge, 87, 90, 168, 175
Kirchoff's law, 10
Kirlian photography, 26

Larmor frequency, 140
Lasers, 86, 103, 213
Lateral inhibition, 206
Lead-zirconate-titanate (PZT), 101
Lichtenberg figures, 26-27
Light valve, 48, 61
Logarithmic response, 196
LogEtronics method, 202-203
Low frequencies, 6-8

Magnet, superconducting, 148
Magnetic tape recorders, 97
Magnetocardiogram, 7
Magnetogyric ratio, 140, 148
Mass attenuation coefficient, 88
Matched filter, 47, 258
Medical images:
  computed tomography, 53-82
  digital methods, 225-231
  displays, 232-258
  image formation, 31-52
  image processing, 197-224
  meaning of, 1-2
  motion measurement and flow, 180-196

nuclear magnetic resonance, 138-165
spectral information, 166-179
spectrum, 3-30
ultrasound, 100-137
X-rays, 83-99
Meson, cosmic rays, 24
Metallic appearances, 253
Microscopy, X-ray, 84
M-mode displays, 219
M-mode scan, 111
Moiré pattern, 15
Monochromatic X-ray beam, 169
Motion and flow, 180-196
Doppler shift, 180, 181, 183, 185, 190
Motion pictures, 253, 254-256

Nausea, 71
Near field, 27-29, 101
Negative resistance, 225
Nuclear magnetic resonance (NMR), 9-10
contrast agents, 163-165
effect and observation, 138-158
hazards, 161-163
speed of image formation, 158-161
Nuclear medicine, radioactivity and, 17-24

Optimum frequency, NMR, 148

Parallel processing, 31-37
Paramagnetic materials, NMR observation, 157, 165
Particle beams, 90
Phased array, 119, 121
Photographic film, 197-200, 215
Piezoelectric effect, 100-101
Pinhole, 31, 34, 47
Pixel, 66, 75, 82, 156, 215, 227
Planck's constant, 4, 140
Point spread function, 50, 57, 210
Poisson distribution, 71
Position emission tomography (PET), 23-24, 35, 57
Positrons, 21
Post mortem effects, ultrasound, 129-130
Potentiometers, 116
Protons, 90, 94, 140
Pseudo three dimensional images, 238-240
Pulfrich effect, 254
Pulsed Doppler, 150, 190

Quantitative analysis, image processing, 219-224
Quantum efficiency, 42, 65
Quantum fluctuations, 42, 45, 50
eye properties, 37-39

Quarter wavelength, matching, 107

Radio, 8-9, 79, 148, 161, 258
Radioactivity, 10, 17-24, 41
Radon, 66-67
Real-time imaging, 118
Reflection inversion, 110
Relaxation time, NMR, 140, 152
Roentgen, 25, 83
Roentgen rays, 16
Roentgen unit, 95
Ross filters, 173

Saturation recovery, 154, 160-161
Scan, 31-37
Scan converter tubes, 124
Scanning X-ray tubes, 73
Scanogram, 37, 68
Scan types, 110-125
Schlieren apparatus, 104
Scintillation detector, 96
Side lobes, 102, 120
Signal-to-noise ratio, 39-41, 65, 71, 160
Slice thickness, 72
Smoothing images, 43, 215
Sound and ultrasound, 25-26
Spatial filtering, 213
Spatial frequency, 49, 51, 209
Speckle, 177-178
Spectral information, 166-179
Spin echo, 155
Spin warp imaging, 152
Steady current, 4-6
Steady-state free precession, 154
Synchrotron radiation, 85
Synthetic aperture, 122-123

Television, 74, 80, 200, 205, 219-220, 227, 255
Third generation scanner, 64
Thermoluminescence, 96
Three-dimensional image, 77, 241-254
Three-dimensional viewing exercise, 250
Tomographic concepts, 53. *See also* Computed tomography
Tracer, 18

Ultrasound, 79, 100-107, 177, 258
bubbles, 130-137
contrast agents, 127-129
hazard, 125-127
interactions, 107-110
post mortem effects, 129-130
scan types, 110-125
sources, 100-107

Ultraviolet, 15-16, 30, 176
Unsharp mask method, 202

Vectorcardiogram, 6-7
Ventricular fibrillation, 162
Videodensitometry, 195
Video magnetic tape recorder, 231
Visible light, 14-15
Voxel, 75

Washout curves, 137, 195
Wiggler, 85

Xerox process, 44, 206
X-rays, 16-17, 29, 30, 33, 35, 42, 83-99
   biological hazard and dose, 94-97
   contrast agents, tracers, and stains, 97-99
   interactions, 86-91
   sources, 83-86
   speed of examination, 91-94

Zeeman effect, 9
Zeugmatography, 158
Zone plate, 32, 48